SIGNAL PROCESSING
The Modern Approach

McGraw-Hill Series in Electrical Engineering

Consulting Editor
Stephen W. Director, *Carnegie-Mellon University*

CIRCUITS AND SYSTEMS
COMMUNICATIONS AND SIGNAL PROCESSING
CONTROL THEORY
ELECTRONICS AND ELECTRONIC CIRCUITS
POWER AND ENERGY
ELECTROMAGNETICS
COMPUTER ENGINEERING
INTRODUCTORY
RADAR AND ANTENNAS
VLSI

Previous Consulting Editors
Ronald N. Bracewell, Colin Cherry, James F. Gibbons, Willis W. Harman, Hubert Heffner, Edward W. Herold, John G. Linvill, Simon Ramo, Ronald A. Rohrer, Anthony E. Siegman, Charles Susskind, Frederick E. Terman, John G. Truxal, Ernst Weber, and John R. Whinnery

Communications and Signal Processing

Consulting Editor
Stephen W. Director, *Carnegie-Mellon University*

Antoniou: *Digital Filters: Analysis and Design*
Candy: *Signal Processing: The Model-Based Approach*
Candy: *Signal Processing: The Modern Approach*
Carlson: *Communications Systems: An Introduction to Signals and Noise in Electrical Communication*
Cherin: *An Introduction to Optical Fibers*
Cooper and McGillem: *Modern Communications and Spread Spectrum*
Davenport: *Probability and Random Processes: An Introduction for Applied Scientists and Engineers*
Drake: *Fundamentals of Applied Probability Theory*
Guiasu: *Information Theory with New Applications*
Keiser: *Optical Fiber Communications*
Papoulis: *Probability, Random Variables, and Stochastic Processes*
Papoulis: *Signal Analysis*
Papoulis: *The Fourier Integral and Its Applications*
Peebles: *Probability, Random Variables, and Random Signal Principles*
Proakis: *Digital Communications*
Schwartz: *Information Transmission, Modulation, and Noise*
Schwartz and Shaw: *Signal Processing*
Smith: *Modern Communication Circuits*
Taub and Schilling: *Principles of Communication Systems*

SIGNAL
PROCESSING
The Modern Approach

James V. Candy

Lawrence Livermore National Laboratory
and
University of Santa Clara

McGraw-Hill Book Company

New York   St. Louis   San Francisco   Auckland   Bogotá   Caracas
Colorado Springs   Hamburg   Lisbon   London   Madrid   Mexico
Milan   Montreal   New Delhi   Oklahoma City   Panama   Paris
San Juan   São Paulo   Singapore   Sydney   Tokyo   Toronto

This book was set in Times Roman by Publication Services.
The editors were Alar E. Elken and John Morriss;
the cover was designed by Joseph Gillians;
the production supervisor was Denise L. Puryear.
Project supervision was done by Publication Services.
R. R. Donnelley & Sons Company was printer and binder.

SIGNAL PROCESSING
The Modern Approach

2 3 4 5 6 7 8 9 0 DOCDOC 8 9 3 2 1 0 9 8

ISBN 0-07-009751-8

Library of Congress Cataloging-in-Publication Data
Candy, J. V.
    Signal processing: the modern approach / James V. Candy.
        p. cm. — (McGraw-Hill series in electrical engineering.
    Communications and signal processing)
    Bibliography: p.
    Includes index.
    ISBN 0-07-009751-8
    1. Signal processing. I. Title. II. Series.
TK5102.5.C27 1988
321.38'043—dc19   87-16619 CIP

ABOUT THE AUTHOR

Dr. James V. Candy is a Project Engineer for Signal Processing at the Lawrence Livermore National Laboratory and an Adjunct Professor of Electrical Engineering at the University of Santa Clara. Dr. Candy received a commission in the USAF in 1967 and was a Systems Engineer/Test Director from 1967 to 1971. He has been a Research Engineer at the University of California, Lawrence Livermore National Laboratory since 1976. He has held various positions including that of Thrust Area Leader for Signal and Control Engineering. Recently he worked at the NATO SACLANT ASW Research Center in La Spezia, Italy and he is currently on sabbatical leave from Lawrence Livermore at Ecole Superieure D'Electricite, Laboratoire Des Signaux et Systèmes Gif-sur-Yvette, France.

His research interests include estimation, identification, spatial estimation, and array signal processing techniques. Dr. Candy is a registered Control System Engineer in the state of California. He received his B.S.E.E. degree from the University of Cincinnati, and his M.S.E. and Ph.D. degrees in electrical engineering from the University of Florida, Gainesville in 1972 and 1976 respectively. He is the author of *Signal Processing: The Model-Based Approach* (McGraw-Hill, 1986).

To Pat, my love, and to the Lord—my life!

CONTENTS

Preface   xi

1 Introduction   1

1.1   Discrete-Time Signals   1
1.2   Representations of Discrete Signals   6
1.3   Representations of Random Discrete Signals   9
1.4   Summary   14
      References   14
      SIG Notes   14
      Exercises   14

2 Discrete Signal System Representations: The Z-Transform   16

2.1   Z-Transform   16
2.2   Discrete Linear Systems   23
2.3   Z-Transform Method of Solving Difference Equations   27
2.4   Frequency Response of Discrete Signals and Systems   29
2.5   Discrete Representations of Continuous Signals   33
2.6   Fundamentals of Digital Filters   41
2.7   Summary   46
      References   48
      SIG Notes   49
      Exercises   49

3 Discrete Signal Analysis   55

3.1   Periodic Signal Analysis: Discrete Fourier Transforms   55
3.2   Aperiodic Signal Analysis: Discrete-Time Fourier Transforms   59
3.3   Transform Equivalence: Z, DtFT, CtFT, DFT, and DFS   64
3.4   Practical Aspects of the DFT   68
3.5   Summary   80

References   80
SIG Notes   80
Exercises   80

4   Discrete Random Signals   83

4.1   Introduction   83
4.2   Stochastic Processes   87
4.3   Spectral Representation of Random Signals   99
4.4   Linear Systems with Random Inputs   111
4.5   Parametric Representations of Stochastic Processes: ARMAX Models   115
4.6   Parametric Representations of Stochastic Processes: Lattice Models   125
4.7   Parametric Representations of Stochastic Processes: State-Space Models   132
4.8   Equivalence of Linear Models: ARMAX, Lattice Transformations
       to State-Space   137
4.9   Summary   142
       References   143
       SIG Notes   144
       Exercises   144

5   Random Signal Analysis: Classical Approach   149

5.1   Estimation Concepts   149
5.2   Minimum-Variance/Maximum-Likelihood Estimation   152
5.3   Covariance Estimation   156
5.4   Nonparametric Methods of Spectral Estimation   160
5.5   Coherence Analysis   171
5.6   Summary   174
       References   174
       SIG Notes   175
       Exercises   175

6   Random Signal Analysis: Modern Approach   178

6.1   Parametric Methods of Spectral Estimation   178
6.2   Autoregressive (All-Pole) Spectral Estimation   180
6.3   Moving-Average (All-Zero) Spectral Estimation   188
6.4   Autoregressive Moving-Average or Rational (Pole-Zero) Spectral
       Estimation   191
6.5   Lattice Spectral Estimation   196
6.6   Minimum-Variance, Distortionless Response Spectral Estimation   200
6.7   Harmonic Methods of Spectral Estimation   204
6.8   Summary   211
       References   211
       SIG Notes   212
       Exercises   212

7   Parametric Signal Processing   214

7.1   Introduction   214
7.2   Optimal Estimation   215

7.3   Levinson (All-Pole) Filters   220
7.4   Levinson (All-Zero) Filters   224
7.5   Lattice Filters   231
7.6   Prediction-Error Filters   236
7.7   Order Estimation   241
7.8   Case Study: Electromagnetic Signal Processing   244
7.9   Summary   257
      References   257
      SIG Notes   258
      Exercises   259

8   Adaptive Signal Processing   263

8.1   Introduction   263
8.2   Adaptation Algorithms   264
8.3   All-Zero Adaptive Filters   267
8.4   Pole-Zero Adaptive Filters   283
8.5   Lattice Adaptive Filters   286
8.6   Applications   294
8.7   Case Study: Plasma-Pulse Estimation   302
8.8   Summary   307
      References   307
      SIG Notes   308
      Exercises   308

9   Model-Based Signal Processing   314

9.1   Introduction   314
9.2   Model-Based Signal Processing   316
9.3   Model-Based Processors: State-Space (Kalman) Filters   321
9.4   Kalman Filter Identifier   331
9.5   Kalman Filter Deconvolver   334
9.6   The Kalman/Wiener Filter Equivalence   337
9.7   Summary   340
      References   340
      SSPACK Notes   341
      Exercises   341

Appendixes

A   Fast Fourier Transform Algorithm   345
B   Maximum Entropy Spectral Estimation   350
C   Recursive Prediction-Error Method (RPEM)   354
D   SIG: A General-Purpose Signal-Processing Package   359
E   SSPACK: An Interactive State-Space, Model-Based Processing Software
    Package   368

Index   375

PREFACE

This text is designed primarily for the advanced senior, entering graduate student, or practicing engineer. The main goal of this text is to bridge the gap between classical and modern methods of signal processing with emphasis toward the random signal cases, somewhat in the same spirit as Schwartz and Shaw [1]. The text incorporates digital signal-processing methods which employ the now classical Fast Fourier transform (FFT), through modern parametric and adaptive signal processing techniques, up to and including an introduction to the sophisticated model-based processing methods. We concentrate on presenting digital signal processing from a very simplistic perspective—*the signal processor is given a set of data records (sometimes noisy) and is asked to extract the required signal information.* Wherever possible, we have included computer simulated examples using SIG™,† the general-purpose signal-processing package, and SSPACK™,‡ the interactive state-space, model-based, signal processing package, to give the reader a feel for the "practice" of signal processing. The required background is a basic course in digital signal processing (discrete systems, transforms, etc.) and a basic course in probability and stochastic processes.

The first four chapters present the fundamentals necessary to comprehend the modern techniques. The reader is led from linear systems and processing of deterministic signals using FFT techniques, to discrete random signals. We avoid most details, and direct the reader to other fundamental texts [2,3,4]. Once the fundamentals are in hand, we show the similarity of deterministic and random signal analysis, with the signal/spectrum being replaced (when possible) by the covariance/power spectrum. Classical estimators are discussed in terms of non-parametric methods, while the modern parametric methods are introduced to es-

---

†SIG is a trademark of the Lawrence Livermore National Laboratory (see Appendix D)
‡SSPACK is a trademark of Techni-Soft (see Appendix E)

timate the power spectrum. The theory and development of modern parametric and adaptive signal processors are developed in some detail, with various techniques, applications, and examples discussed. Finally, the advanced topic of model-based signal processing is discussed at an introductory level. More specialized texts should be consulted for further details on the various processors discussed here [5,6,7,8,9,10].

The first chapter is introductory in nature; we develop the concept of a discrete signal and its various representations. We then introduce discrete random signals and the development of signal estimators.

In Chapter 2, we discuss the basic concepts underlying discrete systems and transforms. We briefly develop the Z-transform and discrete systems, and review the solution of difference equations. Next, we develop the idea of frequency response and show how these concepts can be used to specify and analyze discrete systems. We develop the idea of sampling a continuous signal and representing it as a discrete or "sampled" signal. Finally, we briefly review the fundamentals of digital filters from the signal processor's, rather than the filter designer's, perspective.

Chapter 3 is concerned with analyzing discrete (deterministic) signals. We first develop the basic discrete Fourier transform and show how it can be used to construct other representations (e.g., Z-transforms, discrete-time Fourier, continuous Fourier transforms). Finally, we discuss the practical aspects of the DFT, including its pitfalls and the nagging idea of "frequency resolution."

In Chapter 4, we develop the fundamental concept of a discrete random signal, along with some of its accompanying probabilistic and statistical concepts, by showing its similarity to the deterministic case. Next, we develop the idea of spectral representations of stochastic processes and prove the fundamental Wiener-Khintchine Theorem. Once this is established, we develop the fundamental relations for linear systems excited by random signals. With this in mind, we develop various parametric representations of stochastic processes which are used to simulate signals with specified statistics. We investigate the ARMAX, lattice, and Gauss-Markov (state-space) models and show their equivalence.

Chapter 5 is concerned with the analysis of discrete random signals. The basic concepts of estimation are first developed, followed by popular estimator constructs—minimum-variance estimators. After discussing various techniques of estimating covariances from data, we develop the classical FFT-based methods of spectral estimation. Finally, we develop the related coherence function, which can be used to analyze the quality of measured data.

In Chapter 6, we continue the discussion of random signal analysis from the modern perspective. We discuss the modern approach to spectral analysis using the parametric methods. We develop the AR, MA, ARMA, and lattice techniques of spectral estimation as well as the MVDR and harmonic approaches (Pisarenko and Prony).

In Chapter 7, we discuss some of the popular parametric signal-processing techniques. First, we discuss the basic idea of optimal estimation and then develop the Levinson all-pole and all-zero filters. Next, we develop the lattice and prediction-error filters evolving from system identification. Finally, we develop the crucial notion of order estimation and discuss a case study applying many of these techniques to process electromagnetic signals.

Chapter 8 is concerned with the processing of signals with unknown or nonstationary statistics—adaptive processors. After introducing the basic ideas, we discuss various gradient-based adaptation algorithms. Next, we develop the all-zero, pole-zero, all-pole, and lattice adaptive filters and show how these techniques can be applied to solve various problems: noise cancelling, adaptive prediction, and line enhancement. Finally, we discuss the development of noise-cancelling filters to estimate a plasma pulse from noisy measurement data.

In Chapter 9, we discuss the basic ideas of model-based signal processing. We first introduce the concepts and develop the state-space (Kalman filter) processor using the innovations approach. Next, we show how the processor can be applied to solve the identification and deconvolution problems. Finally, we show the equivalence of the Kalman and Wiener processors.

Simulations are used throughout the text as a teaching aid and should be of interest to the practicing engineer or scientist. At the end of each chapter we include "SIG Notes" and "SSPACK Notes" to inform the signal processor of the capability available in each of these packages.

REFERENCES

1. M. Schwartz and L. Shaw, *Signal Processing: Discrete-Analysis, Detection, and Estimation* (New York: McGraw-Hill, 1976).
2. A. Oppenheim and R. Shafer, *Digital Signal Processing* (Englewood Cliffs, N.J.: Prentice-Hall, 1975).
3. A. Oppenheim, A. Willsky, and I. Young, *Signals and Systems* (Englewood Cliffs, N.J.: Prentice-Hall, 1983).
4. S. Tretter, *Introduction to Discrete-Time Signal Processing* (New York: Wiley, 1976).
5. L. Ljung and T. Soderstrom, *Theory and Practice of Recursive Identification* (Boston: MIT Press, 1983).
6. G. Goodwin and K. Sin, *Adaptive Filtering, Prediction, and Control* (Englewood Cliffs, N.J.: Prentice-Hall, 1984).
7. B. Widrow and S. Stearns, *Adaptive Signal Processing* (Englewood Cliffs, N.J.: Prentice-Hall, 1984).
8. S. Haykin, *Introduction to Adaptive Filtering* (New York: Macmillan, 1984).
9. S. Orfanidis, *Optimum Signal Processing* (New York: Macmillan, 1985).
10. J. Candy, *Signal Processing: The Model-Based Approach* (New York: McGraw-Hill, 1986).
11. D. Lager and S. Azevedo, "SIG—A General Purpose Signal Processing Code," *Proc. IEEE*, 1987.
12. S. Azevedo, J. Candy, and D. Lager, "SSPACK—An Interactive Multi-channel, Model-Based Signal Processing Package," *Proc. IEEE Conf. Circuits, Systems*, 1986.

ACKNOWLEDGEMENTS

I would like to express my gratitude to those who motivated this manuscript. First, I thank my colleague and friend Dr. D. Siljak for his timely advice and continued moral support. I would also like to thank Ikram Abdou, University of

Delaware; Alan Bovik, University of Texas—Austin; Haluk Derin, University of Massachusetts; Dr. M. Hodzig, Paul Prucnal, Columbia University, and Dr. E. Sullivan for their review of the manuscript and suggestions for its improvement. I thank Dr. S. Parker for his encouragement in leading me to the "lattice processors." I also thank the management at Lawrence Livermore National Laboratory, for enabling me to teach internal courses in signal processing which helped to "debug" this manuscript. My thanks go to E. Lafranchi, S. Weissenberger, G. Clark, M. Portnoff, R. Ziolkowski, and especially to P. Phelps, for his continued support. I would also like to thank my colleagues and friends at Lawrence Livermore National Laboratory whose penetrating questions led to a deeper understanding of signal-processing concepts: M. Axelrod, S. Azevedo, F. Barnes, D. Dunn, P. Fitch, F. Followill, W. Gersch, J. Hernandez, and D. Lager. Also, I thank the many students at Santa Clara University who helped in reviewing the manuscript as well as with its organization. I again thank the Lawrence Livermore National Laboratory for the use of SIG, the general-purpose signal-processing package, and Techni-Soft for the use of SSPACK, the state-space model-based processing package. Finally, I would like to thank Ms. Deborah Payne for her skillful "type-setting" of this manuscript using T_EX.

James V. Candy

CHAPTER
1

INTRODUCTION

In this chapter and throughout the text, we introduce the basic concepts involved in signal processing from a practical viewpoint. Our basic premise is that the signal processor is given a set of data records (sometimes noisy) and is asked to extract the required signal information. We first introduce the concept of a signal; then we introduce its various representations. Finally, we discuss various approaches for processing random signals.

1.1 DISCRETE-TIME SIGNALS

Signals occur in many different phenomena. When the space shuttle blasts off into orbit, the astronauts communicate to the control center by transmitting and receiving radio waves through the atmosphere. An earthquake rumbling through the earth's crust is transmitted by an acoustic wave and recorded on sensitive seismometers. The reactor in a nuclear power plant is monitored for malfunction through detectors and sensors to prevent possible melt-down. Aircraft at a busy airport are guided by air traffic controllers who use sophisticated tracking radars. All these examples are applications of the measurement and conditioning of data or signals that carry information that is vital to perform a given function. Radio signals are transmitted and received through antennas in order to keep both astronaut and ground controller informed of the status of the given mission. Seismic signals are used to analyze and predict the occurrence of earthquakes. Transduced electrical signals are used to control the cooling of the reactor core and predict potential failure. Electromagnetic signals reflecting off approaching aircraft are discerned by radar systems and displayed to avoid collisions (see Fig. 1.1–1). So, we see in a heuristic sense that a signal can be thought of as data or information generated by physical phenomenology, like earthquakes or radar,

1

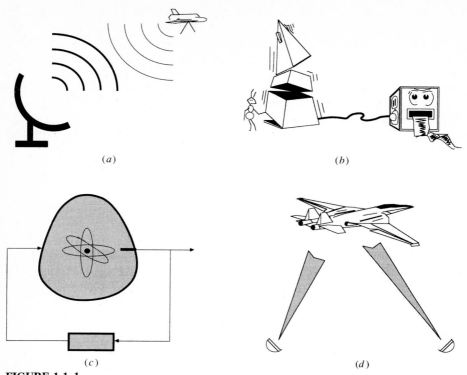

FIGURE 1.1–1
Typical signals (*a*) Radio signals (*b*) Seismic signals (*c*) Transduced electrical signals (*d*) Electromagnetic signals

and collected for current or future use. These are just a few of the typical signals used in various applications.

In a less exotic sense, suppose we are interested in understanding why our electric bills are so high. We begin our analysis by recording the monthly bills to produce a sequence of dollar costs. We can consider this sequence of values recorded every month as a discrete signal. Another simple example of a signal is the daily balance in our checkbooks. If we record the balance every day for a year, then we can consider this sequence a discrete signal as well. A signal can be thought of as an ordered set (usually by time) of values, or more precisely, a set of values indexed by a set of numbers. In these examples the values are the dollar cost of electricity or the dollar balance of our checkbook, but the index sets are different. In the first case the bills are recorded monthly, whereas in the second the checkbook is balanced daily. In any case, the concept of a signal is identical. So, we see that signals occur naturally (earthquakes) or are created (radio). Most applications rely on the basic notion of measuring a signal that contains information necessary to perform a function, ranging from the fighter pilot engaged in a high-speed dogfight who depends on timely displays for survival, to the homemaker who depends on a clothes dryer to sense that its contents are dry to yield clothes for the next day's activities.

With the concept of a signal as an ordered set of values containing information about the process under investigation, we can define *signal processing* (simply) as the process of extracting useful information and discarding the extraneous information from measured data. This text is concerned with the development of signal-processing techniques to analyze, extract, and predict signals from measured data.

Mathematically, a signal can be represented by a scalar or vector function of one or more variables constituting the index set. Typically, a scalar *discrete signal* is defined by the function

$$x(t) \quad \text{for } t \in \mathbf{I}$$

where t is the discrete-time index which is a member of the set of integers $\mathbf{I}$. *Continuous signals* are represented similarly by

$$x_c(\tau) \quad \text{for } \tau \in \mathbf{R}$$

where τ is the continuous-time index which is a member of the set of real numbers $\mathbf{R}$.

Signals are classified according to some unique characteristic which can usually be represented mathematically. A signal is classified as *deterministic* if it is repeatable; that is, when we measure its value over a specified period of time and duplicate the measurement some time later, the same signal values occur. If a signal does not achieve the same values, then it is no longer repeatable and considered *random*. A deterministic signal is defined as *periodic* with period M if

$$x(t) = x(t + M) \quad \forall \, t \qquad (1.1-1)$$

A signal that is not periodic is called *aperiodic*. A discrete signal is called a *sampled* signal if it is the result of sampling a continuous time signal; that is, x_s is a sampled signal if

$$x_s(\tau) = x_c(\tau)\big|_{\tau = k\Delta T} \qquad (1.1-2)$$

where ΔT is the specified sampling interval and k is the discrete index. Finally, a discrete signal is defined as *random* if it can be represented by the process

$$x(t, \xi) \quad t \in \mathbf{T}, \ \xi \in \Xi$$

where t is a member of the index set $\mathbf{T}$ and ξ is a member of the sample space Ξ. We shall discuss random signals in Sec. 1.3 and concentrate on deterministic signals here.

Fundamental deterministic signals are useful for analysis. The *unit impulse* $\delta(t)$ is defined by

$$x(t) = \delta(t) := \begin{cases} 1 & t = 0 \\ 0 & t \neq 0 \end{cases} \qquad (1.1-3)$$

The *unit step* $\mu(t)$ is defined by

$$x(t) = \mu(t) := \begin{cases} 0 & t < 0 \\ 1 & t \geq 0 \end{cases} \tag{1.1-4}$$

Just as with continuous-time signals, the unit impulse and step are related. The unit step can be synthesized by a sum of impulses analogous to integration in continuous time:

$$\mu(t) = \sum_{t=0}^{\infty} \delta(t) \tag{1.1-5}$$

Inversely, the unit impulse can be synthesized by taking the first difference of the unit step analogous to differentiation in continuous time:

$$\delta(t) = \mu(t) - \mu(t-1) \tag{1.1-6}$$

Two other basic discrete functions of interest are the *unit ramp*, given by

$$x(t) = t \quad \text{for } t \geq 0 \tag{1.1-7}$$

and the discrete *complex exponential*, defined by

$$x(t) = \alpha^t \quad \text{for } \alpha \in \mathbf{C} \tag{1.1-8}$$

and $\mathbf{C}$ is the set of complex numbers.
   We can think of α as

$$\alpha = e^{\Theta}$$

where Θ can be real or complex, analogous to continuous-time signals. If α is real, then x is an increasing or decreasing exponential for $\alpha > 1$ or $\alpha < 1$, respectively. Note that with $\alpha < 0$, the sign of $x(t)$ alternates. For the case of Θ being purely imaginary, we have the discrete exponential

$$x(t) = e^{j\Omega_k t} \tag{1.1-9}$$

where Ω_k is the discrete frequency.
   Similar to the continuous case, exponentials are related to sinusoidal signals of the form

$$x(t) = A \cos(\Omega_k t + \phi) \tag{1.1-10}$$

where both Ω and ϕ are in units of radians, through the *Euler* relations, that is,

$$e^{j\Omega_k t} = \cos \Omega_k t + j \sin \Omega_k t \tag{1.1-11}$$

and
$$A \cos(\Omega_k t + \phi) = \frac{A}{2} e^{j\phi} e^{j\Omega_k t} + \frac{A}{2} e^{-j\phi} e^{-j\Omega_k t} \tag{1.1-12}$$

where $j := \sqrt{-1}$.
   We note in passing that, unlike its continuous counterpart, the discrete

exponential possesses unique properties. Consider a complex exponential with frequency $\Omega_k + 2\pi$,

$$e^{j(\Omega_k + 2\pi)t} = e^{j2\pi t}e^{j\Omega_k t} = 1 \times e^{j\Omega_k t} = e^{j\Omega_k t}$$

which implies that the discrete exponential is *periodic* with period 2π. Continuous exponentials are all distinct for distinct values of frequency; that is, values for $e^{j\omega_0 \tau}$ are distinct for each value of ω_0. Since signals at Ω_k, $(\Omega_k \pm 2\pi)$, $(\Omega_k \pm 4\pi)$, $\ldots$, $(\Omega_k \pm m\pi)$ are identical, discrete exponentials are not distinct; therefore, they require only investigation in an interval of length 2π. Note that the discrete exponential does not have a continuously increasing rate of oscillation as Ω_k is increased in magnitude. Rather, oscillation rates increase from 0 until $\Omega_k = \pi$ and then decrease until we reach 2π. High frequencies are reached near $\pm \pi$.

Also from the periodicity property, we have [1]

$$e^{j\Omega_k M} = 1$$

or equivalently,

$$\Omega_k M = 2\pi k$$

or

$$\Omega_k = \frac{2\pi k}{M} \tag{1.1--13}$$

which implies that Ω_k is periodic only if $\Omega/2\pi$ or k/M is a rational number. Here we call M the *fundamental period* of the complex exponential or sinusoid. Each of these basic discrete signals is shown in Fig. 1.1–2. It will be shown later that these signals possess certain unique properties useful in analyzing discrete signals and systems.

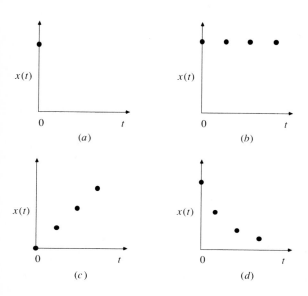

FIGURE 1.1–2
Fundamental discrete signals (*a*) Unit impulse (*b*) Unit step (*c*) Unit ramp (*d*) Exponential

1.2   REPRESENTATIONS OF DISCRETE SIGNALS

In this section we discuss various representations of discrete signals under investigation. From the signal processor's viewpoint, once we assume a particular class of signal, then the analysis that follows must be consistent with that assumption. For instance, if we assume a signal is discrete and periodic, then we know that it can be represented by a Fourier series [2]. The *discrete Fourier series* (DFS) representation of a signal is given by the pair

$$x(t) = \sum_{m=<M>} a_m e^{j\Omega_m t} \qquad (1.2-1)$$

and

$$a_m = \frac{1}{M} \sum_{t=<M>} x(t) e^{-j\Omega_m t} \qquad (1.2-2)$$

where $\Omega_m = 2\pi m / M$ and $<M>$ means any interval of M samples. These equations are called the *synthesis* and *analysis* relations, respectively, and the a_m variables are called the *discrete Fourier coefficients*. These relations imply that the discrete signal under investigation can be decomposed into a unique set of harmonics. The following example illustrates these relations.

> **Example 1.2–1.** Suppose we assume that a signal under investigation is periodic, with Ω_0 an integer multiple of M and we would like to investigate its properties. We model the periodic signal as
>
> $$x(t) = \cos \Omega_0 t$$
>
> From the Euler relations we know that
>
> $$\cos \Omega_0 t = \frac{e^{j\Omega_0 t} + e^{-j\Omega_0 t}}{2}$$
>
> which implies that the DFS representation is given by
>
> $$x(t) = \cos \Omega_0 t = \frac{1}{2} e^{j\Omega_0 t} + \frac{1}{2} e^{-j\Omega_0 t}$$
>
> or
>
> $$a_m = \begin{cases} \frac{1}{2} & m = \pm 1 \\ 0 & \text{elsewhere} \end{cases}$$
>
> The signal and its Fourier series representation is shown in Fig. 1.2–1 where we show the coefficients over one period.

We see that a periodic signal can be represented by a discrete harmonic series, and the corresponding coefficients can be considered an equivalent representation of the information in the given signal. Note that we construct or synthesize the signal using the harmonic functions of Eq. (1.2–1) and analyze its spectral content using Eq. (1.2–2). We shall discuss the DFS representations of signals in more detail in Chapter 3. Next, we consider similar representations for aperiodic signals.

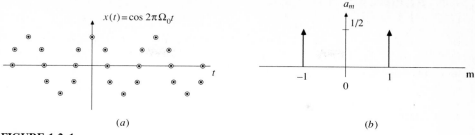

FIGURE 1.2–1
Discrete Fourier series of a sinusoid (*a*) Function (*b*) Spectral coefficients

Analogous to continuous-time processes, we use transform representations as an alternative way of analyzing the information in a signal. For instance, if we believe a signal has periodic components, then it is more convenient to represent or transform the signal to the frequency domain where sinusoids appear (in theory) as spectral lines. Transformations to the frequency domain exist for discrete time signals, just as they do for continuous-time signals. In the continuous domain, we use the Laplace transform and its inverse to transform the signal to the frequency domain, with $s = \sigma \pm j\omega$ as the complex variable. The Laplace transform provides nice convergence properties and possesses the *Fourier transform* as a special case when the region of convergence contains the $j\omega$-axis; that is, when $s = j\omega$. In discrete time, we have an analogous situation: the Z-transform plays the role of its continuous-time counterpart, the Laplace transform, and possesses similar convergence properties. It has the so-called *discrete-time Fourier transform* (DtFT) as a special case when the region of convergence contains the *unit circle*, that is, $z = e^{j\Omega}$. To be more precise, we define the Z-transform of a signal as

$$X(z) = Z[x(t)] := \sum_{t=-\infty}^{\infty} x(t)z^{-t} \qquad (1.2\text{–}3)$$

and the corresponding *inverse* Z-transform as

$$x(t) = \frac{1}{2\pi j} \oint X(z)z^{t-1}dz \qquad (1.2\text{–}4)$$

The Z-transform representation of a discrete signal enables us to analyze the spectral content of aperiodic signals and therefore determine such quantifiers as bandwidth and spectral shape. We will discuss the Z-transform and its properties more fully in the next chapter. From the signal processor's perspective, the Z-transform representation offers a convenient domain to analyze spectral properties of a discrete signal, as shown in the following example.

Example 1.2–2. Consider the sinusoid of the previous example multiplied by a unit step function,

$$x(t) = \cos \Omega_0 t \times \mu(t)$$

Then the corresponding Z-transform is given by

$$X(z) = Z[\cos \Omega_0 t \ \mu(t)] = \sum_{t=0}^{\infty} \cos \Omega_0 t \ z^{-t} \qquad |z| > 1$$

or, from the Euler relations, we obtain

$$X(z) = \frac{1}{2} \sum_{t=0}^{\infty} e^{j\Omega_0 t} z^{-t} + \frac{1}{2} \sum_{t=0}^{\infty} e^{-j\Omega_0 t} z^{-t}$$

or

$$X(z) = \frac{1}{2} \sum_{t=0}^{\infty} (e^{j\Omega_0} z^{-1})^t + \frac{1}{2} \sum_{t=0}^{\infty} (e^{-j\Omega_0} z^{-1})^t$$

Using the relations for the geometric series,† we have

$$X(z) = \frac{1}{2}\left(\frac{1}{1 - e^{j\Omega_0} z^{-1}}\right) + \frac{1}{2}\left(\frac{1}{1 - e^{-j\Omega_0} z^{-1}}\right)$$

Combining under a common denominator and using the Euler relations, we obtain

$$X(z) = \frac{1 - \cos \Omega_0 z^{-1}}{1 - 2 \cos \Omega_0 z^{-1} + z^{-2}} \qquad \text{for } |z| > 1$$

We see that the Z-transform offers an alternate means of representing a discrete time aperiodic signal.

The discrete-time Fourier transform (DtFT) is defined as

$$X(\Omega) = \text{DtFT} \ [x(t)] := \sum_{t=-\infty}^{\infty} x(t)e^{-j\Omega t} \qquad (1.2\text{–}5)$$

and the corresponding inverse transform is given by

$$x(t) = \frac{1}{2\pi} \int_{2\pi} X(\Omega)e^{j\Omega t} d\Omega \qquad (1.2\text{–}6)$$

We discuss the DtFT in more detail in Chapter 3.

The DtFT is useful in determining the spectral- or frequency-domain representation of discrete signals. The frequency response for a discrete signal can be determined from its Z-transform by letting $z = e^{j\Omega}$ and varying Ω; that is, the *frequency response* of $x(t)$ is given by

$$X(z)\big|_{z=e^{j\Omega}} = |X(e^{j\Omega})| \angle X(e^{j\Omega}) \qquad (1.2\text{–}7)$$

We discuss this representation of a discrete signal in the next chapter and show its relationship to sampled continuous-time signals.

We can also view a signal as the output of some system whose action has

---

†Recall that the *geometric series* is given by $\sum_{n=0}^{\infty} \alpha^n = 1/(1 - \alpha)$ for $|\alpha| < 1$.

changed its properties. For instance, a digital filter will pass certain portions of the signal spectrum while attenuating others. Thus, a filter can be thought of as a system. Using the *convolution* relations of linear systems theory, we can represent the output signal that results from the filter with impulse response $h_f(t)$ as

$$y(t) = h_f(t) * x(t) = \sum_{k=0}^{\infty} h_f(k)x(t - k) \qquad (1.2\text{--}8)$$

In this instance, the signal $y(t)$ is the output of the system characterized by a digital filter. From the signal processor's viewpoint, it is very important to know how the filter will alter the characteristics of the input signal.

Finally, we can represent a signal in a model-based sense [3]. Here we assume that the signal is the output of a system or model excited by a specified input or noise or both. The model can range from a simple set of difference equations like

$$y(t) = a_1 y(t - 1) + a_2 y(t - 2) + u(t) \qquad (1.2\text{--}9)$$

to the exotic vector-matrix state-space model given by

$$x(t) = Ax(t - 1) + u(t - 1) \quad \text{and} \quad y(t) = Cx(t) \qquad (1.2\text{--}10)$$

From the signal processor's perspective, the model-based approach incorporates the underlying physical phenomenology in a mathematical model to aid in understanding, analyzing, and decomposing the measured data. We will discuss the model-based approach in Chapter 9.

Thus, we see that the signal processor must be aware of the various representations available for a signal whether performing analysis or designing a sophisticated predictor for processing.

1.3 REPRESENTATIONS OF RANDOM DISCRETE SIGNALS

Naturally, we would like to extend our knowledge of fundamental deterministic signals and spectra to the random case. Random signals usually occur when we make measurements of a process. Measurements are typically contaminated with noise, either from the sensors, measurement instrumentation, the surrounding environment, or even the process itself. In practice, the signal processor will encounter random signals almost exclusively. In this section, we introduce the notion of the random signal and show how the concepts of linear systems theory apply to this case.

As mentioned previously, a discrete random signal can be characterized by the stochastic process

$$x(t, \xi) \quad t \in \mathbf{T}, \ \xi \in \Xi \qquad (1.3\text{--}1)$$

where $\mathbf{T}$ is the index set of integers for the discrete case and Ξ is a sample space. Because of this characterization, the random signal and subsequent processing become enveloped in the complex mathematics of probability theory and stochastic processes. However, if we think of transforming the random signal to a deterministic equivalent, then we can apply all of the concepts of linear systems theory to analyze and extract the required information. This is the approach we take in this text.

To accomplish this transformation, we must introduce two basic functions: the covariance and power-spectral density. Mathematically, if we define $x(t, \xi_i)$ or, more succinctly, $x(t)$ as a discrete random signal, then we can define the corresponding *covariance* function as being, under certain conditions,

$$R_{xx}(k) := E\{x(t)x(t + k)\} \qquad \text{for lag } k \qquad (1.3-2)$$

and the *power-spectral density* function as

$$S_{xx}(\Omega) := E\{X(\Omega)X^*(\Omega)\} \qquad (1.3-3)$$

where E is the expectation operator. We can heuristically think of the expectation as removing the randomness from the signal. Because of this expectation operation, these functions are *deterministic* and therefore can be considered the result of transforming the random signal. It will be shown in Chapter 4 that the covariance and power spectrum are a discrete-time Fourier transform pair. Therefore, we see that for random signals under certain conditions, the covariance replaces the signal and the power-spectral density replaces the spectrum. This transformation is depicted in Fig. 1.3–1.

Once this transformation is accomplished, then the concepts of linear systems theory are directly applicable. For instance, the response of a linear system excited by a deterministic input is characterized by convolution as

$$y(t) = h(t) * x(t)$$

whereas if $x(t)$ is random, then $y(t)$ is random and must be replaced by its corresponding covariance $R_{yy}(k)$. The convolutional relation in this case is given by

$$R_{yy}(k) = g(k) * R_{xx}(k) \qquad (1.3-4)$$

where the random excitation $x(t)$ has been replaced by its corresponding covariance and $g(k) = h(k) * h(-k)$. So we see that random signals can still be analyzed like deterministic signals when we are able to transform them according to the covariance-power spectrum relations. In fact, the first step in analyzing a random signal is to convert it (if possible) to a covariance and/or power-spectral den-

FIGURE 1.3–1
Deterministic to random transformations: signal to covariance—spectrum to power spectral density

sity function to extract the pertinent information. The problem of estimating the covariance and power spectrum of a random signal from noisy data is discussed in Chapters 5 and 6.

Once these relations are established, then concepts similar to filtering deterministic signals can be established. In fact, a filter designed for random signals is called an *estimation filter* because of its origins in statistical estimation theory. An estimation filter is usually designed to produce estimates in the past, present, or future through procedures called smoothing, filtering, and prediction, respectively. More precisely, the output of an estimation filter is the *signal estimate* $\hat{x}(T|t)$ which is the "best" estimate of the random signal $x(T)$ given the measurement data up to time t. Each of these conditions (see Fig. 1.3–2) can be expressed in terms of T:

1. For $T = t - \Delta$: The fixed-lag *smoothed* estimate $\hat{x}(t - \Delta|t)$ results
2. For $T = t$: The *filtered* estimate $\hat{x}(t|t)$ results
3. For $T = t + l$: The *l*-step *predicted* estimate $\hat{x}(t + l|t)$ results

Consider the output of the estimation filter designed to eliminate random noise from a transient signal. The estimator is a function of the signal $x(t)$ and noise $n(t)$:

$$\hat{y}(t|t) = f(x(t), n(t))$$

The noisy and processed data are shown in Fig. 1.3–3. Here we see how the filtered response has attenuated the random noise. We discuss the concepts of signal estimation using the modern parametric design methods in Chapter 7.

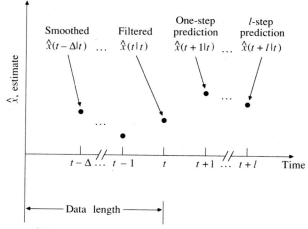

FIGURE 1.3–2
Signal smoothing, filtering, and prediction estimates

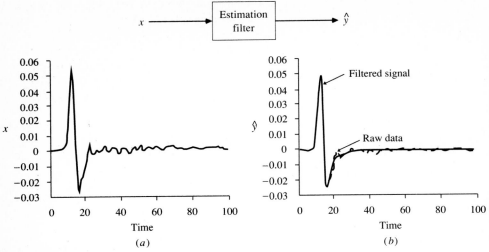

FIGURE 1.3–3
Signal estimator for noisy transient signal: (*a*) Noisy measurement (*b*) Signal estimate

If the estimation filter, or more commonly, the *estimator* employs a model of phenomenology or process under investigation, then it is a model-based processor. For example, suppose we use the so-called *Gauss-Markov* model (see Chapter 4) in our estimator design:

$$x(t) = Ax(t-1) + w(t-1)$$
$$y(t) = Cx(t) + v(t)$$

(1.3–5)

Then the resulting signal estimator,

$$\hat{x}(t|t) = A\hat{x}(t-1|t-1) + K(t)e(t)$$

(1.3–6)

is called a model-based signal processor—in this case, a *Kalman filter*. Model-based signal processing is discussed more fully in Chapter 9. Consider the following simple example of a model-based processor.

Example 1.3–1. Suppose we are asked to design a simple RC circuit, as shown in Fig. 1.3–4. The measurement is contaminated with random measurement noise, as well as with uncertainties in the values of the resistor and capacitor. Writing the Kirchoff node equations, we have

$$I_{\text{in}} - \frac{e}{R} - C\frac{de}{dt} = 0$$

and the measurement given by the voltmeter is

$$e_{\text{out}} = K_e e$$

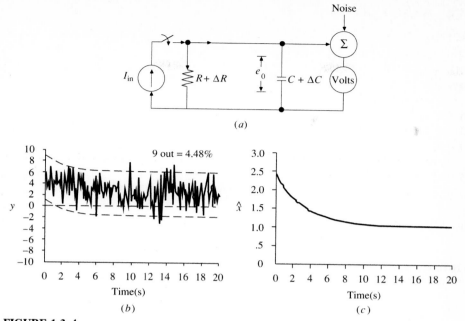

FIGURE 1.3–4
Model-based processor design (Kalman Filter) for an RC circuit (*a*) Circuit (*b*) Noisy voltage measurement (*c*) Estimated output voltage

Discretizing this equation using first differences and including white gaussian noise sources for the parameter and measurement uncertainties, we develop the following discrete Gauss-Markov model:

$$e(t) = \left(1 - \frac{\Delta T}{RC}\right) e(t - 1) + \frac{\Delta T}{C} I_{in}(t - 1) + w(t - 1)$$

$$e_{out}(t) = K_e e(t) + v(t)$$

The resulting estimator (Kalman filter) is given by

$$\hat{e}(t|t) = \left(1 - \frac{\Delta T}{RC}\right) \hat{e}(t - 1|t - 1) + K(t)e(t)$$

For $R = 3.3$ kΩ and $C = 1000$ μF, $\Delta T = 100$ ms, $K_e = 2$, the noisy measurement and corresponding model-based, random signal estimate is shown in Fig. 1.3–4, using SSPACK [5].

We shall see that random signals can be characterized by stochastic processes (Chapter 4), transformed to equivalent deterministic representations covariance and power spectrum (Chapters 5 and 6), and processed by parametric and model-based processors (Chapters 7, 8 and 9) in much the same way as a deterministic signal.

1.4   SUMMARY

In summary, discrete signals can be represented according to their unique characteristics. The signal can be deterministic or random, periodic or aperiodic, or transient. We introduced various representations of discrete signals, including the discrete Fourier series for periodic signals, the Z-transform, and the discrete-time Fourier transform for aperiodic signals. We then showed how these signals can be represented as outputs of linear systems and introduced various models. We presented the concept of frequency response of discrete signals. Finally, we introduced the notion of discrete random signals and showed how they can be processed similar to deterministic signals after the covariance–power-spectrum transformation has occurred. We also introduced the idea of a model-based processor and investigated a simple RC circuit example using a Kalman filter.

REFERENCES

1. A. Oppenheim and R. Shafer, *Digital Signal Processing* (Englewood Cliffs, N. J.: Prentice-Hall, 1975).
2. A. Oppenheim, A. Wilsky, and I. Young, *Signals and Systems* (Englewood Cliffs, N.J.: Prentice-Hall, 1983).
3. J. Candy, *Signal Processing: The Model-Based Approach* (New York: McGraw-Hill, 1986).
4. D. Lager and S. Azevedo, "SIG—A General Purpose Signal Processing Code," *Proc. IEEE,* 1987.
5. S. Azevedo, J. Candy, and D. Lager, "SSPACK—An Interactive Multi-channel, Model-Based Signal Processing Package," *Proc. IEEE Confr. Circuits Systems,* 1986.

SIG NOTES

SIG [4] can be used to simulate varous basic functions. For instance, the unit impulse can be simulated with the KRONECKER Command (SHARE), whereas the step and ramp are simulated with the CONSTANT and RAMP commands. Sinusoidal signals, including FM, can be simulated using the CHIRP (changing frequency) command. The discrete Fourier transform can be calculated using the FFT and IFFT commands and various forms of the DFT (e.g., exponential, sinusoidal) are selected by typing "MENU FSP" and selecting the desired representation. Linear systems are simulated with the LSS**** commands depending on the desired structure. SIG does *not* simulate state-space systems directly (see SSPACK [5]). SIG has a correlation/covariance estimator called CORRELATE and a suite of spectral estimators SD****.

EXERCISES

1.1. Find the (DFS) representation of the sinusoidal signal with Ω_0 an integer multiple of M

$$x(t) = A \sin(\Omega_0 t + \phi)$$

1.2. Derive the DFS representation of the periodic pulse train given by

$$x(t) = \begin{cases} 1 & 0 \le t \le 2N_1 \\ 0 & 2N_1 < t \le N \end{cases}$$

1.3. Suppose that N is very large for the pulse train signal of Exercise 1.2. Use limiting arguments ($\lim_{N\to\infty}(2\pi/N) \to \Omega$) to derive the DtFT representation.

1.4. Assume the pulse of Exercise 1.2 is aperiodic, that is,

$$x(t) = \begin{cases} 1 & 0 \le t \le 2N_1 \\ 0 & \text{elsewhere} \end{cases}$$

Calculate the corresponding Z-transform and compare this result to the DtFT of Exercise 1.3.

1.5. Derive the Z-transform of the sinusoidal signal

$$x(t) = A \sin 2\pi f_0 t \mu(t)$$

1.6. Calculate (numerically) the response of a linear time-invariant system (LTI) (see Eq. (1.2–8)) with unit impulse response

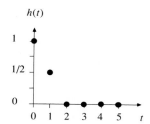

when excited by

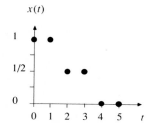

CHAPTER

2

DISCRETE
SIGNAL/SYSTEM
REPRESENTATIONS:
THE Z-TRANSFORM

In this chapter we briefly review the characterization of a discrete signal or system by its corresponding Z-transform. First, we define the Z-transform and its properties and show how it can be used to represent basic discrete signals. Next we extend this representation to describe a discrete linear system characterized by difference equations and show how to obtain solutions for those equations. We then develop the idea of frequency response and discuss the transformations between the continuous (S-plane) and discrete (Z-plane) which prove useful for designing digital filters.

2.1 Z-TRANSFORM

The Z-transform is the fundamental mechanism used to represent discrete signals and systems. The Z-transform of the signal $x(t)$ is defined by

$$X(z) := Z[x(t)] = \sum_{t=-\infty}^{\infty} x(t)z^{-t} \qquad \text{for } z \in \mathbf{C} \qquad (2.1\text{--}1)$$

and the inverse Z-transform

$$x(t) = Z^{-1}[X(z)] = \frac{1}{2\pi j} \oint_C X(z)z^{t-1}dz \qquad (2.1\text{--}2)$$

We see that the direct transform of $x(t)$ is an infinite power series in z^{-1} with coefficients $\{x(t)\}$, whereas the inverse transform is given by the contour integral which encircles the origin of the Z-plane. The basis of the Z-transform is the Laurent series of complex variable theory (see [1] for details). The Z-transform converges for those values of z that lie in the so-called *region of convergence* (ROC). In general, Eq. (2.1–1) will converge in an annular region of the Z-plane bounded by

$$R_- < |z| < R_+$$

Thus, we see that the Z-transform represents an analytic function at every point inside the ROC; therefore, the transform and all its derivatives must be continuous functions of $x(t)$ within the ROC.

Example 2.1–1. Suppose we have a discrete signal given by

$$x(t) = \alpha^t \mu(t)$$

and we would like to find its Z-transform. From the definition, we have

$$X(z) = Z[\alpha^t \mu(t)] = \sum_{t=-\infty}^{\infty} (\alpha^t \mu(t))z^{-t} = \sum_{t=0}^{\infty} (\alpha z^{-1})^t$$

Using properties of the geometric series,

$$\sum_{t=0}^{\infty} \beta^t = \frac{1}{1-\beta} \qquad \text{for } |\beta| < 1 \qquad (2.1\text{–}3)$$

we have, with $\beta = (\alpha z^{-1})$,

$$X(z) = \frac{1}{1-\alpha z^{-1}} \qquad \text{for } |\alpha z^{-1}| < 1$$

or equivalently $\qquad X(z) = \dfrac{z}{z-\alpha} \qquad \text{for } |z| > \alpha$

The ROC is shown in Fig. 2.1–1 as the region strictly greater than α. From this example we see that the properties of the signal determine the ROC. We shall discuss this point shortly. In Table 2.1–1 we list the Z-transform pairs of important discrete functions which can easily be derived in a similar manner.

An important class of Z-transforms are those for which $X(z)$ is a ratio of two polynomials, that is, a *rational function*. Recall that for $X(z)$ rational,

$$X(z) = \frac{B(z)}{A(z)} = \frac{\prod_{i=1}^{N_b}(z-z_i)}{\prod_{i=1}^{N_a}(z-p_i)} \qquad (2.1\text{–}4)$$

then the *zeros* of $X(z)$ are those roots of the numerator polynomial for which $X(z)$ is null:

$$Zero: \qquad X(z)\big|_{z=z_i} = 0 \qquad (2.1\text{–}5)$$

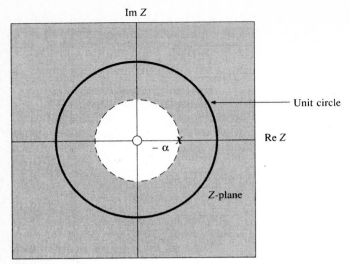

FIGURE 2.1–1
Z-transform region of convergence for $Z[\alpha^t \mu(t)]$

Similarly the *poles* of $X(z)$ are the roots of the denominator polynomial for which $X(z)$ is infinite, that is,

$$\text{Pole:} \quad X(z)\big|_{z=p_i} = \infty \qquad (2.1\text{–}6)$$

It can be shown [2] that a discrete linear time-invariant system is *stable* if and only if its poles lie within the unit circle. We saw in the previous example that $X(z)$ had a zero at $z = 0$ and pole at $z = \alpha$, as shown in Fig. 2.1–1.

With these representations in mind, let us review the properties of a signal that determine the corresponding ROC. We list those properties (see [2] for details) which determine the region of convergence of the Z-transform:

Property 1: The ROC of $X(z)$ consists of an *annular ring* in the Z-plane centered about the origin.

Property 2: The ROC does *not* contain any *poles*.

Property 3: If $x(t)$ is of *finite length*, then the ROC is the entire Z-plane except if $z = 0$ and/or $z = \infty$.

Property 4: If $x(t)$ is a *right-sided* signal ($x(t) = 0$, $t < t_0$) and if the circle $|z| = \rho$ is in the ROC, then all finite values of z for which $|z| > \rho$ will also be in the ROC.

Property 5: If $x(t)$ is a *left-sided* signal ($x(t) = 0$, $t > t_0$) and if the circle $|z| = \rho$ is in the ROC, then all values of z for which $0 < |z| < \rho$ will also be in the ROC.

Property 6: If $x(t)$ is a *two-sided* signal ($-\infty < t < \infty$) and if the circle $|z| = \rho$ is in the ROC, then the ROC consists of a ring in the Z-plane containing $|z| = \rho$.

TABLE 2.1–1
Common Z-transform pairs

Signal	Transform	ROC
$\delta(t)$	1	All z
$\mu(t)$	$\dfrac{1}{1 - z^{-1}}$	$|z| > 1$
$\delta(t - m)$	z^{-m}	All z except 0 for $m > 0$ ∞ for $m < 0$
$\alpha^t \mu(t)$	$\dfrac{1}{1 - \alpha z^{-1}}$	$|z| > \alpha$
$t\alpha^t \mu(t)$	$\dfrac{\alpha z^{-1}}{(1 - \alpha z^{-1})^2}$	$|z| > \alpha$
$[\cos \Omega_0 t]\mu(t)$	$\dfrac{1 - [\cos \Omega_0]z^{-1}}{1 - [2\cos \Omega_0]z^{-1} + z^{-2}}$	$|z| > 1$
$[\sin \Omega_0 t]\mu(t)$	$\dfrac{[\sin \Omega_0]z^{-1}}{1 - [2\cos \Omega_0]z^{-1} + z^{-2}}$	$|z| > 1$
$[r^t \cos \Omega_0 t]\mu(t)$	$\dfrac{1 - [r\cos \Omega_0]z^{-1}}{1 - [2r\cos \Omega_0]z^{-1} + r^2 z^{-2}}$	$|z| > r$
$[r^t \sin \Omega_0 t]\mu(t)$	$\dfrac{[r\sin \Omega_0]z^{-1}}{1 - [2r\cos \Omega_0]z^{-1} + r^2 z^{-2}}$	$|z| > r$

Property 1 is a restatement of the basic convergence property of the Z-transform (see [2] for proof), while Property 2 is a consequence of the fact that, by definition, the Z-transform at a pole is infinite and does not converge. Property 3 simply requires that $|x(t)|$ is bounded ($< \infty$) for the prescribed values of t, z can then assume all values except possibly at zero and infinity. Properties 4 through 6 all may be proven (see [2]) in the same manner: assume a circle of radius ρ which lies within the ROC, then for this value of z it is shown that the transform converges. Note that a right-sided signal is *causal* ($x(t) = 0$ for $t < 0$) and therefore the ROC is a region greater than a circle of radius ρ. The causal signal depicted in Fig. 2.1–1 illustrates this property. The properties of a two-sided signal become important when we analyze the covariance and power spectra of random signals in Chapter 4. So, it is important when listing the Z-transform of a signal to include the ROC as well.

The Z-transform possesses many important properties which make it useful for representing discrete-time signals and systems. The Z-transform is a linear operator; that is, the property of *linearity* holds.

If
$$x_1(t) \leftrightarrow X_1(z) \qquad \text{ROC} = R_1$$
$$x_2(t) \leftrightarrow X_2(z) \qquad \text{ROC} = R_2$$

then
$$Z[ax_1(t) + bx_2(t)] = aZ[x_1(t)] + bZ[x_2(t)]$$

or
$$Z[ax_1(t) + bx_2(t)] \leftrightarrow aX_1(z) + bX_2(z) \qquad \text{ROC contains } R_1 \cap R_2$$

The *time-shift* property which is important for solving difference equations is given by

$$x(t) \leftrightarrow X(z) \qquad \text{ROC} = R_x$$

Then

$$Z[x(t - t_0)] = \sum_{t=-\infty}^{\infty} x(t - t_0)z^{-t} = z^{-t_0} \sum_{t=-\infty}^{\infty} x(t - t_0)z^{-(t-t_0)} = z^{-t_0}X(z)$$

and the ROC $= R_x$, except for possibly at the origin or infinity.

The *frequency scaling* properties all take the same basic form:

$$x(t) \leftrightarrow X(z) \qquad\qquad\qquad \text{ROC} = R_x$$

$$Z[a^t x(t)] = \sum_{t=-\infty}^{\infty} x(t)(a^{-1}z)^{-t} = X(a^{-1}z) \qquad \text{ROC} = aR_x$$

whereas the *time reversal* property is easily shown to be

$$Z[x(-t)] = \sum_{t=-\infty}^{\infty} x(-t)z^{-t} = \sum_{\beta=-\infty}^{\infty} x(\beta)(z^{-1})^{-\beta} = X(z^{-1}) \qquad \text{ROC} = \frac{1}{R_x}$$

In performing operations on discrete-time systems it is crucial that the *convolution* property holds:

$$Z\left[x_1(t) * x_2(t)\right] = Z\left[\sum_{k=-\infty}^{\infty} x_1(k)x_2(t - k)\right] = \sum_{t=-\infty}^{\infty} \left(\sum_{k=-\infty}^{\infty} x_1(k)x_2(t - k)\right)z^{-t}$$

Reversing the order of summation and performing a change of variable $m = t - k$, we have

$$Z[x_1(t) * x_2(t)] = \sum_{k=-\infty}^{\infty} x_1(k)\left[\sum_{m=-\infty}^{\infty} x_2(m)z^{-m}\right]z^{-k} = X_1(z)X_2(z)$$

$$\text{ROC contains } R_1 \cap R_2$$

Its inverse property, or complex *convolution* in the Z-plane, leads to multiplication in the time domain (see [2] for details):

$$Z[x_1(t)x_2(t)] = \frac{1}{2\pi j} X_1(z) * X_2(z) \qquad \text{ROC} = R_{x_1}R_{x_2}$$

Derivations of other properties follow directly as a consequence of the previously stated properties and definition of the Z-transform. For instance, a periodic signal is a consequence of the time-shifting property; that is, by defining $x_M(t)$ as

$$x_M(t) = \begin{cases} x(t) & 0 \leq t \leq M - 1 \\ 0 & \text{elsewhere} \end{cases}$$

then $\qquad Z[x(t + M)] = Z[x_M(t) + x_M(t - M) + x_M(t - 2M) + \ldots]$

From the linearity and time-shift properties we have

$$Z[x(t + M)] = X_M(z) + z^{-M}X_M(z) + \ldots = X_M(z)(1 + z^{-M} + z^{-2M} + \ldots)$$

and therefore

$$Z[x(t + M)] = \frac{1}{1 - z^{-M}}X_M(z)$$

We summarize these and other important properties of the Z-transform in Table 2.1–2.

In order for the Z-transform to be useful we must be able to use it to solve discrete-time problems. The inverse Z-transform plays a key role, just as in the continuous case using Laplace transforms. The inverse transform is given by the contour integral of Eq. (2.1–2) where C is a closed contour in the Z-plane encircling the origin and lying within the ROC. Three popular techniques of inverting rational transforms are the power-series, the partial-fraction, and the inversion-integral methods (see [3] for more details).

If the Z-transform is available in power-series form, $x(t)$ is the coefficient of the term involving z^{-t} in the series

$$X(z) = \sum_{t=-\infty}^{\infty} x(t)z^{-t}$$

TABLE 2.1–2
Properties of the Z-transform

Property	Signal	Transform
Linearity	$ax_1(t) + bx_2(t)$	$aX_1(z) + bX_2(z)$
Time shift	$x(t - t_0)$	$z^{-t_0}X(z)$
Frequency scale	$e^{j\Omega_0 t}x(t)$	$X(e^{-j\Omega_0}z)$
	$z_0^t x(t)$	$X(z/z_0)$
	$a^t x(t)$	$X(a^{-1}z)$
Time reversal	$x(-t)$	$X(z^{-1})$
Convolution	$x_1(t) * x_2(t)$	$X_1(z)X_2(z)$
Multiplication	$x_1(t)x_2(t)$	$\dfrac{1}{2\pi j}\oint_C X_1(\alpha)X_2(z/\alpha)\alpha^{-1}d\alpha$
Derivative	$tx(t)$	$-z\dfrac{dX(z)}{dz}$
Finite sum	$\sum_{t=-\infty}^{n} x(t)$	$\dfrac{1}{1 - z^{-1}}X(z)$
Periodic	$x(t + M)$	$\dfrac{1}{1 - z^{-M}}X_M(z)$ (for $X_M(z)$ over one period)
Initial value	$\lim_{t\to 0} x(t)$	$\lim_{z\to\infty} X(z)$
Final value	$\lim_{t\to\infty} x(t)$	$\lim_{z\to 1}[1 - z^{-1}X(z)]$
Parseval's theorem	$\sum_{t=-\infty}^{\infty} \lvert x(t)\rvert^2$	$\dfrac{1}{2\pi j}\oint X(\alpha)X^*(1/\alpha^*)\alpha^{-1}d\alpha$

The *power-series method* first obtains the Z-transform as an infinite series in z^{-t} and then seeks to find a closed-form representation using summation relations (e.g., geometric series). If $X(z)$ is given as a ratio of two polynomials, then the power-series method can be derived by long division.

For rational Z-transforms, the *partial-fraction method* is obtained by converting the polynomials to the partial-fraction form:†

$$X(z) = \frac{N(z)}{D(z)} = \sum_{i=1}^{N_d} \frac{R_i}{z - p_i} \tag{2.1-7}$$

where p_i is a pole or root of the denominator polynomial $D(z)$ and R_i is a residue at the pole, given by

$$R_i = (z - p_i)X(z)\big|_{z = p_i}$$

Finally, the so-called *inversion-integral method*, which is based on the residue theorem of complex variable theory [1], states that the inverse transform given by the inversion integral is equal to the sum of the residues at poles inside the closed contour. That is,

$$x(t) = \frac{1}{2\pi j} \oint_c X(z)z^{t-1}dz = \sum_i [\text{Residues of } X(z)z^{t-1} \text{at } z = p_i] \tag{2.1-8}$$

where for simple poles we have

$$\text{Res } [X(z)z^{t-1} \text{ at } z = p_i] = (z - p_i)X(z)z^{t-1}\big|_{z=p_i}$$

The following example demonstrates these principles.

Example 2.1–2. Suppose we have a discrete signal transform $X(z)$ given by

$$X(z) = \frac{1}{(1 - z^{-1})(1 - \frac{1}{2}z^{-1})} \qquad |z| > 1$$

The poles of $X(z)$ are at $p_1 = 1$ and $p_2 = 1/2$. Now we find $x(t)$ using each of the methods discussed above:

(i) Power-series method:

$$
1 - \tfrac{3}{2}z^{-1} + \tfrac{1}{2}z^{-2} \overline{\big)\, 1 \qquad\qquad 1 + \tfrac{3}{2}z^{-1} + \tfrac{7}{4}z^{-2} + \ldots}
$$

$$
\begin{array}{r}
1 - \tfrac{3}{2}z^{-1} + \tfrac{1}{2}z^{-2} \\ \hline
\tfrac{3}{2}z^{-1} - \tfrac{1}{2}z^{-2} \\
\tfrac{3}{2}z^{-1} - \tfrac{9}{4}z^{-2} + \tfrac{3}{4}z^{-3} \\ \hline
\tfrac{7}{4}z^{-2} - \tfrac{3}{4}z^{-3}
\end{array}
$$

or $X(z) = \sum_{t=0}^{\infty} x(t)z^{-t} \Rightarrow x(0) = 1, \ x(1) = 3/2, x(2) = 7/4, \ldots$

---

†We tacitly assume only simple (no multiplicity) poles; see [2] and [4] for multiple poles.

(ii) Partial-fraction method:

$$X(z) = \frac{R_1}{1 - z^{-1}} + \frac{R_2}{1 - \frac{1}{2}z^{-1}}$$

$$R_1 = (1 - z^{-1})X(z)\Big|_{z=1} = \frac{1}{1 - \frac{1}{2}z^{-1}}\Big|_{z=1} = 2$$

$$R_2 = \left(1 - \frac{1}{2}z^{-1}\right)X(z)\Big|_{z=1/2} = \frac{1}{1 - z^{-1}}\Big|_{z=1/2} = -1$$

or

$$X(z) = \frac{2}{1 - z^{-1}} - \frac{1}{1 - \frac{1}{2}z^{-1}}$$

Taking inverse transforms, we have

$$x(t) = 2\mu(t) - (1/2)^t \qquad t \geq 0$$

and therefore

$$x(0) = 2 - 1 = 1$$

$$x(1) = 2 - 1/2 = 3/2$$

$$x(2) = 2 - 1/4 = 7/4$$

$$\vdots$$

(iii) Inversion-integral method:

$$x(t) = \text{Res}\,[X(z)z^{t-1}, p_1 = 1] + \text{Res}\,[X(z)z^{t-1}, p_2 = 1/2]$$

where

$$X(z)z^{t-1} = \frac{z^{t-1}}{(1 - z^{-1})(1 - \frac{1}{2}z^{-1})} = \frac{z^{t+1}}{(z - 1)(z - 1/2)} \qquad \text{for } t \geq -1$$

$$\text{Res}\,[X(z)z^{t-1}, p_1 = 1] = (z - 1)X(z)z^{t-1}\Big|_{z=1} = \frac{z^{t+1}}{z - 1/2}\Big|_{z=1} = \frac{1}{1/2} = 2$$

$$\text{Res}\,[X(z)z^{t-1}, p_2 = 1/2] = (z - 1/2)X(z)z^{t-1}\Big|_{z=1/2} = -(1/2)^t$$

or

$$x(t) = 2 - (1/2)^t \qquad \text{for } t \geq -1$$

It can be shown that for $t < -1$, $x(t) = 0$.

2.2 DISCRETE LINEAR SYSTEMS

In the previous chapter we defined a discrete signal as an ordered set of values. We showed how it can be represented or transformed to its equivalent representation in the frequency domain to extract spectral information. Through the use of inverse transformations we are able to recover the original signal as well. In this section we develop the concept of a discrete-time system analogous to that of a signal; that is, we define a *discrete-time system* as a transformation of a discrete-time input signal to a discrete-time output signal. We are primarily interested in a *linear discrete-time system*, which implies that the system "transformation" is constrained to be linear; that is, the properties of superposition (homogeneity and additivity) hold. A discrete linear system is usually defined in terms of its (unit) *impulse response*, $h(\cdot)$, or the output of a system when its input is a unit impulse:

$$y(t) = T[x(t)] = \sum_k h(k, t)x(t) \qquad (2.2\text{--}1)$$

such that for $x(t) = ax_1(t) + bx_2(t)$

$$y(t) = aT[x_1(t)] + bT[x_2(t)] = a \sum_k h(k, t)x_1(t) + b \sum_k h(k, t)x_2(t) = ay_1(t) + by_2(t)$$

With this in mind, we can define further properties of discrete systems. If we constrain the system to be *time-invariant*, then for any time shift of the input we must have a corresponding shift of the output; that is, $x(t - m) \rightarrow y(t - m)$. A *linear-time invariant* (LTI) discrete system can be completely characterized by its impulse response as

$$y(t) = h(t) * x(t): = \sum_k h(k)x(t - k) = \sum_k x(k)h(t - k)$$

where $*$ is the convolution operator.

If we constrain the impulse response to be null for negative values of k, then the LTI system is said to be *causal*:

$$y(t) = \sum_{k=0}^{\infty} h(k)x(t - k) \qquad (2.2\text{--}2)$$

We also mention that the LTI system is *stable* if, for a bounded input, its output is bounded. This requires that $\sum_t |h(t)| < \infty \; \forall t$ (see [2,3] for details).

Discrete LTI systems are powerful representations when coupled with the convolution relation of Eq. (2.2–2) which enables us to determine the response of the system to *any* input given the impulse response. This is best illustrated by the following example.

Example 2.2–1. Suppose we have a causal discrete LTI system with unit impulse response

$$h(t) = a^t + b^t \qquad \text{for } a, b < 1$$

and we would like to determine its response to a unit step. From Eq. (2.2–2) we have

$$y_{\text{STEP}}(t) = y(t) = \sum_{k=0}^{\infty} h(k)\mu(t - k)$$

$$y(0) = h(0) = (a^0 + b^0)$$

$$y(1) = h(1) + h(0) = (a^1 + b^1) + (a^0 + b^0)$$

$$y(2) = h(2) + h(1) + h(0) = (a^2 + b^2) + (a^1 + b^1) + (a^0 + b^0)$$

$$\vdots$$

$$y(N) = \sum_{k=0}^{N} a^k + b^k$$

We can calculate the response of the discrete LTI system for any response if we know $h(t)$. Using SIG [5] we can simulate a discrete LTI system if we let $a = 1/2$ and $b = 1/4$; then

$$h(t) = \left(\frac{1}{2}\right)^t + \left(\frac{1}{4}\right)^t$$

and for a step input we see that

$$y(t) = \sum_{k=0}^{t} \left(\frac{1}{2}\right)^k + \left(\frac{1}{4}\right)^k$$

The results of this simulation are shown in Fig. 2.2–1.

An important class of discrete LTI systems is that for which the input and output are related through linear constant-coefficient difference equations. We define an N_a-th order linear constant-coefficient *difference equation* by

$$y(t) + a_1 y(t-1) + \cdots + a_{N_a} y(t - N_a) = b_0 x(t) + b_1 x(t-1) + \cdots + b_{N_b} x(t - N_b)$$

or $$\sum_{i=0}^{N_a} a_i y(t - i) = \sum_{i=0}^{N_b} b_i x(t - i) \qquad \text{for } N_b \le N_a, \, a_0 = 1 \qquad (2.2\text{–}3)$$

We can also write this equation in polynomial operator form. If we define q^{-k}, the *backward shift operator*, as

$$q^{-k} y(t) = y(t - k)$$

then we can write Eq. (2.2–3) as

$$A(q^{-1}) y(t) = B(q^{-1}) x(t) \qquad (2.2\text{–}4)$$

where $$A(q^{-1}) = 1 + a_1 q^{-1} + \cdots + a_{N_a} q^{-N_a}$$

$$B(q^{-1}) = b_0 + b_1 q^{-1} + b_2 q^{-2} + \cdots + b_{N_b} q^{-N_b}$$

A discrete system characterized by this form of the difference equation is called *recursive* because its present output $y(t)$ is determined from past outputs $\{y(t - i)\}$ and past inputs $\{x(t - i)\}$. This representation can be extended to the random or stochastic case, leading to the ARMAX model (see Chapter 4). This form is also called infinite-impulse response (IIR) because the impulse response consists of an infinite number of coefficients. Similarly, when $A(q^{-1}) = 1$, Eq. (2.2–4) becomes

$$y(t) = B(q^{-1}) x(t) = \sum_{i=0}^{N_b} b_i x(t - i) \qquad (2.2\text{–}5)$$

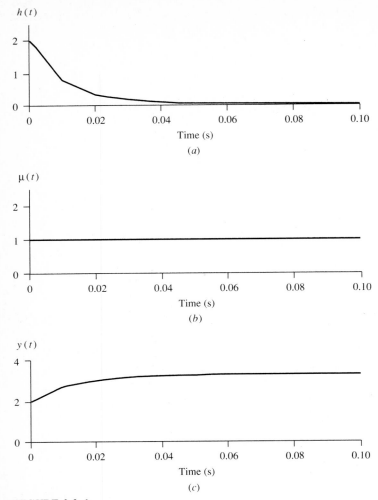

FIGURE 2.2–1
Discrete LTI system simulation (*a*) Impulse response (*b*) Step excitation (*c*) System output or step response

which is *nonrecursive* because its present output $y(t)$ depends only on past inputs $\{x(t - i)\}$. If we identify this relation with Eq. (2.2–2), we note that this is a finite sum of impulse-response weights ($h(i) \rightarrow b_i$). Thus, this form is called *finite-impulse response* (FIR) as well. We will call systems characterized by linear-constant coefficient difference equations *parametric* and those characterized by impulse-response weights *nonparametric*.

The Z-transform can be used to replace a difference equation by polynomials in the complex variable z, which will eventually lead to algebraic solutions. Using the time-shifting property of Table 2.1–2 we can represent the recursive difference equation of Eq. (2.2–3) as

$$(1 + a_1 z^{-1} + \cdots + a_{N_a} z^{-N_a})Y(z) = (b_0 + b_1 z^{-1} + \cdots + b_{N_b} z^{-N_b})X(z)$$

by taking Z-transforms. We can also represent the system under investigation in rational form by

$$H(z) = \frac{Y(z)}{X(z)} = \frac{B(z)}{A(z)} = \frac{\sum_{i=0}^{N_b} b_i z^{-i}}{1 + \sum_{i=1}^{N_a} a_i z^{-i}} = \frac{\prod_{i=1}^{N_b}(z - z_i)}{\prod_{i=1}^{N_a}(z - p_i)} \qquad (2.2\text{--}6)$$

We define the *transfer function* of the system as the rational function

$$H(z) = \left.\frac{Y(z)}{X(z)}\right|_{IC=0} = \frac{B(z)}{A(z)} = \sum_{t=0}^{\infty} h(t)z^{-t} \qquad (2.2\text{--}7)$$

resulting from the transformation of the recursive difference equation with *zero initial conditions*. We now see more clearly why a rational system is termed infinite-impulse response. Similarly, a finite-impulse response system defined by

$$H(z) = B(z) = z^{-N_b}(b_0 z^{N_b} + \cdots + b_{N_b}) = \frac{\prod_{i=1}^{N_b}(z - z_i)}{z^{N_b}} \qquad (2.2\text{--}8)$$

has N_b-poles at the origin and is therefore always stable (with poles inside the unit circle).

Discrete LTI systems thus can be characterized nonparametrically through the convolution sum or parametrically through difference equations.

2.3   Z-TRANSFORM METHOD OF SOLVING DIFFERENCE EQUATIONS

Now that we have specified both the Z-transform representation of discrete LTI systems and various techniques which can be used to invert them, we are ready to examine the solution of difference equations. The Z-transform method of solving difference equations is analogous to the Laplace transform method of solving differential equations. That is, the Z-transform converts the difference equation to a rational representation; then we can employ any of the inverse Z-transform methods discussed previously to obtain a solution.

Recall that a discrete LTI system can be described by the constant-coefficient difference equation as

$$\sum_{i=0}^{N_a} a_i y(t - i) = \sum_{i=0}^{N_b} b_i x(t - i) \qquad N_a \geq N_b, a_0 = 1 \qquad (2.3\text{--}1)$$

If we constrain the signals of interest to be causal, then the *one-sided* Z-transform must be used with the corresponding time-shift property:

$$Z[x(t - m)] = z^{-m}X(z) + \sum_{k=1}^{m} x(-k)z^{k-m} \qquad (2.3\text{--}2)$$

Taking the Z-transform of the difference equation and using this property, we have

$$\sum_{i=0}^{N_a} a_i \left(z^{-i} Y(z) + \sum_{k=1}^{i} y(-k)z^{k-i} \right) = \sum_{i=0}^{N_b} b_i \left(z^{-i} X(z) + \sum_{k=1}^{i} x(-k)z^{k-i} \right)$$

or

$$A(z)Y(z) + \sum_{i=0}^{N_a} \sum_{k=1}^{i} a_i z^{k-i} y(-k) = B(z)X(z) + \sum_{i=0}^{N_b} \sum_{k=1}^{i} b_i z^{k-i} x(-k) \quad (2.3\text{--}3)$$

which shows that for zero initial conditions the transfer function is obtained as before

$$H(z) = \frac{Y(z)}{X(z)} = \frac{B(z)}{A(z)} \quad (2.3\text{--}4)$$

Thus, using the one-sided Z-transform relation we can solve difference equations quite easily.

Example 2.3–1. Reconsider the simple RC circuit of Example 1.3–1. As before, after applying first differences to the Kirchoff node equations, we have the difference equation (ignoring the noise and measurement system).

$$e(t) = \left(1 - \frac{\Delta T}{RC} \right) e(t-1) + \frac{\Delta T}{C} I_{in}(t-1)$$

For $R = 3.3 \text{ k}\Omega$, $C = 1000 \text{ }\mu\text{F}$, $\Delta T = 0.1$ sec, we have

$$e(t) = 0.97e(t-1) + 100 I_{in}(t-1)$$

Taking Z-transforms using Eq. (2.3–2) we have

$$E(z) = 0.97 \left[z^{-1}E(z) + e(-1) \right] + 100 \left[z^{-1}I_{in}(z) + I_{in}(-1) \right]$$

$$(1 - 0.97z^{-1})E(z) = 100z^{-1}I_{in}(z) + 0.97e(-1) + 100I_{in}(-1)$$

or

$$E(z) = \frac{100z^{-1}}{1 - 0.97z^{-1}} I_{in}(z) + \frac{0.97e(-1) + 100I_{in}(-1)}{1 - 0.97z^{-1}}$$

The capacitor is initially charged at 2.5 V, and the current is a step of 300 μA with $I_{in}(-1) = 0$. Thus, we have

$$I_{in}(z) = \frac{0.0003}{1 - z^{-1}}$$

and therefore

$$E(z) = \frac{100z^{-1}}{1 - 0.97z^{-1}} \left(\frac{0.0003}{1 - z^{-1}} \right) + \left(\frac{0.97(2.5)}{1 - 0.97z^{-1}} \right)$$

$$= \frac{0.03z^{-1}}{(1 - 0.97z^{-1})(1 - z^{-1})} + \frac{2.425}{1 - 0.97z^{-1}}$$

Using partial fractions, we obtain

$$E(z) = -\frac{1}{1 - 0.97z^{-1}} + \frac{1}{1 - z^{-1}} + \frac{2.425}{1 - 0.97z^{-1}}$$

$$e(t) = 1.425(0.97)^t + \mu(t)$$

The voltage response is shown in Fig. 1.3–4.

We see that the Z-transform method of solving difference equations is analogous to the Laplace transform method; that is, both techniques first transform equations to polynomial representations in the corresponding complex variable and then invert them to obtain a closed-form solution.

2.4 FREQUENCY RESPONSE OF DISCRETE SIGNALS AND SYSTEMS

In this section, we discuss the idea of spectral representations of both discrete signals and systems. The spectral or frequency-domain representation of a signal or system enables the signal processor to quickly extract the frequency information available in the temporal signal. It is crucial in the design and analysis of processing techniques.

Discrete frequency response is based on the discrete-time Fourier representation given by the following definitions. If the ROC of the Z-transform contains the unit circle ($|z| = 1$), then we can define the discrete-time Fourier transform (DtFT) by substituting $z = e^{j\Omega}$ in the Z-transform pair of Eqs. (2.1–1,2). That is, the DtFT is given by

$$X(\Omega) := \text{DtFT}[x(t)] = \sum_{t=-\infty}^{\infty} x(t)e^{-j\Omega t} \tag{2.4–1}$$

Restricting the contour of integration to the unit circle with $dz = je^{j\Omega}d\Omega$, we obtain the *inverse discrete-time Fourier transform* (IDtFT) as

$$x(t) = \text{IDtFT}[X(\Omega)] = \frac{1}{2\pi} \int_{2\pi} X(\Omega)e^{j\Omega}d\Omega \tag{2.4–2}$$

An important feature of the DtFT is that it is periodic with period 2π, which follows intuitively, since each revolution of the unit circle leads to a periodic repetition of the transform. We will discuss the DtFT in more detail in the next chapter.

The notion of the frequency response of a discrete signal evolves from linear-systems theory and is generated in a manner similar to the generation of continuous-time signals. Recall in the continuous case we let $s = j\omega$ and vary ω along the $j\omega$-axis, calculating the corresponding magnitude and phase. Similarly, for a discrete case we let $z = e^{j\Omega}$ and vary Ω along the unit circle; that is, the frequency response of a discrete signal is given by

$$X(z)\big|_{z=e^{j\Omega}} = \left|X(e^{j\Omega})\right| \angle X(e^{j\Omega}) \qquad (2.4\text{--}3)$$

A typical frequency response mapping is depicted in Fig. 2.4–1. Here we see that each selected value of the frequency Ω_i corresponds to a unique value on the unit circle. We also note that the discrete-time frequency response is periodic as shown in the figure. Each period of the spectrum corresponds to a revolution around the unit circle. In reality, the frequency response is generated from the continuum of values of Ω that lie on the unit circle, so in fact, the discrete frequency response is continuous (shown by the solid line in the figure). Mathematically, we know this is true because we are calculating the DtFT when $z = e^{j\Omega}$. We also note that selecting discrete values of frequency ($\Omega = \Omega_i$) corresponds to the discrete Fourier transform (DFT) calculation. We will discuss these representations and transforms more fully in Chapter 3.

The frequency response of discrete-time systems is also important for purposes of both analysis and synthesis. The frequency response of a discrete LTI system can be determined from its impulse-response (nonparametric) or rational transfer function (parametric) representation.

The nonparametric impulse response is simply transformed using the DtFT, that is,

$$\text{DtFT } [h(t)] \rightarrow H(e^{j\Omega})$$

where H is a function of the complex variable

$$H(e^{j\Omega}) = \text{Re } H(e^{j\Omega}) + j \text{ Im } H(e^{j\Omega}) = \left|H(e^{j\Omega})\right| \angle H(e^{j\Omega}) \qquad (2.4\text{--}4)$$

The magnitude and phase are determined in the usual manner:

$$\left|H(e^{j\Omega})\right| = \sqrt{[\text{Re } H(e^{j\Omega})]^2 + [\text{Im } H(e^{j\Omega})]^2} \qquad (2.4\text{--}5)$$

$$\angle H(e^{j\Omega}) = \arctan\left(\frac{\text{Im } H(e^{j\Omega})}{\text{Re } H(e^{j\Omega})}\right) \qquad (2.4\text{--}6)$$

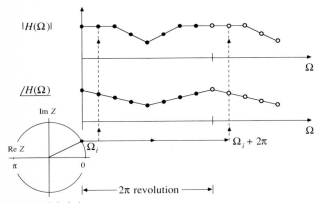

FIGURE 2.4–1
Frequency response of a discrete signal mapped by selecting discrete frequencies on the unit circle

So we see that the frequency response of a discrete LTI system can be determined directly from its impulse response, using the DtFT.

If the system is represented by a rational transfer function in the Z-domain,

$$H(z) = \frac{B(z)}{A(z)} \tag{2.4-7}$$

then the frequency response of the system is given by

$$H(e^{j\Omega}) = H(z)\big|_{z=e^{j\Omega}} = \frac{B(e^{j\Omega})}{A(e^{j\Omega})} = \left|\frac{B(e^{j\Omega})}{A(e^{j\Omega})}\right| \angle \frac{B(e^{j\Omega})}{A(e^{j\Omega})} \tag{2.4-8}$$

It is important to note that the frequency response of a discrete LTI system is periodic because of the periodic property of the discrete-time exponential (see Eq. (1.1–13)):

$$H(e^{j\Omega}) = H(e^{j(\Omega+2\pi)}) \tag{2.4-9}$$

This follows immediately from the DtFT. The following example illustrates how the frequency response may be determined for a discrete LTI system described by difference equations.

Example 2.4–1. Suppose we would like to analyze the performance of a simple two-point averager; that is,

$$y(t) = \frac{1}{2} \sum_{i=0}^{1} x(t-i) = \frac{1}{2}x(t) + \frac{1}{2}x(t-1)$$

To calculate the frequency response for this system, we first determine the transfer function using Z-transforms,

$$Y(z) = \frac{1}{2}X(z) + \frac{1}{2}z^{-1}X(z) = \frac{1}{2}(1 + z^{-1})X(z)$$

or

$$H(z) = \frac{Y(z)}{X(z)} = \frac{1}{2}(1 + z^{-1})$$

Next we obtain the DtFT by substituting $z = e^{j\Omega}$:

$$H(e^{j\Omega}) = \frac{1}{2}(1 + e^{-j\Omega}) = \frac{1}{2}[(1 + \cos\Omega) - j\sin\Omega]$$

The squared magnitude is given by

$$|H(e^{j\Omega})|^2 = \frac{1}{4}(1 + 2\cos\Omega + \cos^2\Omega + \sin^2\Omega) = \frac{1}{2}(1 + \cos\Omega)$$

and the phase is

$$\angle H(e^{j\Omega}) = \arctan\left(\frac{-\sin\Omega}{1 + \cos\Omega}\right) = -\frac{\Omega}{2}$$

Using SIG [5], we simulate the system; the results are shown in Fig. 2.4–2. Here we show the impulse response as well as both magnitude and phase spectra.

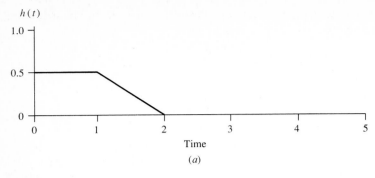

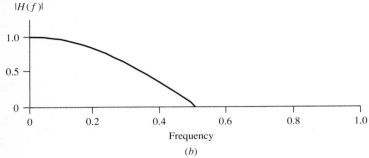

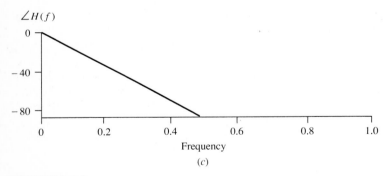

FIGURE 2.4–2
Frequency response of two-point averager: (*a*) Impulse response (*b*) Magnitude spectrum (*c*) Phase spectrum

Just as in the continuous-time case, the frequency response of a discrete LTI system is determined by its poles and zeros. Recall that the rational transfer function can be expressed in terms of its poles and zeros, or in *pole-zero form* as

$$H(z) = \frac{B(z)}{A(z)} = \frac{K\prod_{i=1}^{N_b}(z - z_i)}{\prod_{i=1}^{N_a}(z - p_i)} \tag{2.4–10}$$

where z_i are the zeros and p_i the poles of $H(z)$.

If we have the transfer function in pole-zero form, then the frequency response is easily determined by

$$|H(z)| = \frac{|K|\prod_{i=1}^{N_b}|z - z_i|}{\prod_{i=1}^{N_a}|z - p_i|} = \frac{|K||z - z_1|\cdots|z - z_{N_b}|}{|z - p_1|\cdots|z - p_{N_a}|} \qquad (2.4\text{--}11)$$

and
$$\angle H(z) = \sum_{i=1}^{N_b}\arctan\left(\frac{\operatorname{Im} z_i}{\operatorname{Re} z_i}\right) - \sum_{i=1}^{N_a}\arctan\left(\frac{\operatorname{Im} p_i}{\operatorname{Re} p_i}\right) \qquad (2.4\text{--}12)$$

The location of poles and zeros in the Z-plane can be used to predict the behavior of the discrete LTI system under investigation (see [2] for more details). For instance, if the system is stable then all of its poles must lie within the unit circle.

2.5 DISCRETE REPRESENTATIONS OF CONTINUOUS SIGNALS

Even though we live in a digital world created by advances in microchips and computers, most physical phenomenology is characterized by continuous-time or *analog* signals. The majority of information that the signal processor extracts occurs in natural signals and is continuous; therefore, we must understand the underlying theory that relates the continuous-time signal and domain to their corresponding discrete counterparts. This theory is culminated in the *sampling theorem* which tells us how fast we must sample an analog (continuous) signal in order to be able to reconstruct it exactly from its samples (discrete signal).

Although much effort has concentrated on continuous-time signals and systems, we must be able to transcend both domains with relative ease. The relation between the two domains is depicted in Fig. 2.5–1. First, for continuous signals we go from the Laplace-to-Fourier domain using the $j\omega$-axis through $s = j\omega$, whereas for discrete signals we use the unit circle through $z = e^{j\Omega}$ to go from the Z-to-discrete Fourier domain. The key to representing continuous signals in terms of discrete signals lies in the relations between the Laplace-to-Z or, equivalently, the mapping from the S $\rightarrow$ Z planes. If we use the corresponding

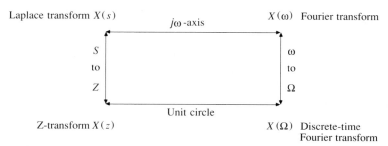

FIGURE 2.5–1
Continuous and discrete signal transform domains

Fourier transforms, then the key lies in the mapping of $\omega \rightarrow \Omega$. Next, let us examine the sampling process and see how these transforms are employed.

We can represent the sampling process mathematically by first defining the sampled signal with sampling interval ΔT as

$$x_s(\tau) = x_c(\tau)\big|_{\tau = k\Delta T} \qquad (2.5\text{--}1)$$

If we assume that we have a perfect impulse sampler,

$$x_s(\tau) = x_c(\tau)p(\tau) \qquad (2.5\text{--}2)$$

where the sampling function is given by

$$p(\tau) = \sum_{k=-\infty}^{\infty} \delta(\tau - k\Delta T)$$

then from continuous Fourier transform theory,[†] we have the convolution relation

$$X_s(\omega) = \frac{1}{2\pi}\big[X(\omega) * P(\omega)\big] \qquad (2.5\text{--}3)$$

Since the Fourier transform of a periodic impulse train is also a periodic impulse train with *sampling frequency* ω_s, where

$$P(\omega) = \frac{2\pi}{\Delta T} \sum_{k=-\infty}^{\infty} \delta(\omega - k\omega_s) \qquad (2.5\text{--}4)$$

then the sampling operation in the frequency domain becomes

$$X_s(\omega) = \frac{1}{2\pi}\big[X(\omega) * P(\omega)\big] = \frac{1}{\Delta T} \sum_{k=-\infty}^{\infty} X(\omega - k\omega_s) \qquad (2.5\text{--}5)$$

Thus, $X_s(\omega)$ is a periodic function of shifted replicas of the continuous spectrum, $X(\omega)$ scaled by $1/\Delta T$. If we assume that $x_c(\tau)$ is bandlimited, that is,

$$X_c(\omega) = \begin{cases} 1 & \omega \leq \omega_B \\ 0 & \omega > \omega_B \end{cases} \qquad (2.5\text{--}6)$$

then various situations result, based on the relationship between the values of ω_s and ω_B. Thus, for ω assuming values such that

$$\omega = \begin{cases} \omega_s > 2\omega_B & (\text{or} \quad \omega_B < \omega_s - \omega_B) \\ \omega_s < 2\omega_B & (\text{or} \quad \omega_B > \omega_s - \omega_B) \end{cases} \qquad (2.5\text{--}7)$$

[†]Recall that the *continuous-time Fourier transform* of $x(\tau)$ is defined by

$$X(\omega) := \int_{-\infty}^{\infty} x(\tau)e^{-j\omega\tau}d\tau$$

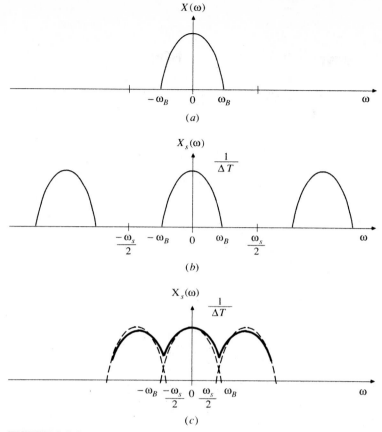

FIGURE 2.5–2
Frequency-domain representation of sampled continuous signals (*a*) Continuous signal spectrum (*b*) Correctly sampled ($\omega_s > 2\omega_B$) signal spectrum (*c*) Undersampled or aliased spectrum ($\omega_s < 2\omega_B$)

we see from Fig. 2.5–2 that the first-case results do not overlap in any frequency between the sampled replicas, whereas in the second there is frequency overlap; that is, the terms in Eq. (2.5–5) overlap. In the first case, if we low-pass filter the sampled spectrum and scale by ΔT, we can recover the continuous signal exactly. However, if $\omega_s < 2\omega_B$, then replica overlap in frequency occurs, and the continuous signal and spectrum are no longer exactly reproducible from their samples. This *undersampling* in time produces the frequency overlap called aliasing. *Aliasing* is the disguising of higher frequencies as lower frequencies caused by undersampling in time (second case of Eq. (2.5–7)), resulting in spectral overlap. We summarize these results in the sampling theorem.

Sampling Theorem: Let $x_c(\tau)$ be a bandlimited signal such that $X(\omega) = 0$ for $|\omega| > \omega_B$. Then $x_c(\tau)$ is uniquely determined by its samples $x_c(k\Delta T)$ if the sampling frequency ω_s satisfies

$$\omega_s > 2\omega_B$$

where
$$\omega_s = \frac{2\pi}{\Delta T}$$

and
$$x_c(\tau) = \sum_{k=-\infty}^{\infty} x_c(k\Delta T) \frac{\sin \omega_B(\tau - k\Delta T)}{\omega_B(\tau - k\Delta T)} \qquad (2.5\text{–}8)$$

This theorem can be derived from Eq. (2.5–5) after low-pass filtering and the substitution of the various Fourier relations. The sinc function appearing in the sampling theorem above acts as an exact interpolation function in the reconstruction. It should also be noted that the *folding* or *Nyquist* frequency is defined by

$$f_{\text{NYQUIST}} = \frac{f_s}{2} \qquad (2.5\text{–}9)$$

If a signal is not sampled in accordance with the sampling theorem, then aliasing occurs. The aliased frequencies can be determined from the Nyquist frequency as

$$f_{\text{alias}} = 2kf_{\text{NYQUIST}} \pm f_{\text{true}} \qquad \text{for } k \geq 1 \qquad (2.5\text{–}10)$$

The following example demonstrates the effect of undersampling data.

Example 2.5–1. Suppose we have data containing two sinusoids, one the signal at 10 Hz and one an unknown disturbance at 20 Hz. We sample at $f_s = 25$ Hz, corresponding to a Nyquist frequency of $f_{\text{NYQUIST}} = 12.5$ Hz, which implies from Eq. (2.5–10) that

$$f_{\text{alias}} = 25k \pm 20 = 5 \text{ Hz}$$

In Fig. 2.5–3, we show the results of simulating this problem in SIG [5]. The continuous signal and spectrum, which were created by simulating the signals at a sampling interval of $\Delta T = 0.01$ sec, are shown in (*a*). Since the highest frequency assumed in our data corresponds to the 12.5 Hz, we resample this simulated data at $\Delta T = 0.04$ sec. The result is shown in (*b*). Here we see that the data has been aliased as predicted by the 5 Hz aliasing frequency; that is, the signal energy of the 20 Hz sinusoid has been aliased or folded down to 5 Hz. This is caused by the undersampling. If we low-pass filter the continuous signal at the assumed Nyquist frequency of 12.5 Hz with an analog filter, then aliasing can be minimized, as shown in (*c*). Note that because the anti-aliasing filter does not possess a sharp roll-off, some aliasing still occurs, as shown in (*c*).

One might note from these representations that every discrete frequency-response function is periodic with period 2π. How then can we characterize continuous frequencies using discrete representations? The answer again lies in the sampling theorem. We know that to reconstruct a continuous function exactly, we must sample at a frequency $\omega_s > 2\omega_B$, where ω_B is the highest frequency present in the continuous spectrum. The sampling frequency is also given in terms of the sampling interval as

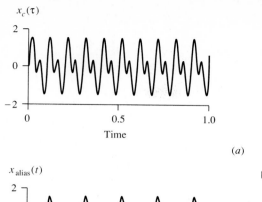

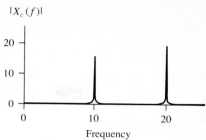

(a)

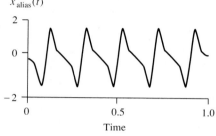

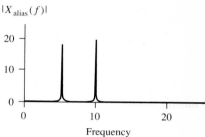

(b)

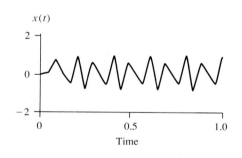

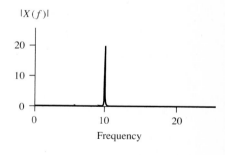

(c)

FIGURE 2.5–3

Undersampling or aliasing data (a) Continuous signal and spectrum (b) Aliased data sampled at 25 Hz (c) Anti-alias filtering

$$\omega_s = 2\pi f_s = \frac{2\pi}{\Delta T} \qquad (2.5\text{–}11)$$

If we investigate the frequency response of a sampled signal, it must coincide with that of the original continuous spectrum. The mapping from the Z- to the S-plane for a sampled signal is given by

$$z = e^{j\omega\Delta T}$$

which maps the $j\omega$-axis into the unit circle, $z = e^{j\Omega}$. Since the discrete frequency response is periodic, it has period $2\pi/\Delta T$, and therefore each strip of width $2\pi/\Delta T$ in the S-plane maps into the unit circle in the Z-plane. This mapping is depicted

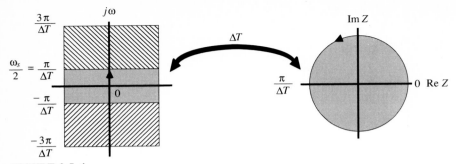

FIGURE 2.5–4
Sampled signal frequency response generation from the $j\omega$-axis to the unit circle

in Fig. 2.5–4. One revolution of the unit circle corresponds to a mapping of each of the $2\pi/\Delta T$ *strips* of the S-plane. However, since all of the continuous-spectral information is contained in the first or *primary* strip, we need only assure ourselves that that first strip is wide enough to include all of the frequencies of the continuous signal. This is assuredly the case in Eq. (2.5–11), since the strip width is determined by $2\pi/\Delta T = 2\pi f_s$; that is, the sampling frequency controls the width and therefore the frequencies of the continuous signals which will be included. If we follow the rules of the sampling theorem, then the sampling rate will be determined by the highest frequencies of the continuous spectrum, and the strip will possess all of this information.

It is also interesting to note that aliasing in this description corresponds to selecting a strip that does not contain the highest frequencies of the continuous signal; that is, the sampling frequency is too small. The aliased frequencies will then be reflected about the Nyquist frequency as shown in Fig. 2.5–3. Using this transformation of continuous to sampled domains we see that the discrete frequencies on the unit circle satisfy the relation

$$z = e^{j\Omega} = e^{j\omega\Delta T} \quad \Rightarrow \quad \Omega = \omega\Delta T \qquad (2.5\text{--}12)$$

which indicates explicitly how a continuous frequency is scaled to the corresponding discrete frequency. Thus, we see how the continuous spectrum $X(\omega)$ is related to the sampled spectrum through Eq. (2.5–5):

$$X_s(\omega) = \frac{1}{\Delta T} \sum_{k=-\infty}^{\infty} X\left(\omega - \frac{2\pi k}{\Delta T}\right) \qquad (2.5\text{--}13)$$

The sampled spectrum is related to the discrete time spectrum as well, through Eq. (2.5–12):

$$X(\Omega) = X_s(\omega)\big|_{\omega=\Omega/\Delta T} \qquad (2.5\text{--}14)$$

This relation can be thought of as the scaling introduced in converting the sampled signal $x_s(\tau)$ to the discrete signal $x(t)$ by a factor of ΔT in time. We see these relations depicted for a bandlimited continuous signal in Fig. 2.5–5. Thus, the

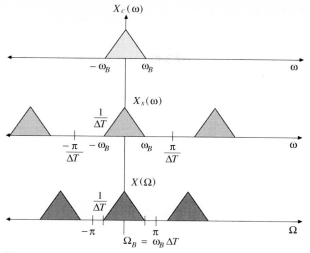

FIGURE 2.5–5
Relationship between continuous, sampled, and discrete spectra

sampling interval ΔT, in conjunction with the sampling theorem, determines the discrete frequency response scale factors relating continuous to discrete spectra.

Before we leave the subject of transformations between the S- and Z-planes, let us consider how the sampling interval ΔT controls the positions of poles and zeros. The transformation from S- to Z-planes is given by

$$z = e^{s\Delta T} = e^{(d\pm j\omega)\Delta T} \tag{2.5-15}$$

or

$$z = |z|\angle z = |e^{d\Delta T}|\arctan\{\tan(\pm\omega\Delta T)\}$$

Any point in the S-plane maps to a unique point in the Z-plane with magnitude $|z| = e^{d\Delta T}$ and angle $\angle z = \omega\Delta T$. We note that the $j\omega$-axis of the S-plane maps into the unit circle of the Z-plane and, therefore, stable poles in the left-half S-plane map into stable poles within the unit circle of the Z-plane. In fact, those points of constant damping in the S-plane map into circles of radius $|z| = e^{d\Delta T}$, while lines of constant frequency in the S-plane correspond to radial lines of angle $\angle z = \omega\Delta T$ in the Z-plane. We summarize these mappings in Fig. 2.5–6.

This mapping of continuous-time poles to the unit circle leads to the *impulse-invariant method* of transforming continuous (Laplace) transfer functions[†] to equivalent discrete (Z) transfer functions (see [6] for more details):

$$H(s) = \sum_{i=1}^{N} \frac{R_i}{s - s_i} \to H(z) = \sum_{i=1}^{N} \frac{R_i}{1 - e^{s_i\Delta T}z^{-1}} \tag{2.5-16}$$

[†]Recall the Laplace transform of $L[h(\tau)] := \int_{-\infty}^{\infty} h(\tau)e^{-s\tau}d\tau$.

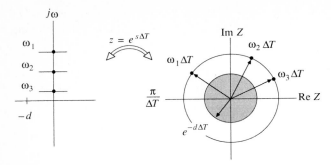

FIGURE 2.5–6
S- to Z-plane mappings

The essence of this method is that the impulse response of the discrete system is precisely the sampled values of the continuous response,

$$h(k\Delta T) = h_c(\tau)\big|_{\tau = k\Delta T}$$

In this transformation the residues $\{R_i\}$ are preserved, and a pole of $H(s)$ at $s = s_i$ is mapped to a pole in $H(z)$ at $z = e^{s_i \Delta T}$. Note that the zeros of $H(s)$ do not correspond to the zeros of $H(z)$. For this transformation to be useful, it is crucial that the continuous system is bandlimited and sampled according to the sampling theorem with $\omega_s > 2\omega_B$, to avoid aliasing of higher-frequency poles into those of lower frequency.

The impulse-invariant transformation not only provides a means to design digital filters (see next section), but also enables us to determine new system coefficients when the sampling interval is changed. We can write the discrete transfer function of a sampled signal as

$$H(z, \Delta T) = \sum_{i=1}^{N_a} \frac{R_i}{1 - a_i(\Delta T)z^{-1}} \qquad (2.5\text{–}17)$$

where $a_i(\Delta T) = e^{s_i \Delta T}$. Suppose that we change the sampling interval to ΔT_1, and that we would like to obtain the new set of transfer function coefficients

$$H(z, \Delta T_1) = \sum_{i=1}^{N_a} \frac{R_i}{1 - a_i(\Delta T_1)z^{-1}} \qquad \text{for } a_i(\Delta T_1) = e^{s_i \Delta T_1} \quad (2.5\text{–}18)$$

We can calculate the new $\{a_i(\Delta T_1)\}$ directly from the previous coefficients $\{a_i(\Delta T)\}$, since

$$s_i = \frac{1}{\Delta T} \ln a_i(\Delta T) \qquad (2.5\text{–}19)$$

and, equating and solving for $a_i(\Delta T_1)$, we have

$$a_i(\Delta T_1) = \exp\left\{ \left(\frac{\Delta T_1}{\Delta T} \right) \ln a_i(\Delta T) \right\} \qquad (2.5\text{–}20)$$

Using the impulse-invariant transformation, we can determine the new set of coefficients of a discrete system subject to the new sample interval. The following example illustrates this idea.

Example 2.5–2. Suppose we are asked to simulate a system at a sampling interval of $\Delta T_1 = 0.1$ sec, but we have developed our original model based on data gathered at $\Delta T = 1$ sec. The transfer function model is given by

$$H(z) = H(z, 1) = \frac{4}{1 - 0.05z^{-1}} \qquad \text{for } a(1) = 0.05$$

Using Eq. (2.5–20), we have

$$a(0.1) = \exp\left\{ \left(\frac{0.1}{1}\right) \ln 0.05 \right\} = \exp\{-0.03\} = 0.74$$

and therefore the new model given by Eq. (2.5–18) is

$$H(z, 0.1) = \frac{4}{1 - 0.74z^{-1}}$$

It should be noted that similar calculations can be made for complex poles as well (see the Exercises). Thus, we see that transforming from the S-plane to the Z-plane can prove quite useful, especially when systems are characterized in rational form.

In passing, it is useful to mention a very popular transformation technique, the *bilinear transformation* given by

$$s = \frac{2}{\Delta T} \left(\frac{1 - z^{-1}}{1 + z^{-1}} \right) \tag{2.5–21}$$

In this transformation, the left-half S-plane (stability region) maps *into* the interior of the unit circle, while the right-half lies outside. The $j\omega$-axis is mapped (nonlinearly) between 0 and $2\pi/\Delta T$ onto the unit circle. In contrast to the impulse-invariant mapping of Eq. (2.5–16), the bilinear transformation *uniquely* maps each distinct value on the $j\omega$-axis onto the unit circle, through frequency warping of

$$\omega = \frac{2}{\Delta T} \tan \frac{\Omega}{2} \tag{2.5–22}$$

which effectively compresses (nonlinearly) the $j\omega$-axis to the unit circle. This transformation is used extensively in the design of digital filters (see [2] for details).

2.6 FUNDAMENTALS OF DIGITAL FILTERS

In this section we briefly discuss special discrete LTI systems—digital filters. We discuss the operation of these filters from the signal processor's rather than from a filter designer's perspective; that is, we are primarily interested in the performance of the filter in magnitude and phase, and in its effect on the signal under investigation. Understanding the fundamentals of digital filters enables us to analyze and predict their effect on incoming signals. First we consider infinite-impulse response (IIR) filters, then we turn to finite-impulse response (FIR) filters.

Recall that an IIR digital filter is simply a discrete LTI system represented by a rational transfer function consisting of poles and zeros. We know from the inverse Z-transform (long division) that

$$H_{\text{IIR}}(z) = \frac{B(z)}{A(z)} = \sum_{t=0}^{\infty} h(t)z^{-t}$$

and that $\{h(t)\}$ is an infinite-length impulse response sequence. The typical method of designing this class of filter consists of two steps: (1) determining a suitable analog filter transfer function meeting the required specifications, and (2) determining the equivalent digital filter through transformations. Ideal filter responses are shown in Fig. 2.6–1, where we see the corresponding magnitude (frequency) response of low-pass, high-pass, bandpass, and band-reject filters. Typical classical analog filter designs follow (see [2] for details):

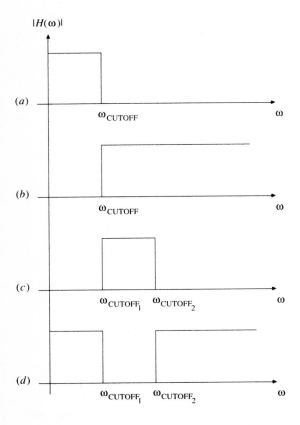

FIGURE 2.6–1
Ideal filter frequency (magnitude) response (a) Low-pass (b) High-pass (c) Bandpass (d) Band-reject

1. Butterworth designs (maximally flat magnitude)
2. Bessel designs (maximally flat delay)
3. Chebyshev I designs (equiripple in passband)
4. Chebyshev II designs (equiripple in stopband)
5. Elliptic designs (equiripple in both bands)

The procedure is called the "analog prototype design method" which consists of [2]

1. Interpreting the design specifications
2. Selecting the type of design, order, and cutoff frequency to meet these specifications
3. Specifying the classical analog filter transfer function $H(s)$
4. Transforming the analog design to digital, that is, $H(s) \rightarrow H(z)$ using the bilinear or impulse-invariant transformations (Sec. 2.5)
5. Implementing the design in hardware

The following example illustrates the design technique.

Example 2.6–1. Suppose we are given a set of design specifications for a low-pass filter. We decide that the corresponding analog filter to meet these specifications should be the Butterworth type, first-order, with a cutoff frequency of 1 r/s, with

$$H(s) = \frac{1}{s + 1}$$

For our digital design, we choose a sampling interval of $\Delta T = 1/2$ sec ($\frac{1}{2}f_{\text{HIGH}}$) and investigate the impulse-invariant and bilinear-tranformation techniques.

(i) Impulse-invariant design:

$$H(z) = \sum_{i=1}^{N} \frac{R_i}{1 - e^{s_i \Delta T} z^{-1}} = \frac{1}{1 - e^{-1/2} z^{-1}}$$

with $N = 1$, $s_1 = -1$, $\Delta T = 1/2$, $R_1 = 1$.

(ii) Bilinear transformation:[†]

$$s = \frac{2}{\Delta T} \left(\frac{1 - z^{-1}}{1 + z^{-1}} \right) = 4 \left(\frac{1 - z^{-1}}{1 + z^{-1}} \right)$$

$$H(z) = H(s) \Big|_{s = 4\left(\frac{1-z^{-1}}{1+z^{-1}}\right)} = \frac{1}{4\left(\dfrac{1 - z^{-1}}{1 + z^{-1}}\right) + 1} = \frac{1/5(1 + z^{-1})}{1 - \frac{3}{5}z^{-1}}$$

---

†Usually the discrete cutoff frequency is selected first, then the corresponding analog cutoff is calculated using the "warping" transformation of Eq. (2.5–22) (see [2] for more details).

So we see that IIR filters are characterized by rational transfer functions

$$H_{\mathrm{IIR}}(z) = \frac{B(z)}{A(z)} = |H(z)| \angle H(z) \qquad (2.6\text{--}1)$$

The main problem with IIR designs are that they can achieve good performance in magnitude (phase) but cause distortion in phase (magnitude). We illustrate this in the following example.

Example 2.6–2. Consider the design of a 6th-order, 10 Hz low-pass filter using the analog prototype design procedure. We use SIG [5] to develop Butterworth and Chebyshev I and II designs (see Fig. 2.6–2). Note in each case the sharp cutoff

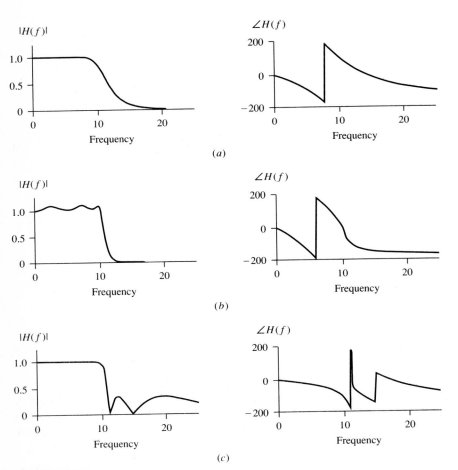

FIGURE 2.6–2
Analog prototype design of an IIR 6th-order, 10 Hz low-pass filter: (*a*) Butterworth (*b*) Chebyshev I (*c*) Chebyshev II

and desirable magnitude response, but with a nonlinear phase response. Note the sawtooth shape of the phase caused by "phase wrapping"; that is, each time the phase response passes through π it resets to zero, because of the definitions of the trigonometric functions on the computer. The nonlinearities are the distorted response characteristics in the figure. So we see from a signal processor's viewpoint that we can achieve acceptable designs with sharp cutoffs in one domain (magnitude or phase), but that we must expect nonlinear behavior in the other domain.

An alternative to IIR designs is the FIR digital filter (see [2,6]), characterized by

$$H_{\text{FIR}}(z) = \sum_{i=0}^{N_h} h(i)z^{-i} = \prod_{i=1}^{N_h} \frac{(z - z_i)}{z^{N_h}} \qquad (2.6\text{–}2)$$

where all of the N_h-poles are located at the origin ($p_i = 0$).
There are three popular methods of FIR filter design:

1. Window design [2,6]
2. Frequency sampling design [6]
3. Optimal (minimax error) design [7]

Window designs evolve to a closed-form solution for filter coefficients, but the other two methods are optimization techniques, requiring iterative methods to meet design objectives. The window method is easy to apply, but it is not as popular as the optimal design techniques.

The design of FIR filters using window functions (see [8] for details) is achieved by truncating the desired infinite-impulse response, using a finite-duration window function,

$$h_d(t) \approx h_{\text{IIR}}(t)W(t) \qquad (2.6\text{–}3)$$

The corresponding frequency response can be thought of as a Fourier series representation, with $h_d(t)$ playing the role of Fourier coefficients:

$$H_d(e^{j\Omega}) = \sum_{t=-\infty}^{\infty} h_d(t)e^{-j\Omega t} \qquad (2.6\text{–}4)$$

Thus, the design of ideal filters by truncating the corresponding impulse response is identical to the study of convergent Fourier series using window functions [2]. A common technique employed in practice is to specify the ideal filter response in the frequency domain and to obtain the corresponding infinite-impulse response using the IDFT,

$$h_d(t) \approx \underbrace{\text{IDFT }[H_{\text{IIR}}(\Omega_m)]}_{h_{\text{IIR}}(t)}W(t) \qquad (2.6\text{–}5)$$

Various window functions are available, with tradeoffs in performance. The typical tradeoff is that the lower the sidelobe (less leakage), the wider the main lobe will be (less resolution). In Table 2.6–1, we list some of the popular windows which can be used for FIR filter designs.

From the signal processor's viewpoint, FIR designs are desirable because they can have precisely linear phase; that is, if we constrain the weights to satisfy

$$h(t) = \pm h(M - t) \tag{2.6--6}$$

then

$$H(e^{j\Omega}) = |H(e^{j\Omega})|e^{-j\Omega M/2} \tag{2.6--7}$$

with $|H(e^{j\Omega})|$ either purely real or imaginary, depending on the $(+)$ or $(-)$ sign, respectively, in Eq. (2.6–6), with

$$\angle H(e^{j\Omega}) = \frac{-\Omega M}{2} \tag{2.6--8}$$

The penalty for achieving both desirable magnitude and phase properties is a large filter order (N_h) for a given performance. The design of a FIR filter is illustrated in the following example.

> **Example 2.6–3.** Consider the design of an FIR filter to meet the specifications of the previous example of low-pass, 10 Hz. We illustrate the Hamming and Kaiser window designs and the Parks-McClellan optimal design procedure [7] in Fig. 2.6–3. We use SIG [5] to develop the filters and analyze the frequency response. In the figure we show the desired (ideal) filter. The windows are then applied to the truncated-impulse response record. The Parks-McClellan algorithm was designed for 128 impulse response weights and for the ideal filter specifications. We can see the sharper cutoff and lower sidelobes of this optimal design along with linear phase. The FIR filter is then implemented using these weights $\{h(t)\}$.

So we see that IIR digital filters can be developed employing classical analog prototype designs using S- to Z-plane transformations, with nonlinear performance in either magnitude or phase response, while FIR filters can be designed with windows or iterative techniques that achieve both acceptable magnitude and phase (linear) response performance, but at the cost of high orders. From the signal processor's viewpoint, it is clear that simple low-order IIR filters designed with sharp cutoffs and using the analog prototypes can be used for real-time applications at the expense of potential phase distortion. FIR designs can achieve similar or even superior cutoff characteristics with linear phase response; however, high-order filters $(N_h \gg N_a)$ are required, which may not make them feasible for real-time applications.

2.7   SUMMARY

In this chapter, we have developed the Z-transform representation of discrete signals and systems. We reviewed the basic concepts in Z-transform theory and showed how they can be used to solve various discrete-time problems. We investigated the frequency response of discrete systems, sampled continuous

TABLE 2.6-1
Window characteristics and performance

Type	Function	Transform	Main lobe width	Side lobe level (dB)
Rectangular	$W_R(t) = \begin{cases} 1 & 0 \le t \le M-1 \\ 0 & \text{elsewhere} \end{cases}$	$W_R(\Omega) = e^{-j\Omega(M-1)/2}\dfrac{\sin(\Omega M/2)}{\sin(\Omega/2)}$	$\dfrac{2\pi}{M}$	-13
Triangular	$W_T(t) = \begin{cases} \dfrac{2t}{M-1} & 0 \le t \le \dfrac{M-1}{2} \\ 2 - \dfrac{2t}{M-1} & \dfrac{M-1}{2} \le t \le M-1 \end{cases}$	$W_T(\Omega) = \dfrac{1}{M}e^{-j\Omega(M-1)/2}\left(\dfrac{\sin(\Omega M/2)}{\sin(\Omega/2)}\right)^2$	$\dfrac{4\pi}{M}$	-27
Hanning	$W_{\text{HAN}}(t) = \dfrac{1}{2}\left[1 - \cos\left(\dfrac{2\pi t}{M-1}\right)\right]$ $0 \le t \le M-1$	$W_{\text{HAN}}(\Omega) = \dfrac{1}{2}W_R(\Omega) + \dfrac{1}{4}W_R\left(\Omega + \dfrac{2\pi}{M}\right)$ $+\dfrac{1}{4}W_R\left(\Omega - \dfrac{2\pi}{M}\right)$	$\dfrac{4\pi}{M}$	-31
Hamming	$W_{\text{HAM}}(t) = 0.54 - 0.46\cos\left(\dfrac{2\pi t}{M-1}\right)$ $0 \le t \le M-1$	$W_{\text{HAM}}(\Omega) = 0.54 W_R(\Omega) + 0.23 W_R\left(\Omega + \dfrac{2\pi}{M}\right)$ $+ 0.23 W_R\left(\Omega - \dfrac{2\pi}{M}\right)$	$\dfrac{4\pi}{M}$	-41
Blackman	$W_B(t) = 0.42 - 0.5\cos\left(\dfrac{2\pi}{M-1}\right) + 0.08\cos\left(\dfrac{4\pi}{M-1}\right)$ $0 \le t \le M-1$	$W_B(\Omega) \approx 0.42 W_R(\Omega) + 0.25\left[W_R\left(\Omega + \dfrac{2\pi}{M}\right) + W_R\left(\Omega - \dfrac{2\pi}{M}\right)\right]$ $+ 0.04\left[W_R\left(\Omega + \dfrac{4\pi}{M}\right) + W_R\left(\Omega - \dfrac{4\pi}{M}\right)\right]$	$\dfrac{6\pi}{M}$	-58
Kaiser	$W_K(t) = \dfrac{I_0(\beta)}{I_0(\alpha)}\quad |t| \le \dfrac{M-1}{2}$ for $I_0(\alpha) = 1 + \displaystyle\sum_{k=1}^{\infty}\left[\dfrac{1}{k!}\left(\dfrac{\alpha}{2}\right)^k\right]^2$ (zeroth order Bessel) $\beta = \alpha\sqrt{1 - \left(\dfrac{2t}{M-1}\right)^2}$	$W_K(\Omega) \approx \dfrac{M}{\alpha I_0(\alpha)}\left[\dfrac{\sin\left[\alpha\sqrt{(\Omega/\Omega_a)^2 - 1}\right]}{\sqrt{(\Omega/\Omega_a)^2 - 1}}\right]$		-46 to -82

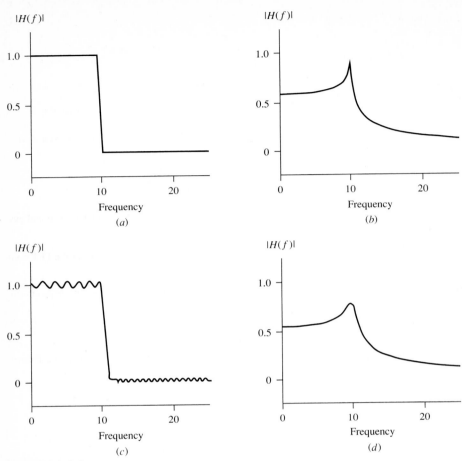

FIGURE 2.6–3
FIR filter design for a 10 Hz, low-pass filter: (*a*) Ideal filter frequency response (*b*) Hamming window (*c*) Parks-McClellan minimax (*d*) Kaiser window

systems, and discussed pole-zero representations and their effect on the time response. Next, we investigated discrete representations of continuous signals and showed how they relate through both the sampling theorem and transformations. We also briefly developed transformation techniques (impulse-invariant and bilinear) from the S- to Z-planes. Finally, we surveyed the design and performance of IIR and FIR digital filters from the signal processor's viewpoint.

REFERENCES

1. R. Churchill, *Complex Variables and Applications* (New York: McGraw-Hill, 1974).
2. A. Oppenheim and R. Shafer, *Digital Signal Processing* (Englewood Cliffs, N.J.: Prentice-Hall, 1975).

3. A. Oppenheim, A. Willsky, and I. Young, *Signals and Systems* (Englewood Cliffs, N.J.: Prentice-Hall, 1983).
4. S. Tretter, *Introduction to Discrete-Time Signal Processing* (New York: Wiley, 1976).
5. D. Lager and S. Azevedo, "SIG—A General Purpose Signal Processing Code," *Proc. IEEE*, 1987.
6. L. Rabiner and B. Gold, *Theory and Application of Digital Signal Processing* (Englewood Cliffs, N.J.: Prentice-Hall, 1975).
7. T. Parks and J. McClellan, "Chebyshev Approximation for Nonrecursive Digital Filters with Linear Phase," *IEEE Trans. Circuit Theory*, 1972.
8. F. Harris, "On the Use of Windows for Harmonic Analysis with the Discrete Fourier Transform," *Proc. IEEE*, 1978.

SIG NOTES

SIG can be used to simulate linear systems. The LSSABONLY command simulates a rational system using difference equations. A special case is that of the impulse response of a rational system obtained using the LSSIMPULSE command. SIG also has a "fast" FFT-based convolution command, CONVOLVE. The frequency response of a discrete system can be determined using the FFT command after determining the impulse response. SIG has two sampling commands: RESAMPLE and DECIMATEINTERP. The RESAMPLE command uses the Wiggins interpolator to obtain the new number of points, whereas the DECIMATEINTERP command employs an FIR filter to eliminate aliasing prior to up- or down-sampling. SIG has a correlation/covariance estimator called CORRELATE and a suite of spectral estimators, SD****. SIG also offers a large choice of digital filter designs. IIR digital filter designs include low-pass, high-pass, band-pass, and band-reject, that is, LP, HP, BP, BR— BESSEL, BUTTERWORTH, CHEBYSHEV I, II. FIR designs include the Parks-McClellan minimax (PMFILTER) and various windows: BLACKMAN, HAMMING, HANNING, KAISER, and TRIANGLE.

EXERCISES

2.1. We are given a causal LTI system with transfer function

$$T(z) = \frac{4}{1 - \frac{5}{2}z^{-1} + z^{-2}}$$

(a) Find the unit impulse response of the system.
(b) Find the unit step response.
(c) Is the system stable?
(d) Find the inverse Z-transform of $T(z)$ by division. Calculate $h(t)$, $t = 0, 1, 2$ from (a) and compare.
(e) Sketch the frequency response (magnitude only) of $T(z)$.

2.2. Find the inverse Z-transform of

$$H(z) = \frac{1 + z^{-1}}{1 - \frac{1}{2}z^{-1} + \frac{1}{18}z^{-1}}$$

for the following ROC:
(a) $|z| > 1/3$
(b) $|z| < 1/6$
(c) $1/6 < |z| < 1/3$

2.3. Find the inverse Z-transform of the LTI system with transfer function

$$H(z) = \frac{1}{(1 - z^{-1})(1 - e^{-T}z^{-1})} \qquad \text{for } |z| > 1$$

using
(a) partial fractions
(b) inversion integral

2.4. We are given the following LTI system with transfer function

$$H(z) = \frac{(7 - 2z^{-1})}{(1 - z^{-1})^2}$$

and asked to find the corresponding impulse response for the following ROC:
(a) $|z| < 1$
(b) $|z| > 1$

2.5. Find the inverse Z-transform of the signal $x(t)$, given its transform representation

$$X(z) = \frac{1 + \frac{1}{8}z^{-1}}{(1 + \frac{1}{4}z^{-1})(1 + \frac{5}{12}z^{-1} + \frac{1}{24}z^{-2})}$$

for the following ROC:
(a) $|z| > 1/4$
(b) $|z| < 1/6$
(c) $1/6 < |z| < 1/4$

2.6. We are given a causal LTI system with transfer function

$$H(z) = \frac{6}{(1 - \frac{1}{2}z^{-1})(1 - \frac{1}{3}z^{-1})}$$

(a) Find the corresponding impulse response.
(b) Compare the results of (a) with $h(t)$, $t = 0, 1, 2$ obtained by long division.

2.7. Suppose we are given two subsystems, S_1 and S_2, connected in series, as shown in this diagram:

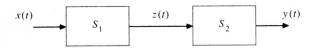

such that

$$S_1: \quad z(t) + \tfrac{1}{2}z(t - 1) = x(t) \qquad\qquad \text{for } z(-1) = 2$$

$$S_2: \quad y(t) - \tfrac{1}{3}y(t - 1) = z(t) - \tfrac{1}{4}z(t - 1) \qquad \text{for } y(-1) = 0$$

(a) Find the transfer function of the connected system, $H(z)$.
(b) Find the response of the overall system to a unit-step excitation, that is,

$$x(t) = \mu(t)$$

(c) Determine the corresponding frequency response $H(\Omega) = |H(\Omega)|\angle H(\Omega)$.

2.8. Suppose we are given the following system:

$$y(t) = e^{-t}x(t + 1)$$

Determine if it is
(*a*) time-invariant
(*b*) causal
(*c*) linear

2.9. Calculate the response of a LTI system with impulse response

$$h(t) = \begin{cases} \alpha^t & 0 \le t \le N - 1 \\ 0 & \text{elsewhere} \end{cases}$$

when excited by the pulse

$$x(t) = \mu(t) - \mu(t - N)$$

2.10. Calculate the response of a LTI system with unit impulse response given by

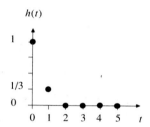

when excited by

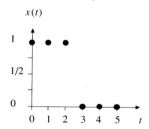

2.11. Suppose we are given a digital filter

$$y(t) = -ay(t - 1) + bx(t) \qquad \text{for } a < 0, \, b > 0$$

and the values of the impulse response

$$h(1) = 5/2, \quad h(2) = 5/4$$

Find the values of a, b, and $h(0)$.

2.12. Find the impulse response of the system

$$y(t) = -4y(t - 1) - \sin \Omega_0 y(t - 2) + \ln \Omega_0 x(t - 1) \qquad \text{for } \Omega_0 > 0$$

2.13. Suppose we have a continuous signal with Fourier spectrum

$$X(\omega) = \begin{cases} 1 & |\omega| < 10\pi \\ 0 & \text{elsewhere} \end{cases}$$

(a) Calculate the appropriate sampling frequency ω_s and sketch the "sampled" spectrum.

(b) Suppose the spectrum is aliased with $\omega_s = 5\pi$; sketch the aliased spectrum.

(c) Simulate the spectra and validate your results.

2.14. Suppose we have a signal composed of three sinusoids at 100, 120, and 140 Hz.

(a) What is the appropriate choice for sampling and Nyquist frequencies?

(b) If we assume the highest frequency is $f_{HIGH} = 100$ Hz, what are the aliased frequencies for this signal, assuming no filtering has occurred?

(c) For this signal, suggest ways of recovering the 100 Hz sinusoid from the aliased data. •

(d) Suppose a sinusoid exists at 1 Hz and that we sample appropriately (anti-alias filter) for $f_{HIGH} = 100$ Hz. Are there enough samples to represent this sinusoid? Discuss your answer.

(e) Simulate these processes at a very small sample interval ($\approx 10^{-5}$ sec) for an approximate continuous signal, then resample according to the problem. Show spectra.

2.15. Reconsider the composite sinusoidal signal of Exercise 1.4, and sketch the S-plane, mapping to the unit circle when $f_{HIGH} = 100$ Hz and the composite signal is

(a) appropriately (anti-alias filtered) sampled

(b) aliased

(c) Show the pole locations of the "aliased" sinusoids.

2.16. We are given an LTI continuous system with differential equation

$$\dot{y}_c(\tau) = -ay_c(\tau) + x_c(\tau) \qquad y_c(0) = 0$$

with bandlimited frequency response

$$H_c(\omega) = \begin{cases} H(\omega) & |\omega| \leq \omega_c \\ 0 & |\omega| > \omega_c \end{cases}$$

and $\omega_s = 4\omega_c$. Find the corresponding discrete-time transfer function, $H(\Omega)$, after sampling and scaling.

2.17. Find the frequency response of the following digital filter,

$$y(t) = 0.5x(t) + 0.5x(t-1)$$

and sketch the corresponding magnitude and phase.

2.18. Calculate the frequency response of the following signal:

$$x(t) = (1/8)^t \mu(t)$$

2.19. Given the two digital filters, F_1 and F_2, characterized by difference equations

$$F_1: \quad y_1(t) - \tfrac{1}{4}y_1(t-1) = x_1(t) + \tfrac{1}{2}x_1(t-1) \qquad y_1(-1) = 4$$

and $\quad F_2: \quad y_2(t) - \tfrac{1}{8}y_2(t-1) = 4x_2(t) \qquad\qquad\qquad y_2(-1) = -8$

find the overall performance of the cascaded filters when they are connected in series.

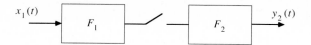

Find the overall transfer function and unit impulse response.

2.20. Suppose we are given an analog (continuous) filter of the form

$$H_c(s) = \frac{s + a}{(s + a + jb)(s + a - jb)}$$

Find the corresponding digital filter for sample interval T using the

(*a*) impulse invariant method

(*b*) bilinear transformation method with $a = 0$

2.21. Suppose we have the following digital filter,

$$H(z) = \frac{2z}{z - 0.005} - \frac{z}{z - 0.135}$$

designed at a sampling interval of $\Delta T = 1.0$ sec, and we wish to redesign it for a new sampling interval of $\Delta T' = 2.0$ sec. Find the new filter.

2.22. We are given the second-order discrete system

$$H(z, \Delta T) = \frac{1 + b_1(\Delta T)z^{-1}}{1 + a_1(\Delta T)z^{-1} + a_2(\Delta T)z^{-2}}$$

where

$$b_1(\Delta T) = -e^{-d\Delta T}(\cos \omega \Delta T - \beta \sin \omega \Delta T)$$

$$a_1(\Delta T) = -2e^{-d\Delta T} \cos \omega \Delta T$$

$$a_2(\Delta T) = e^{-2d\Delta T}$$

and the underlying continuous system is

$$H_c(s) = \frac{\beta}{(s + d + j\omega)(s + d - j\omega)}$$

(*a*) Find d, ω, and β.

(*b*) Suppose we change the sampling interval from $\Delta T \rightarrow \Delta T*$; then find $H(z, \Delta T*)$.

2.23. Given the discrete LTI system characterized by

$$y(t) + \frac{5}{36}y(t - 1) = \frac{1}{36}y(t - 2) = x(t - 1) - \frac{1}{2}x(t - 2)$$

with zero initial conditions, find

(*a*) impulse response

(*b*) frequency response

(*c*) Simulate the system, and plot the corresponding frequency response.

2.24. Suppose we are given the analog (continuous) filter

$$H_c(s) = \frac{s + 2}{s^2 + 7s + 12}$$

and are asked to design a digital filter for a sampling interval of $\Delta T = 0.1$ sec, using the

(*a*) impulse invariant method
(*b*) bilinear transformation method
Give the discrete transfer functions of the corresponding digital filters.

2.25. Design a digital low-pass Butterworth filter from the analog (normalized $\omega_c = 1$ rad/sec) prototype given by

$$H_c(s) = \frac{1}{s^2 + \sqrt{2}s + 1}$$

with a cutoff frequency of $\Omega_c = 50$ Hz when the system sampling frequency is 500 Hz. Use
(*a*) impulse invariance
(*b*) bilinear transformation method
(*c*) Redesign the filter to be high-pass, using the transformation $s \to \lambda/s*$ with a new cutoff of 1 kHz.

2.26. We are investigating the design of an instrument which has a digital filter designed using the impulse invariant method at $\Delta T = 0.1$ sec. If the filter is given by the difference equation

$$y(t) - 0.9512y(t - 1) = 7.3214x(t)$$

(*a*) Find the original analog filter, $H_c(s)$.
(*b*) Redesign the filter to operate at a new sampling interval of $\Delta T = 0.01$ sec.

2.27. A digital filter designed to operate in a microprocessor need not be causal. Suppose we are given the system

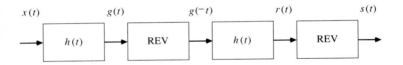

and REV means time reverse the input $x(t) \to x(-t)$.
(*a*) Find $h'(t)$, the overall impulse response.
(*b*) Find the overall frequency response, $H'(\Omega)$.
(This is a technique to remove the phase effect of a filter.)

2.28. Let $h(t)$ be the unit impulse response of a causal FIR filter with $h(t)$ real; then the frequency response can be represented in the form

$$H(\Omega) = \hat{H}(\Omega)e^{j\phi(\Omega)} \qquad \text{for } \hat{H} \text{ real}$$

Find $\phi(\Omega)$ for $0 \le \Omega \le \pi$ when $h(t) = h(N - 1 - t)$, N is even.

2.29. Show that an FIR filter is always stable.

CHAPTER
3

DISCRETE SIGNAL ANALYSIS

In this chapter we develop spectral analysis techniques for deterministic signals. First, we introduce the discrete Fourier transform (DFT) and show how it can be used to obtain spectral representations of periodic discrete signals. Then we show its relationship to the discrete Fourier series. Next, we derive the discrete-time Fourier transform (DtFT) from the DFT and show, from a pragmatic point of view, how they are related. Finally, we discuss the practical considerations that must be accounted for when applying the DFT technique to represent continuous or equivalently sampled signals.

3.1 PERIODIC SIGNAL ANALYSIS: DISCRETE FOURIER TRANSFORMS

In this section we develop the discrete Fourier transform (DFT) as a mechanism to represent periodic discrete signals. We develop its properties and show how it can be used to obtain spectral representations of discrete signals.

Suppose we have a discrete periodic signal of period M, so that

$$x(t) = x(t + kM) \qquad \forall \, k \qquad\qquad (3.1\text{--}1)$$

55

We define the discrete Fourier transform (DFT) pair[†] as

$$X(\Omega_m) := \text{DFT} [x(t)] = \sum_{t=0}^{M-1} x(t)e^{-j\Omega_m t} \tag{3.1-2}$$

$$x(t) = \text{IDFT} [X(\Omega_m)] = \frac{1}{M} \sum_{m=0}^{M-1} X(\Omega_m)e^{j\Omega_m t} \tag{3.1-3}$$

for the periodic signal $x(t)$ where the discrete frequency $\Omega_m = 2\pi m/M$. To be more precise, we can write the DFT pair as

$$X(\Omega_m) = X(e^{j\Omega_m}) = X(e^{j(2\pi/M)m}) = \sum_{t=0}^{M-1} x(t)e^{-j(2\pi/M)mt} \tag{3.1-4}$$

and

$$x(t) = \frac{1}{M} \sum_{m=0}^{M-1} X(e^{j(2\pi/M)m})e^{j(2\pi/M)mt} \tag{3.1-5}$$

emphasizing the idea of sampling the DtFT around the unit circle ($z = e^{j\Omega_m} = e^{j(2\pi/M)m}$), as mentioned in Sec. 2.4. We note from the properties of the discrete exponential of Eq. (1.1–13) that $X(\Omega_m)$ is also periodic with period M.

Thus, we see that the DFT is defined for periodic signals; however, it can be applied to finite-duration aperiodic signals as well, by simply assuming that they represent *one* period of the periodic signal $x_p(t)$,

$$x(t) := \begin{cases} x_p(t) & 0 \le t \le M - 1 \\ 0 & \text{elsewhere} \end{cases}$$

Thus, the DFT coefficients $\{X(\Omega_m)\}$ *uniquely* represent this aperiodic signal. We will discuss this further in the next section. Consider the following example of calculating the DFT.

Example 3.1–1. Determine the DFT of the periodic sequence

$$x(t) = \alpha^t \qquad 0 \le t \le M - 1$$

From the definition of Eq. (3.1–2), we have

$$X(\Omega_m) = \text{DFT} [\alpha^t] = \sum_{t=0}^{M-1} (\alpha^t)e^{-j\Omega_m t} = \sum_{t=0}^{M-1} (\alpha e^{-j\Omega_m})^t$$

Using the geometric series for finite sums[‡] with $\beta = \alpha e^{-j\Omega_m}$, we have

---

[†]We choose this notation because of its similarity to the discrete-time Fourier transform $X(\Omega)$, emphasizing that it is a sampled value; that is, $X(\Omega_m) = X(\Omega)|_{\Omega=\Omega_m}$ or, to be more precise, $X(e^{j\Omega_m}) = X(e^{j\Omega})|_{\Omega=\Omega_m}$.

[‡]Recall that the geometric series for finite sums is given by $\sum_{i=0}^{M-1} \beta^i = (1-\beta^M)/(1-\beta)$ for $|\beta| < 1$.

$$X(\Omega_m) = \frac{1 - (\alpha e^{-j\Omega_m})^M}{1 - \alpha e^{-j\Omega_m}} = \frac{1 - \alpha^M e^{-j(2\pi m/M)M}}{1 - \alpha e^{-j\Omega_m}}$$

or since $e^{-j2\pi m} = 1 \; \forall \, m$, we have

$$X(\Omega_m) = \frac{1 - \alpha^M}{1 - \alpha e^{-j\Omega_m}} \qquad 0 \le m \le M - 1$$

If we choose $\alpha = 0.95$ with $M = 64$, we show the results of the DFT of this sequence simulated using SIG [1]. We see the simulated exponential signal, along with the real and imaginary parts of the corresponding DFT for one period, in Fig. 3.1–1.

We assume that all signals are periodic when we use the DFT; that is, the DFT is actually a representation of a periodic signal. Clearly, it must be related to the discrete Fourier series representation of Eq. (1.2–2). If we choose the interval of the DFS to be $[0, M - 1]$, then

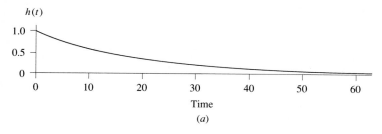

(a)

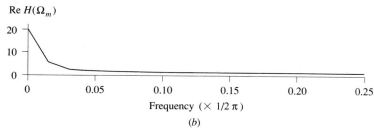

(b)

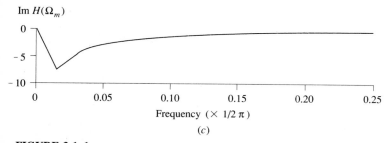

(c)

FIGURE 3.1–1
DFT of a discrete exponential signal: (a) Signal $x(t) = (0.95)^t$ (b) Real part of DFT (c) Imaginary part of DFT

$$\text{DFS}\,[x(t)] = a_m = \frac{1}{M} \sum_{t=0}^{M-1} x(t) e^{-j\Omega_m t} = \frac{1}{M}\,\text{DFT}\,[x(t)] = \frac{X(\Omega_m)}{M}$$

or, equivalently, the DFT and DFS coefficients are related by

$$X(\Omega_m) = M a_m \tag{3.1-6}$$

Alternatively, the DFT representations can be viewed as "modulo M"; that is, the time indices of the signals must be interpreted as modulo M, since the finite-duration signal $x(t)$ can be expressed as the periodic signal

$$x_p(t) = \sum_{k=-\infty}^{\infty} x(t + kM) = x(t \bmod M) \tag{3.1-7}$$

Like the Z-transform, the DFT has some very useful properties which can be employed in the analyses of discrete signals and systems. Following the same

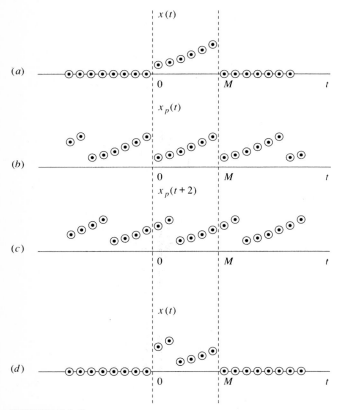

FIGURE 3.1–2
Circular (periodic) shift of a finite-duration signal (a) Original finite-duration signal (b) Periodic extension of signal (c) Circular shift of two samples (d) Shifted finite-duration signal

TABLE 3.1–1
Properties of the DFT

Property	Signal (mod M)	Transform
Linearity	$ax_1(t) + bx_2(t)$	$aX_1(\Omega_m) + bX_2(\Omega_m)$
Time shift	$x(t - t_0)$	$e^{-j\Omega_m t_0}X(\Omega_m)$
Frequency scale	$e^{j\Omega_0 t}x(t)$	$X(\Omega_m - \Omega_0)$
Convolution	$x_1(t) * x_2(t)$	$X_1(\Omega_m)X_2(\Omega_m)$
Multiplication	$x_1(t)x_2(t)$	$\dfrac{1}{M}X_1(\Omega_m) * X_2(\Omega_m)$
Parseval's theorem	$\sum_{t=0}^{M-1} |x(t)|^2$	$\dfrac{1}{M}\sum_{m=0}^{M-1} |X(\Omega_m)|^2$

approach as in the previous chapter, we can show that the DFT possesses proper-ties similar to those of the Z-transform, except for modulo M. Most properties can be derived from the transform pair of Eqs. (3.1–2,3). For instance, the property of *linearity* follows

$$x_1(t) \leftrightarrow X_1(\Omega_m)$$

if

$$x_2(t) \leftrightarrow X_2(\Omega_m)$$

then

$$\text{DFT}\,[ax_1(t) + bx_2(t)] = \sum_{t=0}^{M-1} [ax_1(t) + bx_2(t)]e^{-j\Omega_m t}$$

$$= a\sum_{t=0}^{M-1} x_1(t)e^{-j\Omega_m t} + b\sum_{t=0}^{M-1} x_2(t)e^{-j\Omega_m t}$$

or

$$\text{DFT}\,[ax_1(t) + bx_2(t)] = aX_1(\Omega_m) + bX_2(\Omega_m)$$

which proves the property. We summarize other useful properties of the DFT in Table 3.1–1 (see [2] for more details). Note that these properties are interpreted as modulo M because of the periodic nature of the signal; that is, a periodic or *circular shift*, rather than a linear shift, is inherited from the periodic signals. This shift is illustrated in Fig. 3.1–2.

3.2 APERIODIC SIGNAL ANALYSIS: DISCRETE-TIME FOURIER TRANSFORMS

In this section we develop the discrete-time Fourier transform (DtFT) directly from the DFT, rather than as a special case of the Z-transform. We do this in order to provide the basis for using the DFT to approximate the DtFT in analyzing discrete and sampled signals.

Suppose we have an aperiodic signal of duration $2M_1 + 1$,

$$\tilde{x}(t) = 0 \qquad |t| > M_1 \tag{3.2–1}$$

and we construct a periodic signal of period M such that

$$x(t) = \tilde{x}(t) \text{ over } M, \text{ which includes } |t| \leq M_1$$

The DFT of $x(t)$ is

$$X(\Omega_m) = \sum_{t=-M_1}^{M_1} x(t)e^{-j\Omega_m t} = \sum_{t=-\infty}^{\infty} x(t)e^{-j(2\pi/M)mt} \qquad (3.2\text{--}2)$$

We define the *frequency interval* associated with the DFT by

$$\Delta f := \frac{1}{M}$$

As we let the period $M \gg M_1$ become large, then the frequency interval becomes small or equivalently

$$\Omega_m = \left(\frac{2\pi}{M}\right) m = (2\pi\Delta f)\, m \to 0 \qquad (3.2\text{--}3)$$

As $M \to \infty$, we have

$$\lim_{M\to\infty} \Omega_m \to \Omega = 2\pi f$$

or

$$\lim_{M\to\infty} X(e^{j(2\pi/M)m}) \to X(e^{j\Omega})$$

or

$$\lim_{M\to\infty} \text{DFT}\,[x(t)] \to \text{DtFT}\,[x(t)]$$

which gives the DtFT relation of Eq. (2.4–1):

$$X(\Omega) = \lim_{M\to\infty} X(\Omega_m)\bigg|_{\Omega_m=(2\pi/M)m} = \sum_{t=-\infty}^{\infty} x(t)e^{-j\Omega t} \qquad (3.2\text{--}4)$$

Thus, we see that the coefficients of the DFT are identical to values of the DtFT sampled at integer multiples $\Omega_m = (2\pi/M)m$ around the unit circle. From the IDFT of Eq. (3.1–3), we have

$$x(t) = \frac{1}{M} \sum_{m=0}^{M-1} X(e^{j(2\pi/M)m})e^{j(2\pi/M)mt}$$

or, as $M \to \infty$, we have $\Omega_m \to \Omega$, and the sum approaches the integral (see [2] for details), to give the IDtFT relation of Eq. (2.4–2):

$$x(t) = \frac{1}{2\pi} \int_{2\pi} X(\Omega)e^{j\Omega t}d\Omega \qquad (3.2\text{--}5)$$

In summary, as we allow the period of the discrete signal to become larger, the frequency interval Δf of the DFT becomes smaller, and in the limit, DFT $\to$ DtFT. This will be important when we investigate the processing of sampled signals. So, in contrast to the Z-transform approach of the previous chapter, the DtFT can be obtained from the DFT.

Example 3.2–1. Suppose we have a discrete rectangular pulse given by

$$x(t) = \begin{cases} 1 & |t| \le M_1 \\ 0 & |t| > M_1 \end{cases}$$

We would like to calculate the DFT and DtFT and compare them. The DFT is simply

$$X(\Omega_m) = \sum_{t=-\infty}^{\infty} x(t)e^{-j\Omega_m t} = \sum_{t=-M_1}^{M_1} e^{-j\Omega_m t}$$

assuming that $x(t)$ is periodic with period M. Performing a change of variable with $n = t + M_1$, we have

$$X(\Omega_m) = \sum_{n=0}^{2M_1} e^{-j\Omega_m(n-M_1)} = e^{j\Omega_m M_1} \sum_{n=0}^{2M_1} (e^{-j\Omega_m})^n$$

Applying the geometric-series relation for finite sums with $\beta = e^{-j\Omega_m}$, we obtain

$$X(\Omega_m) = e^{j\Omega_m M_1}\left(\frac{1 - e^{-j\Omega_m(2M_1+1)}}{1 - e^{-j\Omega_m}} \right)$$

or $$X(\Omega_m) = e^{j\Omega_m M_1}\left(\frac{e^{-j(\Omega_m/2)(2M_1+1)}}{e^{-j\Omega_m/2}} \right)\left(\frac{e^{j(\Omega_m/2)(2M_1+1)} - e^{-j(\Omega_m/2)(2M_1+1)}}{e^{j\Omega_m/2} - e^{-j\Omega_m/2}} \right)$$

or, using the Euler relations (Eq. (1.1–11)), we have the discrete sinc function,

$$X(\Omega_m) = \text{DFT}\,[x(t)] = \left.\frac{\sin(\Omega_m/2)(2M_1+1)}{\sin(\Omega_m/2)}\right|_{\Omega_m = (2\pi/M)m}$$

Next, let us calculate the DtFT of $x(t)$:

$$X(\Omega) = \sum_{t=-\infty}^{\infty} x(t)e^{-j\Omega t} = \sum_{t=-M_1}^{M_1} e^{-j\Omega t}$$

Making the same substitutions as before, we have

$$X(\Omega) = e^{j\Omega M_1} \sum_{t=0}^{2M_1} (e^{-j\Omega})^t$$

Again using the finite-sum geometric series with $\beta = e^{-j\Omega}$, we obtain

$$X(\Omega) = \frac{e^{j(\Omega/2)(2M_1+1)} - e^{-j(\Omega/2)(2M_1+1)}}{e^{j\Omega/2} - e^{-j\Omega/2}}$$

or $$X(\Omega) = \text{DtFT}\,[x(t)] = \frac{\sin(\Omega/2)(2M_1+1)}{\sin(\Omega/2)}$$

Clearly, as $\Delta f \to 0$, then

$$\lim_{M\to\infty} X(\Omega_m)\big|_{\Omega_m=(2\pi/M)m} \to X(\Omega)$$

or $$\lim_{M\to\infty} \text{DFT}\,[x(t)] \to \text{DtFT}\,[x(t)]$$

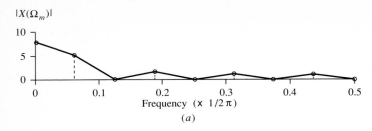

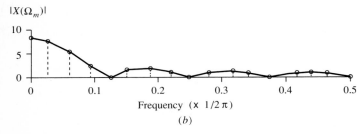

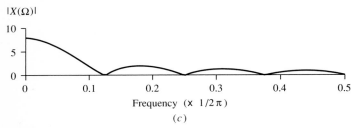

FIGURE 3.2–1
DFT and DtFT of a rectangular pulse: (a) DFT for period $M = 16$ (b) DFT for period $M = 32$ (c)
DFT $\rightarrow$ DtFT for period $M = 1024$

as developed previously. We illustrate this using SIG [1]. Let $M_1 = 16$ and let $M = 16$, 32, and 1024; then as M increases we expect $\Omega_m \rightarrow \Omega$ and $X(\Omega_m) \rightarrow X(\Omega)$. These results are confirmed in Fig. 3.2–1. We see from the figure that as M is increased, the discrete spectral lines approach the continuum of the DtFT.

The DtFT is used primarily to evaluate the frequency response of discrete aperiodic signals as well as discrete LTI systems. All of the theorems and properties developed in the previous chapter are valid for the DtFT as well, as long as $z = e^{j\Omega}$ is contained within the ROC. Thus, DtFT transform pairs for aperiodic signals can be obtained from the common Z-transform pairs of Table 2.1–1 by letting $z = e^{j\Omega}$. Note that all of the methods of inverting the Z-transform discussed in Sec. 2.1 apply by merely substituting $z = e^{j\Omega}$ into each technique, similar to substituting $s = j\omega$ into the Laplace transform relations to obtain the continuous Fourier transform. So we see the DtFT can be used to solve difference equations and characterize discrete-time systems, as long as the ROC contains the unit circle.

The DtFT for a periodic signal is also quite useful in analysis of spectra. The basic transform pairs are derived from the periodic property of the discrete exponential signal

$$x(t) = e^{j\Omega_0 t} = e^{j(\Omega_0 + 2\pi k)t} \qquad \forall k \in \mathbf{I} \tag{3.2-6}$$

which has corresponding DtFT

$$X(\Omega) = 2\pi \sum_{k=-\infty}^{\infty} \delta(\Omega - \Omega_0 - 2\pi k) \tag{3.2-7}$$

This is easily shown by substituting Eq. (3.2–7) into the IDtFT of Eq. (3.2–5):

$$x(t) = \frac{1}{2\pi} \int_{2\pi} \left[2\pi \sum_{k=-\infty}^{\infty} \delta(\Omega - \Omega_0 - 2\pi k) \right] e^{j\Omega t} d\Omega = e^{j(\Omega_0 + 2\pi k)t} = e^{j\Omega_0 t}$$

since any interval of length 2π includes precisely one impulse in the summation. Using this transform pair, we can derive the DtFT of many useful periodic functions. We summarize some of the more common DtFT pairs in Table 3.2–1.

As with the Z-transform, properties of the DtFT are easily derived. For instance, consider the time-shift property useful in solving difference equations:

$$\text{DtFT}\,[x(t - t_0)] = \sum_{t=-\infty}^{\infty} x(t - t_0)e^{-j\Omega t} = e^{-j\Omega t_0} \sum_{t=-\infty}^{\infty} x(t - t_0)e^{-j\Omega(t - t_0)}$$

TABLE 3.2–1
Common DtFT pairs

Signal	Transform
$\sum_{t=<M>} a_t e^{j\Omega mt}$	$2\pi \sum_{m=-\infty}^{\infty} a_m \delta(\Omega - \Omega_m)$
$e^{j\Omega_0 t}$	$2\pi \sum_{k=-\infty}^{\infty} \delta(\Omega - \Omega_0 - 2\pi k)$
$\cos \Omega_0 t$	$\pi \sum_{k=-\infty}^{\infty} [\delta(\Omega - \Omega_0 - 2\pi k) + \delta(\Omega + \Omega_0 - 2\pi k)]$
$\sin \Omega_0 t$	$\frac{\pi}{j} \sum_{k=-\infty}^{\infty} [\delta(\Omega - \Omega_0 - 2\pi k) - \delta(\Omega + \Omega_0 - 2\pi k)]$
1	$2\pi \sum_{k=-\infty}^{\infty} \delta(\Omega - 2\pi k)$
$\sum_{k=-\infty}^{\infty} \delta(t - kM)$	$\frac{2\pi}{M} \sum_{m=-\infty}^{\infty} \delta(\Omega - \Omega_m)$
$\alpha^t \mu(t) \qquad \|\alpha\| < 1$	$\dfrac{1}{1 - \alpha e^{-j\Omega}}$
$x(t) = \begin{cases} 1 & \|t\| \le M \\ 0 & \text{elsewhere} \end{cases}$	$\dfrac{\sin[M + (1/2)]\Omega}{\sin(\Omega/2)}$
$\dfrac{\sin kt}{\pi t} \qquad 0 < k < \pi$	$X(\Omega) = \begin{cases} 1 & 0 \le \|\Omega\| \le k \\ 0 & k < \|\Omega\| \le \pi \end{cases}$
$\delta(t)$	1
$\mu(t)$	$\dfrac{1}{1 - e^{-j\Omega}} + \sum_{k=-\infty}^{\infty} \pi \delta(\Omega - 2\pi k)$
$\delta(t - t_0)$	$e^{-j\Omega t_0}$

TABLE 3.2–2
Properties of the DtFT

Property	Signal	Transform
Linearity	$ax_1(t) + bx_2(t)$	$aX_1(\Omega) + bX_2(\Omega)$
Time shift	$x(t - t_0)$	$e^{-j\Omega t_0}X(\Omega)$
Frequency scale	$e^{j\Omega_0 t}x(t)$	$X(\Omega - \Omega_0)$
Time reversal	$x(-t)$	$X(-\Omega)$
Convolution	$x_1(t) * x_2(t)$	$X_1(\Omega)X_2(\Omega)$
Multiplication	$x_1(t)x_2(t)$	$\dfrac{1}{2\pi j}\displaystyle\int_\pi X_1(\alpha)X_2(\Omega - \alpha)d\alpha$
Derivative	$tx(t)$	$j\dfrac{dX(\Omega)}{d\Omega}$
Finite sum	$\sum_{t=-\infty}^{n} x(t)$	$\dfrac{1}{1 - e^{-j\Omega}}X(\Omega) + \pi X(0)\sum_{k=-\infty}^{\infty} \delta(\Omega - 2\pi k)$
Parseval's theorem	$\sum_{t=-\infty}^{\infty}|x(t)|^2$	$\dfrac{1}{2\pi}\displaystyle\int_{2\pi}|X(\Omega)|^2 e^{-j\Omega}d\Omega$

or
$$\text{DtFT}\,[x(t - t_0)] = e^{-j\Omega t_0}X(\Omega) \qquad (3.2\text{–}8)$$

Similarly, the frequency-scaling property follows as

$$\text{DtFT}\,[e^{j\Omega_0 t}x(t)] = \sum_{t=-\infty}^{\infty} e^{j\Omega_0 t}x(t)e^{-j\Omega t} = \sum_{t=-\infty}^{\infty} x(t)e^{-j(\Omega - \Omega_0)t}$$

or
$$\text{DtFT}\,[e^{j\Omega_0 t}x(t)] = X(\Omega - \Omega_0) \qquad (3.2\text{–}9)$$

We summarize these and other properties of the DtFT in Table 3.2–2.

3.3   TRANSFORM EQUIVALENCE: Z, DtFT, CtFT, DFT, and DFS

In this section we discuss the equivalence of the various transforms with one goal in mind—to calculate DFT and then transform it to the other representations. We start with the Z-transform and show its equivalence to the DtFT and, therefore, its equivalence to the DFT.

Suppose we have a finite-duration sequence $x(t)$, $0 \le t \le M - 1$. The Z-transform is then given by

$$X(z) = \sum_{t=0}^{M-1} x(t)z^{-t}$$

Let us replace $x(t)$ with the IDFT, that is

$$X(z) = \sum_{t=0}^{M-1}\left[\frac{1}{M}\sum_{m=0}^{M-1} X(\Omega_m)e^{j\Omega_m t}\right]z^{-t}$$

Interchanging the order of the summations and grouping, we have

$$X(z) = \frac{1}{M} \sum_{m=0}^{M-1} X(\Omega_m) \sum_{t=0}^{M-1} (e^{j\Omega_m} z^{-1})^t$$

or, using the finite sum geometric series,

$$X(z) = \frac{1}{M} \sum_{m=0}^{M-1} X(\Omega_m) \left(\frac{1 - z^{-M}}{1 - e^{j\Omega_m} z^{-1}} \right)$$

or
$$X(z) = \frac{1 - z^{-M}}{M} \sum_{m=0}^{M-1} \frac{X(\Omega_m)}{1 - e^{j\Omega_m} z^{-1}} \qquad (3.3\text{--}1)$$

     This shows that the Z-transform of the finite-duration signal can be obtained from the DFT coefficients, as indicated in Eq. (3.3–1). If this equation is evaluated on the unit circle, $z = e^{j\Omega}$, then the DtFT is obtained, and we have

$$X(e^{j\Omega}) = \frac{1 - e^{-j\Omega M}}{M} \sum_{m=0}^{M-1} \frac{X(\Omega_m)}{1 - e^{j\Omega_m} e^{-j\Omega}}$$

or
$$X(e^{j\Omega}) = \frac{e^{-j\Omega M/2}}{M e^{j(\Omega_m - \Omega)/2}} \sum_{m=0}^{M-1} \frac{X(\Omega_m) \sin(\Omega M/2)}{\sin[(\Omega - \Omega_m)/2]}$$

or
$$X(e^{j\Omega}) = \sum_{m=0}^{M-1} \frac{X(\Omega_m)}{M} \left[\frac{e^{-j\Omega(M-1/2)} \sin(\Omega M/2)}{e^{j\Omega M/2} \sin[(\Omega - \Omega_m)/2]} \right] \qquad (3.3\text{--}2)$$

where the sinusoidal functions act as interpolation functions for the DFT coefficients $X(\Omega_m)$.[†] In fact, from the DtFT derivation of the previous section, we see that by increasing the period so that $X(\Omega_m) \to X(\Omega)$, we are actually interpolating between values of the DFT, using these sinusoidal interpolators. If we are interested in the continuous-time Fourier transform (CtFT), then we can use the rectangular integration approximation:

$$X(\omega) := \text{CtFT}\,[x_c(\tau)] = \int_{-\infty}^{\infty} x(\tau) e^{-j\omega\tau} d\tau \approx \Delta T \sum_{t=-\infty}^{\infty} x(t) e^{-j\Omega t} \qquad (3.3\text{--}3)$$

or simply
$$\text{CtFT}\,[x_c(\tau)] \approx \Delta T\, \text{DtFT}\,[x(t)] \qquad (3.3\text{--}4)$$

We also note that the DFT is simply the DtFT when $\Omega \approx \Omega_m = (2\pi/M)m$, that is, when the DtFT is evaluated uniformly around the unit circle

---

[†]Note that the DFS coefficients are $a_m = X(\Omega_m)/M$, so we can think of interpolating between the spectral lines of the DFS as well.

TABLE 3.3.1
Transform equivalence

	DFT Relation	Approximation
Z	$X(z) = \displaystyle\sum_{m=0}^{m-1} \dfrac{X(\Omega_m)}{M}\left[\dfrac{1-z^{-M}}{1-e^{j\Omega_m}z^{-1}}\right]$	$\lim_{M\to\infty} \text{DFT}\,[x(t)]$ $(\Delta f \to 0)$
DtFT	$X(\Omega) = \displaystyle\sum_{m=0}^{M-1} \dfrac{X(\Omega_m)}{M}\left[\dfrac{e^{j\Omega[(M-1)/2]}\sin(\Omega M/2)}{e^{-j\Omega M/2}\sin[(\Omega-\Omega_m)/2]}\right]$	$\lim_{M\to\infty} \text{DFT}[x(t)]$ $(\Delta f \to 0)$
CtFT	$X(\omega) \approx \Delta T X(\Omega)$	$\Delta T \lim_{M\to\infty} \text{DFT}\,[x(t)]$
DFS	$a_m = \dfrac{X(\Omega_m)}{M}$	$\dfrac{1}{M}\text{DFT}\,[x(t)]$

	Z	DtFT	CtFT	DFT	DFS
Z	1			$M \to \infty$	
DtFT	$e^{j\Omega}$	1	$\dfrac{1}{\Delta T}$	$M \to \infty$	
CtFT		ΔT	1		
DFT	$e^{j\Omega_m}$	$\Omega = \Omega_m$		1	M
DFS				$\dfrac{1}{M}$	1

$$X(\Omega_m) = X(\Omega)\big|_{\Omega=(2\pi/M)m} = \sum_{t=0}^{M-1} x(t)e^{-j(2\pi/M)mt} \qquad (3.3\text{--}5)$$

or $$\text{DFT}\,[x(t)] = \text{DtFT}\,[x(t)]\big|_{\Omega=\Omega_m} \qquad (3.3\text{--}6)$$

In summary, we see that the Z, DtFT, DFT and DFS† are related simply by

$$X(z) \xrightarrow{e^{j\Omega}} X(\Omega) \xrightarrow{(2\pi/M)m} X(\Omega_m) \xrightarrow{1/M} a_m$$

---

†We derived the relationship between the DFT and DFS in Sec. 3.1, Eq. (3.1–6).

We summarize these relations in Table 3.3–1. The following example illustrates these ideas.

Example 3.3–1. Suppose we have the following finite-duration signal

$$x(t) = \alpha^t \qquad \text{for } 0 \le t \le M - 1, \ |a| < 1$$

and we would like to find the Z, DtFT, CtFT, DFT, and DFS representations. From the definition of Z-transform and geometric series, we have

$$X(z) = \sum_{t=0}^{M-1} \alpha^t z^{-t} = \sum_{t=0}^{M-1} (\alpha z^{-1})^t = \frac{1 - \alpha^M z^{-M}}{1 - \alpha z^{-1}}$$

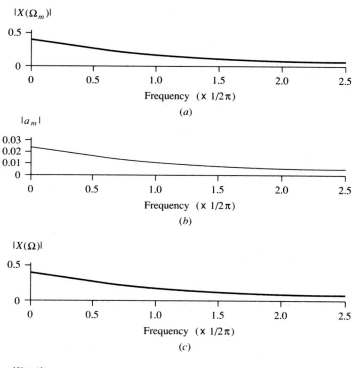

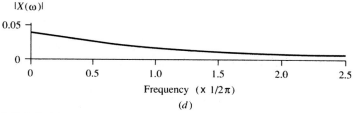

FIGURE 3.3–1
Transform equivalence of $x(t) = (3/4)^t$: (a) DFT (b) DFS (c) DtFT (d) CtFT

The DtFT is obtained by directly substituting $z = e^{j\Omega}$

$$X(\Omega) = \frac{1 - \alpha^M e^{-j\Omega M}}{1 - \alpha e^{-j\Omega}}$$

The CtFT is simply a scaling by ΔT

$$X(\omega) \approx \Delta T X(\Omega) = \Delta T \left(\frac{1 - \alpha^M e^{-j\Omega M}}{1 - \alpha e^{-j\Omega}} \right)$$

The DFT can be obtained directly from the DtFT by substituting $\Omega = \Omega_m = (2\pi/M)m$ and assuming $x(t)$ is periodic with period M:

$$X(\Omega_m) = \frac{1 - \alpha^M e^{-j\Omega_m M}}{1 - \alpha e^{-j\Omega_m}} = \frac{1 - \alpha^M}{1 - \alpha e^{-j(2\pi/M)m}}$$

and the DFS is simply

$$a_m = \frac{X(\Omega_m)}{M} = \frac{1}{M} \left(\frac{1 - \alpha^M}{1 - \alpha e^{-j(2\pi/M)m}} \right)$$

For this problem, we select $\Delta T = 0.1$ sec, $\alpha = 3/4$, $M = 16$ and use SIG [1] to simulate the various transforms. The results are shown in Fig. 3.3–1. On the computer, we calculate the DFT as shown in (a). The DFS is obtained immediately by a $1/M = 0.0625$ scaling as depicted in (b). The DtFT is found by interpolating with sinusoids, that is, letting $M = 1024$ in (c). Finally, the CtFT is found by multiplying the DtFT by ΔT as shown in (d).

3.4   PRACTICAL ASPECTS OF THE DFT

In this section, we show how the DFT can be used as a tool to analyze deterministic signals and transform them to more desirable representations. Before we employ the transformations developed in the previous section (see Table 3.3–1), let us consider the various aspects involved in the DFT calculation for a sampled signal.

In reality, most signals we deal with in nature are continuous; however, through the sampling theorem discussed in Chapter 2, we are able to reconstruct the continuous signal from its samples. Therefore, let us consider the calculation of the DFT from a general continuous-time signal and dissect the various aspects of the operations to evaluate their effect on the final DFT spectrum. In essence, the process involves

1. Continuous signal $[x_c(\tau), X_c(\omega)]$
2. Impulse sampler $[p(\tau), P(\omega)]$
3. Sampling process $[x_s(\tau), X_s(\omega)]$
4. Finite data window $[w(\tau), W(\omega)]$
5. Windowed data $[x_w(\tau), X_w(\omega)]$
6. Frequency sampler $[f(\tau), F(\omega)]$
7. Frequency sampling $[x(t), X(\Omega_m)]$

The separate aspects of the DFT calculation are depicted in Fig. 3.4–1. Here we see a general continuous signal and corresponding bandlimited spectrum. Next we note the idealized impulse sampler and spectrum (see Sec. 2.5) where the spectra are amplitude-scaled by $(2\pi/\Delta T)$ and located at integer multiples of the sampling frequency ω_s. Since the sampling process involves the modulation (multiplication) of the sampler with the signal, the resulting sampled spectrum is given as before by

$$X_s(\omega) = \frac{1}{\Delta T} \sum_k X(\omega - k\omega_s) \tag{3.4–1}$$

resulting in a periodic spectrum amplitude scaled by $1/\Delta T$ (see Fig. 3.4–1(c)). This phenomenon is a direct result of sampling a continuous signal.

Next we note that our data records are of finite duration (T_d in length); this is equivalent to "windowing" the data in the time domain with a rectangular window function. Note that the Fourier transform of the (continuous) rectangular window is a sinc function given by and shown in Fig. 3.4–1(d)

$$W(\omega) = T_d e^{-j(\omega T_d/2)} \frac{\sin(\omega T_d/2)}{\omega T_d/2} \tag{3.4–2}$$

This function is multiplied in time or convolved in frequency by the sampled spectrum. The spectrum is given by

$$X_w(\omega) = X_s(\omega) * W(\omega) \tag{3.4–3}$$

which results in "leakage" or aliasing caused by the infinite duration of the sinc function. This leakage is depicted by the "ripples" in the resulting windowed spectrum of Fig. 3.4–1(e). Since the DFT is a discrete spectral function, it can be represented by sampling in the frequency domain with an impulse sampler (see Fig. 3.4–1(f)),

$$F(\omega) = \sum_m \delta(\omega - m\omega_d) \qquad \text{with } \omega_d = \frac{2\pi}{T_d} \tag{3.4–4}$$

The frequency sampling operation is a multiplication in the frequency domain, resulting in a discrete spectrum and a convolution in the time domain. The resulting spectrum (see Fig. 3.4–1(g)) is given by

$$X(\Omega_m) = X_w(\omega)F(\omega) \tag{3.4–5}$$

which corresponds to the DFT $[x_c(\tau)]$. So we see that taking the DFT of a continuous signal results in a series of approximations (periodicities, truncation, leakage, etc.) that causes artifacts, such as ripples and aliasing, in the approximated continuous spectrum. The following example illustrates this process further.

Example 3.4–1. Suppose we have a continuous sinusoidal signal, and we want to calculate its Fourier transform (CtFT) using the DFT technique. The signal is

$$x_c(\tau) = \sin 2\pi f_0 \tau \qquad \text{for } f_0 = 0.0156 \text{ Hz}$$

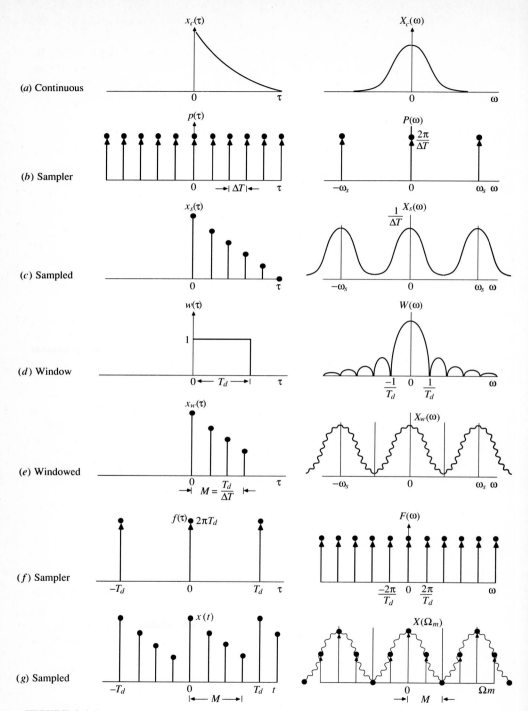

FIGURE 3.4–1
DFT calculation: (*a*) Continuous signal and spectrum (*b*) Impulse sampler and spectrum (*c*) Sampled signal and spectrum (*d*) Window and spectrum (*e*) Windowed sampled signal and spectrum (*f*) Frequency sampler and spectrum (*g*) Discrete signal (IDFT) and DFT spectrum

We sample this signal at $\Delta T = 1$ sec and assume we can only store $M = 256$ data values, so the total duration of our signal is $T_d = M\Delta T = 256$ sec. Using SIG[1] we will investigate two cases of this sinusoid: (1) with the truncated sinusoid ending precisely at a period, and (2) with the sinusoid truncated at some other point. For the sampled signal, we have

$$x_s(\tau) = x_c(\tau)\big|_{\tau=k\Delta T} = \sin 2\pi f_0 k\Delta T = \sin 0.098k$$

The CtFT of the continuous signal is given by

$$X_c(f) = \frac{1}{2j}[\delta(f - f_0) - \delta(f + f_0)]$$

We know that the spectral amplitude of the impulse is scaled by

$$\text{Amp} = \quad \left(\frac{1}{\Delta T}\right) \quad \times \quad (T_d) \quad \times \quad \left(\frac{A}{2}\right) \quad = (1)(256)\left(\frac{1}{2}\right) = 128$$

$$\text{time sampling} \qquad \text{finite duration} \qquad \text{amplitude scaling}$$

Case (1): Truncation interval equal to period.
The results for this case are shown in Fig. 3.4–2(*a*). Here we see the sinusoid truncated precisely at the period and an impulse appearing at $f_0 = 0.0156$ Hz. The sidelobe characteristics do not smear the transform, since the selected truncation interval is equal to a multiple of the period, the zeros of the sinc function (truncation) correspond precisely to the sampling function interval (see [3] for details).

Case (2): Truncation interval *not* equal to period.
The results for this case are shown in Fig. 3.4–2(*b*). The sinusoid is truncated ($M = 240$) and not equal to a period. Since the DFT assumes that the signal is periodic, we see that there is a discontinuity because the next point would be at $x(241) = 0$. We note that a nonzero *dc* value now exists because of this truncation. Smearing has occurred because of the leakage of the sinc function, and the amplitude is scaled using $M = 240$ or Amp = 120 as shown.

If we choose the CtFT representation, then we must multiply the DFT by the sampling interval:

$$\text{CtFT }[x_c(\tau)] \approx \Delta T \text{ DFT }[x(t)]$$

For this example the transforms are identical, since $\Delta T = 1$.

Three major pitfalls associated with the DFT calculation can evolve from the time-sampling, truncation, and frequency-sampling processes [4]. As discussed in Chapter 2, the sampling process (ΔT) controls the ability to reconstruct the continuous signal (spectrum) through the sampling theorem, so we must be assured that $\omega_s > 2\omega_B$ (usually $5\omega_B$ in practice). If we alias the data, then we know that the spectrum will reflect the aliased frequency about the Nyquist or folding frequency

$$\omega_{\text{aliasing}} = 2k\omega_{\text{NYQUIST}} \pm \omega_{\text{TRUE}}$$

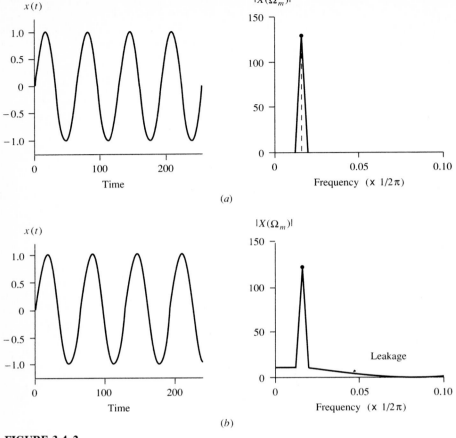

FIGURE 3.4–2
DFT of $\sin 2\pi f_0 t$: (a) Truncation interval ($M = 256$) equal to period: signal and spectrum (b) Truncation interval ($M = 240$) not equal to period: signal and spectrum

Thus, frequencies close to the Nyquist frequency will appear to be less than the Nyquist when aliased.

The leakage problem results when a finite-duration window with infinite-length spectrum (sinc) "leaks" spectral components into adjacent frequency bands due to the sidelobes of the window spectrum. The remedy for this problem is to select windows with smaller sidelobes. We summarized popular window functions in Table 2.6–1 and showed their improvement in sidelobe levels.

The spectral characteristics of the finite-duration window also lead to a problem called the "picket-fence" effect. As indicated in Fig. 3.4–1(f), the DFT calculation involves frequency sampling of the windowed transform. This effect is equivalent to observing the Fourier spectrum through a set of bandpass filters whose response is that of the window spectrum (sinc). As depicted in Fig. 3.4–3 (mainlobes only), if the signal being analyzed has spectral components *only* at

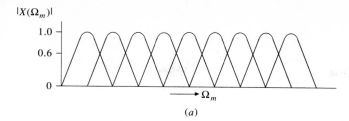

$|X(\Omega_m)|$

1.0

0.6

0

Ω_m

(a)

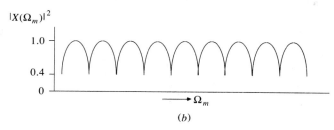

$|X(\Omega_m)|^2$

1.0

0.4

0

Ω_m

(b)

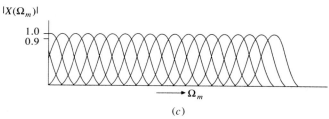

$|X(\Omega_m)|$

1.0
0.9

Ω_m

(c)

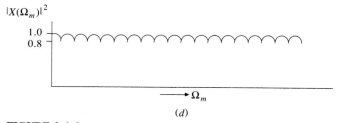

$|X(\Omega_m)|^2$

1.0
0.8

Ω_m

(d)

FIGURE 3.4-3
DFT picket-fence effect: (a) Typical DFT spectrum observed through window spectrum (b) Equivalent power spectrum picket fence (c) Improved DFT spectrum ($N \gg M$) (d) Equivalent power spectrum

integer multiples of $m/T_d = m\Delta f$, no error occurs. However, components exist between these harmonics; therefore, the frequency components are reduced by these bandpass filter spectral windows. The worst case occurs if the component falls halfway between these windows, where the amplitude is attenuated to 0.637 in adjacent windows. This case is similar to viewing the spectrum through a "picket fence," as shown in the corresponding power spectrum in Fig. 3.4-3(b). This effect can be minimized by decreasing the frequency interval ($\Delta f = 1/M\Delta T$) available in the DFT spectrum. Clearly, we can gather more data for $\Delta T' \ll \Delta T$ or, more practically, increase the number of points to $N (N \gg M)$. Then

$$\Delta f' = \frac{1}{N\Delta T} \ll \Delta f$$

and, therefore, the frequency interval will decrease, shifting the bandpass filter center frequencies closer together and minimizing the picket-fence effect, as shown in Fig. 3.4–3(c) and (d). One technique of increasing the number of points from M to N is simply to append zeros to the data record. However, the M-length record must first be windowed, and then zeros must be appended to achieve the desired results. We summarize these concepts in Table 3.4–1 and in the following example.

Example 3.4–2. Suppose we select the discrete exponential signal

$$x(t) = \left(\frac{9}{10}\right)^t$$

with DtFT

$$X(\Omega) = \frac{1}{1 - \frac{9}{10}e^{-j\Omega}}$$

and calculate the DFT for the truncated signal as

$$X(\Omega_m) = \frac{1 - (9/10)^M}{1 - \frac{9}{10}e^{-j\Omega_m}}$$

We use SIG [1] to simulate the various pitfalls and remedies as shown in Fig. 3.4–4. The true signal and spectrum are shown in dotted lines in the figure. When the exponential signal is truncated at sample 5 as shown in (a), the corresponding spectrum shows the effect of a rectangular window and sinc function spectrum. Notice that the first zero-crossing corresponding to $1/M\Delta T = 0.667$. By selecting a Hamming window (half-window for transients), we force the signal to zero, reducing the leakage and sidelobes (see Table 2.5–1). Next, we observe the effect of sampling the exponential lower than the Nyquist rate, causing aliasing in the spectrum. (Note the false spectral lines occurring above 0.5 Hz.) Finally, we observe the picket-fence effect in (d), along with the remedy of appending zeros at the end of the data record to interpolate between values, thereby moving the center frequencies of the bandpass spectra closer together.

When we are performing the analysis of the spectral properties of deterministic signals, it is important to understand to what extent we are able to distin-

TABLE 3.4–1
Practical DFT calculations

Pitfall	Remedy	Effect
Aliasing	Decrease $\Delta T(\omega_s > 2\omega_B)$	$\omega_s = 2\pi/\Delta T \gg \omega_B$
Leakage	Window (low sidelobes)	$W(t) \rightarrow \delta(t)$
Picket fence	Increase M (append zeros)	$\Delta f = 1/M\Delta T \rightarrow 0$

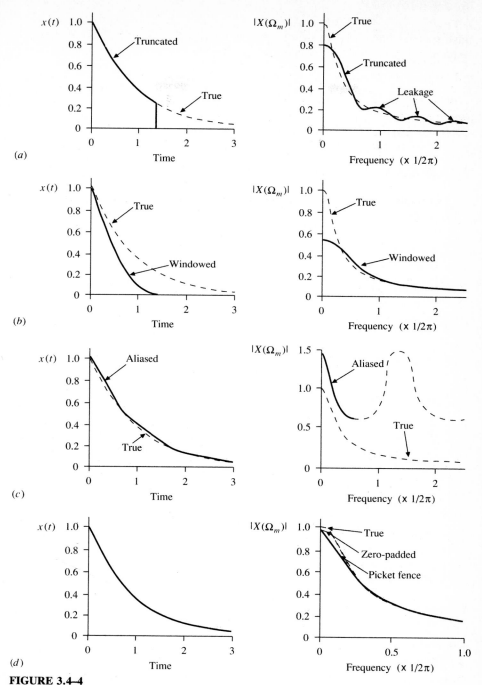

FIGURE 3.4–4
Pitfalls of the DFT for $(0.9)^t$: (a) Truncated signal and spectrum (b) Windowed signal and spectrum (c) Aliased signal and spectrum (d) Picket fence, zero padding, and spectrum

75

guish spectral lines. *Resolution* or *frequency resolution* is defined as the smallest frequency interval or "bin" available in a spectrum. We define the resolution as

$$\Delta f = \frac{1}{T_d} = \frac{1}{M\Delta T} \tag{3.4-6}$$

where T_d is the signal duration time
              M is the total number of samples or period
              ΔT is the sampling interval

The following example illustrates this concept.

Example 3.4–3. Suppose we have decided that the highest expected frequency of our data is $f_{\text{HIGH}} = 2$ Hz, and that we sample at $\Delta T = 0.1$ sec for a data length of $M = 256$ points. The frequency resolution is

$$\Delta f = \frac{1}{M\Delta T} = \frac{1}{T_d} = 0.039 \text{ Hz}$$

Therefore, any spectral lines closer than 0.039 Hz are not distinguishable or cannot be resolved. Suppose we generate sinusoids at 1, 1.02 and 1.07 Hz and calculate the corresponding DFT,

$$x(t) = \sin 2\pi t + \sin 2\pi (1.02)t + \sin 2\pi (1.07)t$$

Theoretically, we should see three spectral lines, but because of this value of Δf we only observe two: the line at 1 Hz and the line at 1.07 Hz. Using SIG [1] we show the signal and spectra in Fig. 3.4–5. Padding the signal with additional zeros ($M' = 4096$) gives an *apparent* resolution of $\Delta f = 0.0024$; however, we are still unable to resolve the 1.02 Hz spectral line (dashed lines). This can be explained by the fact that the underlying assumption of the DFT is that the signal is periodic and resolution is based on M, the assumed period. Increasing M implies that we add more signal information, that is, more cycles; however, padding with zeros is equivalent to multiplying the periodic signal by a rectangular window (note sidelobes), making it more and more aperiodic. Thus, even though padding enhances the lines, through interpolation (see following discussion) it does not increase resolution. If we increase the samples (add more cycles) to $M = 1024$, then the true resolution increases to

$$\Delta f = \frac{1}{M\Delta T} = 0.0098$$

which implies that we will have enough resolution to discern the three sinusoids. This is in fact the case, as shown in the figure. So we see that our resolution can be increased by collecting more data, by choosing a larger time-sampling interval, or by using modern techniques of spectral estimation (see Chapter 6).

It is a common practice to *increase resolution* by padding the signal with zeros, since

$$\Delta f_1 = \frac{1}{(M + M_1)\Delta T} \tag{3.4-7}$$

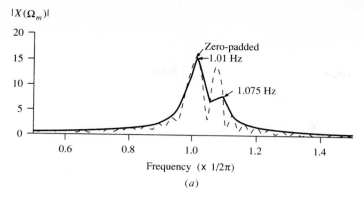

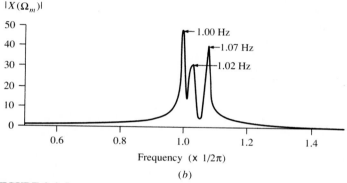

FIGURE 3.4–5

Frequency resolution of composite sinusoids at 1, 1.02, and 1.07 Hz using the DFT (*a*) Spectrum with $\Delta f = 0.039$ (*b*) Spectrum with $\Delta f = 0.0098$

where M_1 is the additional number of zeros (see [5] for details). Clearly,

$$\Delta f_1 < \Delta f$$

but the zeros have not really yielded additional signal information. In fact, we have basically caused an interpolation to take place between the values of the DFT. In the previous example this is precisely the case since padding the composite sinusoids with zeros only enhances the lines at 1 and 1.07 Hz, but does not resolve the line at 1.02 even though $\Delta f_1 \approx 0.002$ Hz, apparently.

To see this, suppose we append M_1 additional zeros to our signal and calculate the DFT

$$X(\Omega_m) = \sum_{m=0}^{(M-1)+M_1} x(t)e^{-j\Omega_m t} = \sum_{m=0}^{M-1} x(t)e^{-j[2\pi/(M+M_1)]mt}$$

since $x(t) = 0$, $M - 1 < t \le M_1$. Note that the resolution or frequency interval has decreased as given in Eq. (3.4–7) because of the appended zeros; however, the effect is to interpolate between values using the function developed in deriving the DtFT from the DFT in Sec. 3.3. Recall that

$$X(\Omega) = \sum_{m=0}^{M-1} \frac{X(\Omega_m)}{M} \left[\frac{e^{-j(M-1)/2} \sin\left[\Omega(M-1)/2\right]}{e^{-j\Omega M/2} \sin\left[\Omega M(\Omega - \Omega_m)/2\right]} \right]$$

where $M \to M + M_1$ in this case. Examples 3.4–3 and 3.4–4 illustrate this idea.

Example 3.4–4. Calculate the DFT coefficients for the case $M = 2$, then append two zeros ($M_1 = 2$) and recalculate the coefficients to investigate the resolution

Case (i): $M = 2$

$$X(\Omega_m) = \sum_{t=0}^{1} x(t)e^{-j\Omega_m t}$$

$m = 0$: $X(\Omega_0) = x(0) + x(1)e^{-j\Omega_0} = x(0) + x(1)e^{-j0} = x(0) + x(1)$

$m = 1$: $X(\Omega_1) = x(0) + x(1)e^{-j\Omega_1} = x(0) + x(1)e^{-j\pi} = x(0) - x(1)$

Case (ii): $M \to M + M_1 = 4$

$$X'(\Omega_m) = \sum_{t=0}^{3} x(t)e^{-j\Omega_m t}$$

$m = 0$: $X'(\Omega_0) = x(0) + x(1)e^{-j\Omega_0} + x(2)e^{-j2\Omega_0} + x(3)e^{-j3\Omega_0}$

$\qquad\qquad = x(0) + x(1) + x(2) + x(3) = x(0) + x(1) = X(\Omega_0)$

$m = 1$: $X'(\Omega_1) = x(0) + x(1)e^{-j\Omega_1} + x(2)e^{-j2\Omega_1} + x(3)e^{-j3\Omega_1}$

$\qquad\qquad = x(0) + x(1)e^{-j\pi/2} + x(2)e^{-j\pi} + x(3)e^{-j3\pi/2} = x(0) - jx(1)$

$m = 2$: $X'(\Omega_2) = x(0) + x(1)e^{-j\Omega_2} + x(2)e^{-j2\Omega_2} + x(3)e^{-j3\Omega_2}$

$\qquad\qquad = x(0) + x(1)e^{-j\pi} + x(2)e^{-j2\pi} + x(3)e^{-j3\pi} = x(0) - x(1)$

$\qquad\qquad = X(\Omega_1)$

$m = 3$: $X'(\Omega_3) = x(0) + x(1)e^{-j\Omega_3} + x(2)e^{-j2\Omega_3} + x(3)e^{-j3\Omega_3}$

$\qquad\qquad = x(0) + x(1)e^{-j3\pi/2} + x(2)e^{-j3\pi} + x(3)e^{-j9\pi/2} = x(0) + jx(1)$

$\qquad\qquad = [X'(\Omega_1)]^*$

So we see that at each of the original values,

$$X'(\Omega_0) = X(\Omega_0)$$
$$X'(\Omega_2) = X(\Omega_1)$$

and in between,

$$X'(\Omega_1) = [X'(\Omega_3)]^* = x(0) - jx(1) = \sqrt{x^2(0) + x^2(1)}\ \arctan\left(\frac{-x(1)}{x(0)}\right)$$

Suppose $x(0) = 1$ and $x(1) = 1/2$; then the magnitude

$$X'(\Omega_0) = X(\Omega_0) = \frac{3}{2}$$

$$X'(\Omega_1) = \sqrt{\frac{5}{2}} = [X'(\Omega)]$$

$$X'(\Omega_2) = X(\Omega_1) = \frac{1}{2}$$

as shown in Fig. 3.4–6.

So we see that appending zeros to a finite-duration signal causes an interpolation of spectral values using discrete sinusoidal functions. Therefore, this technique does not provide a "true increase" in resolution that could be obtained by decreasing the sampling interval.

We have investigated the shortcomings of the DFT caused by the three most prevalent errors: smearing, aliasing, and the picket-fence effect. We have suggested some of the popular remedies which, although simple in concept, are widely accepted in practice. The main point is that the signal processor must be aware of these problems and should not attempt to interpret artifacts caused by faulty transforms as true signal information. Before we leave the subject of the DFT, we must also mention that the DFT is calculated on the computer by a very efficient algorithm called the fast Fourier transform (FFT). The FFT was developed by Cooley and Tukey [6] and takes advantage of the symmetry of calculation on the unit circle. It reduces the N^2 calculations required by the DFT to $N \log N$ calculations. The FFT algorithm is developed in Appendix A.

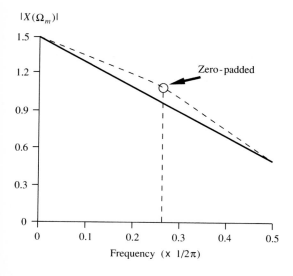

FIGURE 3.4–6
Zero-padding the DFT: $M = 2$;
$M + M_1 = 4$

3.5   SUMMARY

In this chapter we have developed the techniques commonly used in discrete signal analysis. We started with periodic signals and showed how they can be represented by the DFT. We showed that the DFT is a legitimate transform and precisely related to the DFS. Next, we developed the DtFT as the limiting case ($M \rightarrow \infty$) of the DFT in contrast to the development of the previous chapter. We showed the equivalence of the Z, DtFT, CtFT, DFT, and DFS transforms and discussed how each can be obtained from the DFT. Finally, we investigated the practical aspects of the DFT by dissecting the calculation and analyzing its constituent parts. We showed how various artifacts arise during the calculation due to limitations such as truncation or sampling. We examined the common pitfalls encountered in calculating the DFT and discussed the popular remedies. We mentioned the FFT algorithm (see Appendix A for details) used for efficient DFT calculation on the computer.

REFERENCES

1. D. Lager and S. Azevedo, "SIG—A General Purpose Signal Processing Code," *Proc. IEEE*, 1987.
2. A. Oppenheim, A. Willsky, and I. Young, *Signals and Systems* (Englewood Cliffs, N.J.: Prentice-Hall, 1983).
3. E. Brigham, *The Fast Fourier Transform* (Englewood Cliffs, N.J.: Prentice-Hall, 1974).
4. G. Bergland, "A Guided Tour of the Fast Fourier Transform," *IEEE Spectrum*, 1969.
5. L. Rabiner and B. Gold, *Theory and Application of Digital Signal Processing* (Englewood Cliffs, N.J.: Prentice-Hall, 1975).
6. J. Cooley and J. Tukey, "An Algorithm for the Machine Calculation of Complex Fourier Series," *Math. Comput.*, 1965.

SIG NOTES

SIG can be used to analyze discrete signals in the frequency domain, using its suite of Fourier transform, window, and graphics commands. SIG uses the fast Fourier transform techniques to obtain the DFT of a signal and its inverse with the FFT and IFFT command. The resulting spectrum is stored as a complex array with the graphical display selected by menu (MENU FSP) or parameter file commands (PFWRITE *.*). The graphics processor enables the display of magnitude spectrum, phase spectrum, the real part of transform, and the imaginary part of the transform. It also gives the user the choice of continuous (sampled) or discrete spectral lines and display of DC value. The choice of exponential or trigonometric spectra is also available. The command processor will "automatically" append zeros to the closest power of two or allow the user to choose. As mentioned previously, SIG includes a full complement of window functions in the time domain Bartlett (TRIANGLE), BLACKMAN, HAMMING, HANNING, KAISER as well as in the frequency domain (SHARE).

EXERCISES

3.1. Derive the discrete Fourier transform (DFT) of the following finite N-length sequences
(a) $x(t) = \delta(t)$
(b) $x(t) = \delta(t - t_0)$ $0 < t_0 < N$

(c) $x(t) = e^{-\alpha t}$

(d) $x(t) = \mu(t) - \mu(t - 4)$

3.2. Find the DFT $[X(\Omega_m)]$ in terms of $x(t)$.

3.3. Let a LTI causal system be characterized by difference equation

$$y(t) = -\sum_{i=1}^{N_a} a_i y(t - i) + x(t) \qquad \text{for } N_a < N$$

Show how to find $H(e^{j\Omega_m})$ from this equation using the N-point DFT.

3.4. (a) Calculate the DFT of

$$x(t) = \begin{cases} e^{-\alpha t} & 0 < t < N - 1 \\ 0 & \text{elsewhere} \end{cases}$$

(b) Determine the corresponding Z-transform.

(c) When are the DFT and the Z-transform of $x(t)$ identical? (Show your result using (a) and (b).)

(d) Determine the corresponding DtFT of $x(t)$.

(e) When are the DtFT and Z-transform of $x(t)$ identical?

3.5. Find the DtFT of the following signals.

(a) $x(t) = (1/2)^t \mu(t) - 1/4(1/2)^{t-1} \mu(t - 1)$

(b) $y(t) = (1/3)^t \mu(t)$

(c) Using (a) and (b), determine the transfer function of the system with input $x(t)$ and output $y(t)$. Find the corresponding impulse response.

(d) For this transfer function, find the corresponding difference equation using the IDtFT.

(e) $x(t) = \sin 2t + \cos\left[(18\pi/7)t\right]$

(f) $x(t) = \cos^2 \Omega$

3.6. We are given the causal difference equation with zero initial conditions

$$y(t) - e^{-\alpha} y(t - 1) = cx(t) \qquad 0 < \alpha < 1$$

(a) Find the transfer function and corresponding impulse response, $H(z)$, $h(t)$ respectively.

(b) Use (a) to calculate the corresponding DtFT of $h(t)$.

(c) Use (b) to calculate the corresponding DFT of $h(t)$.

3.7. Suppose we are given a continuous linear constant system

$$\ddot{y}_c(\tau) - \alpha y_c(\tau) = x_c(\tau) \qquad \text{for } y(0) = \dot{y}(0) = 0$$

bandlimited such that

$$H_c(\omega) = \begin{cases} H(\omega) & |\omega| < \omega_c \\ 0 & \text{elsewhere} \end{cases}$$

with $\omega_s = 5\omega_c$.

(a) Find the corresponding discrete transfer function, $H(\Omega)$, after ideal sampling.

(b) Repeat with a so-called *zero-order hold* replacing the sampler

$$H_0(\omega) := e^{-j\omega\Delta T/2}\left[\frac{2\sin(\omega\Delta T/2)}{\omega}\right]$$

(c) What is the ideal reconstruction filter for part (b)?

3.8. Suppose we are given a LTI system characterized by the following difference equation

$$y(t) + \frac{3}{4}y(t-1) + \frac{1}{8}y(t-2) = x(t) + \frac{1}{3}x(t-1)$$

(a) Find the frequency response.
(b) Find the impulse response.
(c) Find the response to the excitation

$$x(t) = (1/3)^t \mu(t)$$

and zero initial conditions.

3.9. Suppose we are given the LTI system with impulse response

$$h(t) = (1/6)^t \mu(t)$$

(a) Let $x(t) = (1/3)^t$, and find $y(t)$.
(b) If $x(t) \rightarrow x^*(t-L)$ above, what is $y^*(t)$?

3.10. Suppose we are given the signal

$$x(t) = (0.9)^t \mu(t) + (0.5)^t \mu(t) \qquad t \geq 0$$

Find the Z, DtFT, CtFT, DFT, and DFS representations.

3.11. We are designing a sampling system with a highest expected frequency of 10 Hz. Our buffer storage length is limited to 1024 points. What is our frequency resolution at $f_s = kf_{HIGH}$ for $k = 2, 4, 6$? Suppose we fix our sampling interval to be $\Delta T = 0.05$ sec; how does our resolution change if we increase the buffer length $M = 2048, 4096$?

3.12. Calculate the frequency response of the three-point averager given by the difference equation

$$y(t) = \frac{1}{3} \sum_{i=0}^{2} x(t-i)$$

4

DISCRETE
RANDOM
SIGNALS

In this chapter we develop the idea of a discrete random signal by first drawing on an analogy from deterministic signals, then from the probabilistic concept of a stochastic process. We show how random signals can be characterized by their spectral representations. We then show how random signals can be simulated from these representations. Finally, we develop parametric models of stochastic processes, namely the ARMAX, lattice, and Gauss-Markov (state-space) representations, and we show their equivalence.

4.1 INTRODUCTION

We classify signals as deterministic if they are repeatable, that is, if continued measurements can reproduce the identical signal. However, when a signal is classified as random, it is no longer repeatable, and continued measurements produce different signals each time.

Recall that signal processing is concerned with extracting the useful information from a signal while discarding the extraneous. When a signal is classified as deterministic, then techniques from linear system theory can be applied to extract the required signal information. In analyzing a deterministic signal, we may choose a particular representation to extract specific information. For example, Fourier transformed signals enable us to easily extract spectral content:

$$\text{Transforms:} \quad x(t) \longleftrightarrow X(\Omega)$$

Systems theory plays a fundamental role in our understanding of the operations on deterministic signals. The convolution operation provides the basis for the analysis of signals passed through a linear system, as well as for analysis of its corresponding spectral content,

Convolution: $\quad y(t) \;=\; h(t) * x(t)$

Multiplication: $\quad Y(\Omega) \;=\; H(\Omega)X(\Omega)$

For example, a digital filter is designed to act on a deterministic signal, altering its spectral characteristics in a prescribed manner. That is, a low-pass filter eliminates all of the signal spectral content above its cutoff frequency, and the resulting output is characterized by

Filtering: $\quad Y(\Omega) \;=\; H_f(\Omega)X(\Omega)$

The following example illustrates the application of transforms and linear systems theory in deterministic signal processing.

> **Example 4.1–1.** Suppose we have a measured signal, $x(t)$, consisting of two sinusoids
>
> $$x(t) \;=\; \sin 2\pi(10)t \;+\; \sin 2\pi(20)t$$
>
> $$\qquad\qquad\quad 10\ \text{Hz}\qquad\qquad 20\ \text{Hz}$$
>
> $$Measurement \;=\; signal \qquad +\; disturbance$$
>
> We would like to remove the disturbance at 20 Hz. A low-pass filter $h_f(t) \leftrightarrow H_f(\Omega)$ can easily be designed to remove the disturbance and extract the signal at 10 Hz. Using SIG [1], we simulate this signal; the results are shown in Fig. 4.1–1(a). Here we see the temporal sinusoidal signal and corresponding spectrum showing the signal and disturbance spectral peaks at 10 and 20 Hz, respectively. After designing the parameters of the filter, we know from linear systems theory that the filtered or processed output, $y(t)$, and its corresponding spectrum satisfy the following relations
>
> $$y(t) = h_f(t) * x(t)$$
>
> $$Y(\Omega) = H_f(\Omega)X(\Omega)$$
>
> The results of the processing are shown in Fig. 4.1–1(b). Here we see the operation of the deterministic filter on the measured data. Analyzing the data shows that the processor has extracted the desired signal information at 10 Hz and rejected the disturbance at 20 Hz.

Clearly, we can apply a filter to a random signal, but since its output is still random, we must find a way to eliminate or reduce this randomness in order to employ the powerful techniques available from systems theory. We shall show that techniques from statistics combined with linear systems theory can be applied to extract the desired signal information and reject the disturbance or noise. In this case, the filter is called an estimation filter or simply an *estimator* and it is required to extract the useful information (signal) from noisy or random measurements.

Techniques similar to linear deterministic systems theory hold when it is possible to replace a random signal by its (auto) *covariance sequence* and its Fourier spectrum by the (auto) *power spectrum*,

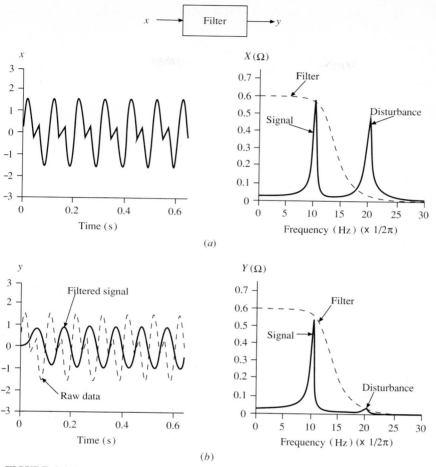

FIGURE 4.1–1

Deterministic signal processing (*a*) Analysis of raw data and spectrum (*b*) Processed data and spectrum

$$
\begin{array}{lll}
\text{(random)} & x(t) \ \rightarrow R_{xx}(k) & \text{(deterministic)} \\
\text{(random)} & X(\Omega) \rightarrow S_{xx}(\Omega) & \text{(deterministic)}
\end{array}
$$

Once this replacement is accomplished, the techniques of linear systems theory can be applied to obtain results similar to deterministic signal processing. In fact, we will show that the covariance sequence and power spectrum are a Fourier transform pair, analogous to a deterministic signal and its corresponding spectrum; that is,

$$
\text{Transform:} \quad R_{xx}(k) \longleftrightarrow S_{xx}(\Omega)
$$

and, as in the deterministic case, we can analyze the spectral content of a random signal by investigating its power spectrum.

The *power spectral density* (PSD) function for a discrete process is defined as

$$S_{xx}(\Omega) = \lim_{N \to \infty} E\left\{ \frac{X(\Omega)X^*(\Omega)}{2N + 1} \right\} \tag{4.1-1}$$

where the expected value operation, $E\{ \cdot \}$, can be thought of simply as "removing" the randomness.

Similarly, the *covariance*[†] is given by

$$R_{xx}(k) = E\{x(t)x(t + k)\} - m_x^2 \tag{4.1-2}$$

Techniques of linear systems theory for random signals are valid, just as in the deterministic case where the covariance at the output of a system excited by a random signal $x(t)$ is given by the convolutional relationship in the temporal or frequency domain as

Convolution: $R_{yy}(k) = h(k) * h(-k) * R_{xx}(k)$
Multiplication: $S_{yy}(\Omega) = H(\Omega)H^*(\Omega)S_{xx}(\Omega)$

Analogously, the filtering operation is performed by an estimation filter, $\hat{H}_f$, designed to shape the output PSD, similar to the deterministic filtering operation,

Filtering: $S_{yy}(\Omega) = |\hat{H}_f(\Omega)|^2 S_{xx}(\Omega)$

where the $\hat{\ }$ notation is used to specify an estimate. The following example analyzes a random signal using covariance and spectral relations.

Example 4.1–2. Suppose we have a measured signal given by

$$x(t) \quad = \quad s(t) \quad + \quad n(t)$$

Measurement = signal + noise

We would like to extract the signal and reject the noise, so we design an estimation filter,

$$y(t) = \hat{h}_f(t) * x(t)$$

$$Y(\Omega) = \hat{H}_f(\Omega)X(\Omega)$$

Using SIG [1], our signal in this case will be the sinusoids at 10 and 20 Hz; the noise is additive random. In Fig. 4.1–2(*a*), we see the random signal and raw (random) Fourier spectrum. Note that the noise severely obscures the sinusoidal signals and that many false peaks could erroneously be selected as sinusoidal signals. Replacing the random signal by its corresponding covariance sequence, and estimating the power spectrum using statistical methods, we can now easily see the sinusoidal signals at the prescribed frequencies, as depicted in Fig. 4.1–2(*b*). So we see that, as in deterministic signal processing techniques, replacing (estimating)

[†]It is also common to use the so-called *correlation* function, which is merely the mean-squared function in Eq. (4.1–2) and is *identical* to the covariance function for a zero-mean ($m_x = 0$) process, in these definitions.

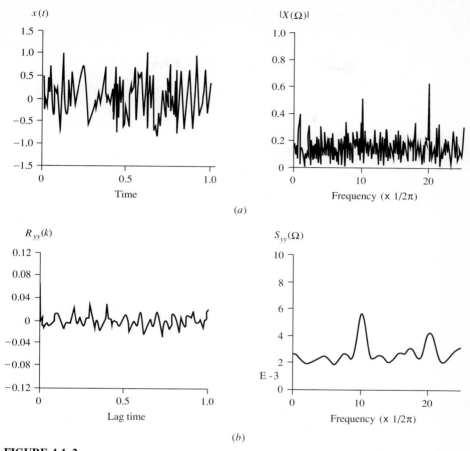

FIGURE 4.1–2
Random signal processing (*a*) Raw data and spectrum (*b*) Covariance data and power spectrum

random signals with covariance and power spectra enables us to apply the powerful methods of linear systems theory to extract useful signal information from noisy data.

In this chapter, we are concerned with the development of methods of analysis for random data and the design of estimators to extract useful signal information from noisy data. The prime impetus, however, is to develop methods to estimate the power spectrum $S(\Omega)$ from noisy data.

4.2 STOCHASTIC PROCESSES

In this section, we review some of the basic ideas of stochastic processes which are essential to characterize a random signal. We limit our discussion to discrete processes and refer the reader to a more specialized text such as [2] for more detailed information.

Stochastic processes find their roots in probability theory, which provides the basis for problem solving. The idea of an underlying sample space is essential to the treatment of stochastic processes. Recall that a *sample space*, Ξ, is a collection of samples corresponding to the set of all possible outcomes, ξ, which are elements of the space. Certain subsets of Ξ, or collections of outcomes, are called *events*. When we attempt to specify how likely it is that a particular event will occur during an experiment, we define the notion of *probability*. A *probability function*, Prob(.), is a function defined on a class of events which assigns a number to the likelihood of a particular event occurring during an experiment. We constrain the events to a set of events, B, on which the probability function is defined and satisfies certain properties [3]. Finally, we define an *experiment* by the triple $\{\Xi, B, P\}$. Let us summarize these ideas with the following example.

> **Example 4.2–1.** Consider the experiment $\{\Xi, B, P\}$ of tossing a fair coin; we then see that
>
> | | | |
> |---|---|---|
> | Sample space: | Ξ | $= \{H, T\}$ |
> | Events: | B | $= \{0, H, T, \Xi\}$ |
> | Probability: | $P(H)$ | $= p$ |
> | | $P(T)$ | $= 1 - p$ |

With the ideas of a sample space, probability function, and experiment in mind, we can start to define the concept of a discrete random signal more precisely.

Recall that a signal is considered *random* if its precise value cannot be predicted. In order to characterize this notion, we recall that a discrete *random variable* is defined as a real function whose value is determined by the outcome of an experiment. Simply, it assigns (or maps) a real number to each point of a sample space Ξ. Notationally, a random variable X is written as

$$X(\xi) = x \qquad \text{for } \xi \in \Xi \tag{4.2–1}$$

Consider the following example of a simple experiment.

> **Example 4.2–2.** We are asked to analyze the experiment of flipping a fair coin. The sample space is shown in Example 4.2–1, and the set of events consists of a head or tail as possible outcomes. That is,
>
> $$B = \left\{ \begin{array}{c} 0 \\ H \\ T \\ \Xi \end{array} \right\} \Rightarrow X(\xi) = x$$
>
> $$\xi = \{H, T\}$$
>
> If we assign a 1 for a head and 0 for a tail, then the random variable X performs the mapping of

$$X(\xi = H) = x(H) = 1$$
$$X(\xi = T) = x(T) = 0$$

where $x(\,\cdot\,)$ is called the sample value or realization of the random variable X.

Once we have a (discrete) random variable, then it is possible to define the *probability mass function* in terms of the random variable,

$$P_X(x_i) = \text{Prob}(X(\xi_i) = x_i) \tag{4.2-2}$$

and the *probability distribution function* by

$$F_X(x_i) = \text{Prob}(X(\xi_i) \le x_i) \tag{4.2-3}$$

These are related by

$$P_X(x_i) = \sum_i F_X(x_i)\delta(x - x_i)$$

$$F_X(x_i) = \sum_i P_X(x_i)\mu(x - x_i) \tag{4.2-4}$$

where δ and μ are the unit impulse and step functions, respectively.

It is easy to show that the distribution function is a monotonically increasing function (see [2] for details) satisfying the following properties:

$$\lim_{x_i \to -\infty} F_X(x_i) = 0$$

and

$$\lim_{x_i \to \infty} F_X(x_i) = 1$$

These properties can be used to show that the mass function satisfies

$$\sum_i P_X(x_i) = 1$$

Either the distribution or probability mass function completely describes the properties of a random variable. Given the distribution or mass function, we can calculate probabilities that the random variable takes on values in any set of events on the real line. To complete our coin-tossing example, if we define the probability of a head occurring as p, then we can calculate the distribution and mass functions as shown in the following example.

Example 4.2–3. Consider the coin-tossing experiment, and calculate the corresponding mass and distribution functions. From the previous example, we have

Sample space:	Ξ	$= \{H, T\}$
Events:	B	$= \{0, H, T, \Xi\}$
Probability:	$P_X(x_1 = H)$	$= p$
	$P_X(x_0 = T)$	$= 1 - p$

Random variable: $X(\xi_1 = H) = x_1 = 1$
$X(\xi_2 = T) = x_2 = 0$

Distribution: $F_X(x_i) = \begin{cases} 1 & x_i \geq 1 \\ 1 - p & 0 \leq x_i \leq 1 \\ 0 & x_i < 0 \end{cases}$

The mass and distribution functions for this example are shown in Fig. 4.2–1. Note that the sum of the mass function value must be 1, and that the maximum value of the distribution function is 1, satisfying the properties mentioned previously.

If we extend the idea that a random variable is now a function of time as well, then we can define a stochastic process. More formally, a random or *stochastic process* is a two-dimensional function of t and ξ:

$$X(t, \xi) \quad \xi \in \Xi, \quad t \in T \tag{4.2–5}$$

where T is a set of index parameters (continuous or discrete), and Ξ is the underlying sample space. There are four possibilities of $X(\cdot, \cdot)$ above:

(i) For any t, ξ varying $\quad - X(t, \xi) \quad -$ random time function
(ii) For fixed t and ξ varying $- X(t_i, \xi) \quad -$ random variable
(iii) For fixed ξ and t varying $- X(t, \xi_i) \quad -$ stochastic process
(iv) For fixed ξ and fixed $t \quad - X(t_i, \xi_i) -$ number

So we see that a discrete random signal can be precisely characterized by

$$x_i(t) := X(t, \xi_i) \tag{4.2–6}$$

This is a *realization* of a stochastic process and in fact *each* distinct value of time can be interpreted as a random variable. Thus, we can consider a stochastic process as simply a sequence of ordered (in time) random variables. A collection of realizations of a stochastic process is called an *ensemble*. In Fig. 4.2–2, we see an example of an ensemble of three realizations of a random process; that is, for each ξ_i, a sequence of random numbers is generated to form the ensemble ($\{X(t, \xi_1)\}, \{X(t, \xi_2)\}, \{X(t, \xi_3)\}$). Here each realization is simulated by selecting a

$$P_x(x) = \begin{cases} 1 - p & x = 0 \\ p & x = 1 \\ 0 & \text{Otherwise} \end{cases}$$

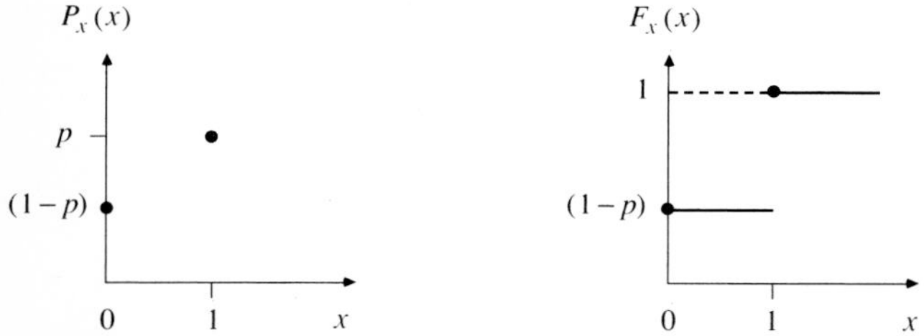

FIGURE 4.2–1
Probability mass and distribution functions for coin tossing experiment

$X(t, \xi_1)$

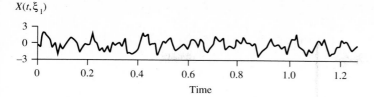

$X(t, \xi_2)$

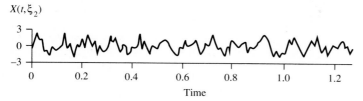

$X(t, \xi_3)$

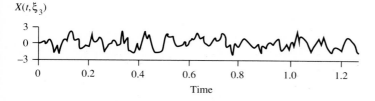

$m_x(t)$

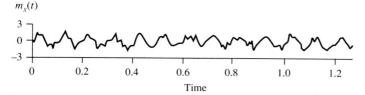

FIGURE 4.2–2
Ensemble of three realizations of a random process and the corresponding ensemble mean

different seed for the random number generator. A more common representation of a random process is given in the next example.

Example 4.2–4. (Random Walk) Suppose we toss a fair coin at each time instant; that is,

$$X(t, \xi_i), \qquad t \in T = \{0, 1, ..., N - 1\}, \qquad \xi \in \Xi = \{H, T\}$$

where

$$X(t, \xi_i) = \begin{cases} K & \xi_1 = H \\ -K & \xi_2 = T \end{cases}$$

We define the corresponding probability mass function as

$$P_X(X(t, \xi_i) = \pm K) = 1/2$$

Here we see that for each $t \in T$, $X(t, \xi)$ is just a random variable.

Now if we define the function

$$y(N, \xi_i) = \sum_{t=0}^{N-1} X(t, \xi_i) \qquad i = 1, 2, ...$$

as the position at t for the ith realization, then for each t, y is a sum of discrete random variables, $X(\,\cdot\,,\xi_i)$, and for each i, $X(t,\xi_i)$ is the ith realization of the process. If for a given trial, we have $\xi_1 = \{HHTHTT\}$ and $\xi_2 = \{THHTTT\}$, then for these realizations we see the ensemble depicted in Fig. 4.2–3. Note that $y(5,\xi_1) = 0$, and $y(5,\xi_2) = -2K$, which can be predicted using the given random variables and realizations.

In summary, we see that a random signal can be represented precisely by a discrete stochastic process, which is a two-dimensional function of time (or an index set) and of samples of the sample space. Since a random signal can be thought of as a sequence of ordered (in time) random variables, standard statistical definitions may be extended.

For a random variable, we can define basic statistics in terms of the probability mass function. The *expected value* or *mean* of a random variable X is given by

$$m_x = E\{X\}$$

and is considered the typical or representative value of a given set of data. For this reason, the mean is called a measure of *central tendency*. The degree to which numerical data tend to spread about the expected value is usually measured by the *variance*, or equivalently, the *auto-covariance*, given by

$$R_{xx} = E\{(X - m_x)^2\}$$

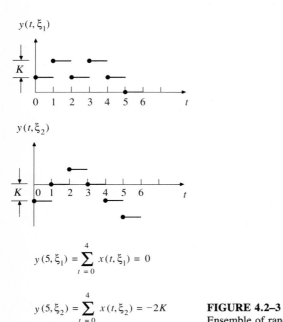

$$y(5,\xi_1) = \sum_{t=0}^{4} x(t,\xi_1) = 0$$

$$y(5,\xi_2) = \sum_{t=0}^{4} x(t,\xi_2) = -2K$$

FIGURE 4.2–3
Ensemble of random walk realizations

The basic statistical measures are called *ensemble statistics*, because they are measured across the ensemble ($i = 1, 2, \ldots$) of data values; that is, the expectation is always assumed over an ensemble of realizations. We summarize these statistics in terms of their mass function as:

$$
\begin{aligned}
\text{Expected value:} && m_x &= E\{X\} = \sum_i X_i P_X(x_i) \\
\text{Mean squared:} && E\{X^2\} &= \sum_i X_i^2 P_X(x_i) \\
\text{Variance:} && R_{xx} &= E\{(X_i - m_x)^2\} = \sum_i (X_i - m_x)^2 P_X(x_i) \\
\text{Covariance:} && R_{xy} &= E\{(X_i - m_x)(Y_j - m_y)\} \\
\text{Standard deviation:} && \sigma_{xx} &= \sqrt{R_{xx}}
\end{aligned}
\tag{4.2--7}
$$

These basic statistics possess various properties that enable them to be useful for analyzing operations on random variables. Some of the more important are the following:

$$
\begin{aligned}
\text{Linearity:} && E\{ax + b\} &= aE\{x\} + b = am_x + b \\
\text{Independence:}^{\dagger} && E\{xy\} &= E\{x\}E\{y\} \\
\text{Variance:} && R_{xx}(ax + b) &= a^2 R_{xx} \\
\text{Covariance:} && & \\
\quad \text{Uncorrelated:} && E\{xy\} &= E\{x\}E\{y\} \qquad \{R_{xy} = 0\} \\
\quad \text{Orthogonal:} && E\{xy\} &= 0
\end{aligned}
\tag{4.2--8}
$$

Since a random signal can be interpreted as a sequence of random variables, then we expect the basic statistics and properties to still apply. In fact, this is the case—the resulting statistic merely becomes a function of the index parameter t (time) in our case. So we see that the basic statistics and properties also hold for stochastic processes as well:

$$
\begin{aligned}
\text{Expected value:} && m_x(t) &= E\{x(t)\} = \sum_i x_i(t) P_X(x_i(t)) \\
\text{Mean squared:} && \Psi_{xx}(t, k) &= E\{x(t)x(k)\} \\
\text{Variance:} && R_{xx}(t, k) &= E\{(x(t) - m_x(t))(x(k) - m_x(k))\} \\
&& &= \Psi_{xx}(t, k) - m_x(t)m_x(k) \\
\text{Covariance:} && R_{xy}(t, k) &= E\{(x(t) - m_x(t))(y(k) - m_y(k))\} \\
\text{Linearity:} && E\{ax(t) + b\} &= aE\{x(t)\} + b = am_x(t) + b \\
\text{Independence:} && E\{x(t)y(t)\} &= E\{x(t)\}E\{y(t)\} \\
\text{Covariance:} && & \\
\quad \text{Uncorrelated:} && E\{x(t)y(t)\} &= E\{x(t)\}E\{y(t)\} \quad \{R_{xy}(t, k) = 0\} \\
\quad \text{Orthogonal:} && E\{x(t)y(t)\} &= 0
\end{aligned}
\tag{4.2--9}
$$

---

†Independence states that the joint mass function can be factored, $p(x, y) = p(x)p(y)$, which leads to these properties.

The following example demonstrates the basic statistical operations on a discrete stochastic process.

Example 4.2–5. Suppose we have a discrete stochastic process given by

$$y(t) = a + bt$$

where a and b are random variables. The mean and variance of this process can then be calculated by

$$m_y(t) = E\{y(t)\} = E\{a + bt\} = E\{a\} + E\{b\}t = m_a + m_b t$$

$$R_{yy}(t, k) = E\{(a + bt)(a + bk)\} - m_y(t)m_y(k)$$

$$= E\{a^2\} + E\{ab\}(t + k) + E\{b^2\}tk - m_y(t)m_y(k)$$

$$R_{yy}(t, k) = \Psi_{aa} + R_{ab}(t + k) + \Psi_{bb}(tk) - m_y(t)m_y(k)$$

Note that the expected value operation implies that for stochastic processes these basic statistics are calculated *across* the ensemble. For example, if we want to calculate the mean of a process,

$$m_x(t) = E\{X(t, \xi_i) = x_i(t)\}$$

we simply take the values of $t = 0, 1, \ldots$ and calculate the mean for each value of time ($i = 1, 2, \ldots$) across the ensemble. In Fig. 4.2–2 we see an ensemble of three random processes and the corresponding expected value. Dealing with stochastic processes is similar to dealing with random variables, except that we must account for the time indices.

Before we consider the case with more than one random variable, let us define some concepts about the probabilistic information contained in a random variable. These probabilistic concepts are applied extensively in communications problems and will prove useful in designing parametric signal processors. We define the information contained in the occurrence of the random variable $X(\xi_i) = x_i$ as

$$I(x_i) = -\log_b P_X(x_i) \tag{4.2–10}$$

b is the base of the logarithm which determines the units of the information measure (base $= 2 \rightarrow$ bits) and the *entropy* or *average information* of $X(\xi_i)$ as

$$H(x_i) = E\{I(x_i)\} = -\sum_i P_X(x_i)\log_b P_X(x_i) \tag{4.2–11}$$

Simply knowing that a random signal is characterized by a stochastic process is not sufficient to make this information useful. We must consider properties of stochastic processes which make them useful, but as before we must first develop the ideas from probability theory and then define them in terms of stochastic processes. Suppose we have more than one random variable; then we define the *joint* mass and distribution functions of an N-dimensional random variable, respectively, as

$$P_X(x_1, \ldots, x_N), \qquad F_X(x_1, \ldots, x_N)$$

All of the basic statistical definitions remain as before, except that we replace the scalar with the joint functions. Clearly, if we think of a stochastic process as a sequence of ordered random variables, then we are dealing with joint probability functions. Suppose we have two random variables, x_1 and x_2, and we know that the latter has already assumed a particular value; then we can define the *conditional* probability mass function of x_1, given that $X(\xi_2) = x_2$ has occurred, by

$$p(x_1 \mid x_2) := P_X(X(\xi_1) \mid X(\xi_2) = x_2) \qquad (4.2\text{–}12)$$

and it can be shown from basic probabilistic axioms (see [2]) that

$$p(x_1 \mid x_2) = \frac{p(x_1, x_2)}{p(x_2)} \qquad (4.2\text{–}13)$$

Note that this equation can also be written as

$$p(x_1, x_2) = p(x_2 \mid x_1)p(x_1) \qquad (4.2\text{–}14)$$

Substituting this equation into Eq. (4.2–13) gives *Bayes' rule*,

$$p(x_1 \mid x_2) = p(x_2 \mid x_1)\frac{p(x_1)}{p(x_2)} \qquad (4.2\text{–}15)$$

If we use the definition of joint mass function and substitute in the previous definitions, then we obtain the *chain rule*,

$$p(x_1, \ldots, x_N) = p(x_1 \mid x_2, \ldots, x_N)p(x_2 \mid x_3, \ldots, x_N) \cdots p(x_{N-1} \mid x_N)p(x_N) \quad (4.2\text{–}16)$$

Along with these definitions follows the idea of *conditional expectation*,

$$E\{x_i \mid x_j\} = \sum_i X_i p(x_i \mid x_j)$$

With these definitions in mind, we can now examine further properties of stochastic processes that will prove useful in characterizing random signals. Note that the conditional expectation is also a linear operator which possesses many useful properties (see Exercise 4.1):

$$E\{ax_i + b \mid x_j\} = aE\{x_i \mid x_j\} + b \qquad (4.2\text{–}17)$$

The concepts of information and entropy can also be extended to the case of more than one random variable. We define the *mutual information* between two random variables, x_i and x_j, as

$$I(x_i, x_j) = \log_b \frac{P_X(x_i \mid x_j)}{P_X(x_i)} \qquad (4.2\text{–}18)$$

and the *average mutual information* between $X(\xi_i)$ and $X(\xi_j)$ as

$$I(X_i; X_j) = E_{x_i x_j}\{I(x_i, x_j)\} = \sum_i \sum_j P_X(x_i, x_j)I(x_i, x_j) \qquad (4.2\text{–}19)$$

which leads to the definition of *joint entropy* as

$$H(X_i; X_j) = -\sum_i \sum_j P_X(x_i, x_j) \log_b P_X(x_i, x_j) \qquad (4.2\text{--}20)$$

A common class of useful stochastic processes is called *Markov processes*. Markov processes are defined in terms of a sequence of random variables $\{X(t)\}$; we say that a process is a *Markov* process if and only if

$$p(X(t+1) \mid X(t), X(t-1), \cdots, X(1)) = p(X(t+1) \mid X(t)) \qquad (4.2\text{--}21)$$

That is, the future $X(t+1)$ depends only on the present $X(t)$ and not the past $\{X(t-1), \cdots, X(1)\}$. Thus, if a process is a Markov process, then the chain rule of Eq. (4.2–14) simplifies considerably:

$$p(x_1, \ldots, x_N) = p(x_1 \mid x_2)p(x_2 \mid x_3) \cdots p(x_{N-1} \mid x_N)p(x_N)$$

Just as in systems theory the concept of a time-invariant system is essential for analytical purposes, so is the equivalent concept of stationarity essential for the analysis of stochastic processes. We say that a stochastic process is *stationary* if the joint mass function is time-invariant, that is,

$$p(X(1)\ldots X(N)) = p(X(1+k)\ldots X(N+k)) \qquad \forall k \qquad (4.2\text{--}22)$$

If the equation is only true for values of $t \le M < N$, then it is said to be *stationary of order M*. Special cases arising are

1. For $M = 1$, $p(X(t)) = p(X(t+k)) = p(X) \ \forall k \Rightarrow m_x(t) = m_x$ a constant (*mean stationary*).
2. For $M = 2$, $p(X(t_1), X(t_2)) = p(X(t_1 + k), X(t_2 + k))$ or for $k = t_2 - t_1$, $p(X(t_2), X(t_2 + k))$ (*covariance stationary*) which implies

$$R_{xx}(t, t+k) = R_{xx}(k) = E\{x(t)x(t+k)\} - m_x^2$$

depends only on the lag k.
3. A process that is both mean and covariance stationary is called *wide-sense stationary* (WSS).

Example 4.2–6. Determine if the following process is WSS:

$$x(t) = A \cos t + B \sin t$$

where A and B are uncorrelated, zero-mean, random variables with variance σ^2.

$$m_x(t) = E\{A\} \cos t + E\{B\} \sin t = 0$$

$$R_{xx}(t, k) = E\{x(t)x(t+k)\} - 0$$

$$= E\{(A \cos t + B \sin t)(A \cos(t+k) + B \sin(t+k))\}$$

$$= E\{A^2\} \cos t \cos(t+k) + E\{B^2\} \sin t \sin(t+k)$$

$$\qquad + E\{AB\} \sin t \cos(t+k) + E\{AB\} \cos t \sin(t+k)$$

or, since A and B are uncorrelated,

$$R_{xx}(t, k) = \sigma^2 \cos(t - t + k) = \sigma^2 \cos k$$

which implies the process is wide-sense stationary.

A process is said to be *ergodic* if its time average is identical to its ensemble average, that is,

$$E\{y(t, \xi_i)\} = E_T\{y(t)\} := \lim_{N \to \infty} \frac{1}{2N + 1} \sum_{t=-N}^{N} y(t) \qquad (4.2\text{-}23)$$

where $E_T\{ \cdot \}$ is the associated *time average*.

This means that all of the statistical information in the ensemble can be obtained from one realization of the process. For instance, reconsider the ensemble in Fig. 4.2–2. In order to calculate the mean of this process, we must average across the ensemble for each instant of time to determine, $m_x(t)$; however, if the process is ergodic, then we may select any member (realization) of the ensemble and calculate its time average, $E_T\{x_i(t)\}$, to obtain the required mean:

$$m_x(t) = m_x$$

Note that an ergodic process must be wide-sense stationary, but that the inverse does not necessarily hold.

Two important processes will be used extensively to model random signals. The first is the *uniformly* distributed process, which is specified by the mass function

$$P_X(x_i) = \frac{1}{b - a} \qquad a < x_i < b \qquad (4.2\text{-}24)$$

or, simply,

$$x \sim U(a, b)$$

where the random variable can assume any value within the specified interval. The corresponding mean and variance of the uniform random variable is given by

$$m_x = \frac{b + a}{2}, \qquad R_{xx} = \frac{(b - a)^2}{12} \qquad (4.2\text{-}25)$$

Secondly, the *gaussian* or *normal process* is defined by its probability mass function

$$p(X(t)) = \frac{1}{\sqrt{2\pi R_{xx}}} \exp\left\{ -\frac{1}{2} \frac{(X(t) - m_x)^2}{R_{xx}} \right\} \qquad (4.2\text{-}26)$$

or, simply,

$$x \sim N(m_x, R_{xx})$$

where m_x and R_{xx} are the respective mean and variance completely characterizing the gaussian process, $X(t)$.

The *central-limit theorem* makes gaussian processes very important, since it states that the sum of a large number of independent random variables tends to be gaussian; that is, for $\{x_i(t)\}$ independent, then $y(t) \sim N(0, 1)$ where

$$y(t) = \sum_{t=1}^{N} \frac{(x_i(t) - m_x(t))}{\sqrt{NR_{xx}(t)}} \tag{4.2-27}$$

Both of these distributions will become very important when we attempt to simulate stochastic processes on the computer. Other useful properties of the gaussian process follow:

1. *Linear transformation.* Linear transformations of gaussian variables are gaussian; that is, if $x \sim N(m_x, R_{xx})$ and $y = ax + b$, then

$$y \sim N(am_x + b, a^2 R_{xx}) \tag{4.2-28}$$

2. *Uncorrelated gaussian variables.* Uncorrelated gaussian variables are independent.

3. *Sums of gaussian variables.* Sums of independent gaussian variables yield a gaussian distributed variable with mean and variance equal to the sums of the respective means and variances; that is,

$$x(i) \sim N(m_x(i), R_{xx}(i)) \qquad \text{and} \qquad y = \sum_i k_i x(i)$$

Then
$$y \sim N\left(\sum_i k_i m_x(i), \sum_i k_i^2 R_{xx}(i) \right) \tag{4.2-29}$$

4. *Conditional gaussian variables.* Conditional gaussian variables are gaussian distributed; that is, if x_i and x_j are jointly gaussian, then

$$E\{x_i \mid x_j\} = m_{x_i} + R_{x_i x_j} R_{x_j x_j}^{-1} (x_j - m_{x_j})$$

and
$$R_{x_i|x_j} = R_{x_i x_i} - R_{x_i x_j} R_{x_j x_j}^{-1} R_{x_j x_i} \tag{4.2-30}$$

5. *Orthogonal errors.* The estimation error $\tilde{x}$ is orthogonal to the conditioning variable; that is, for $\tilde{x} = x - E\{x \mid y\}$, then

$$E\{y\tilde{x}\} = 0$$

6. *Fourth gaussian moments.* The fourth moment of zero-mean gaussian variables is given by

$$E\{x_1 x_2 x_3 x_4\} = E\{x_1 x_2\} E\{x_3 x_4\} + E\{x_1 x_3\} E\{x_2 x_4\} + E\{x_1 x_4\} E\{x_2 x_3\} \tag{4.2-31}$$

In the course of engineering applications, the ideas of sample spaces sometimes get lost in an effort to simplify notation, so it is very important for the reader to be aware that in many engineering texts on random signals, $x(t)$ is defined as a stochastic process. It is assumed that the reader will know that

$$x(t) \rightarrow X(t, \xi_i) \qquad t \in T, \qquad \xi \in \Xi$$

We now have all of the probabilistic ingredients to characterize useful random signals and the corresponding tools to analyze and synthesize them.

4.3 SPECTRAL REPRESENTATION OF RANDOM SIGNALS

Before the design of processing algorithms can proceed, in many engineering problems it is necessary to analyze the measured signal. When possible, the covariance function replaces the random signal and the power spectrum replaces its transform. In this section, we derive the relationship between the covariance function and power spectrum and show how they can be used to characterize fundamental random signals. We also introduce the concept of a coherence function, another tool which can be used to analyze random signals.[†]

First, we begin by defining the power spectrum of a discrete random signal. Recall that the discrete-time Fourier transform DtFT pair is given by

$$x(t) := \text{IDtFT } [X(e^{j\Omega})] = \frac{1}{2\pi} \int_{2\pi} X(e^{j\Omega}) e^{j\Omega t} d\Omega$$

$$X(e^{j\Omega}) = \text{DtFT } [x(t)] = \sum_{t=-\infty}^{\infty} x(t) e^{-j\Omega t} \tag{4.3-1}$$

Recall that the transform converges if $x(t)$ is absolutely summable, that is,

$$\sum_{t=-\infty}^{\infty} |x(t)| < \infty$$

or if the sequence has finite energy

$$\sum_{t=-\infty}^{\infty} |x(t)|^2 < \infty$$

If $x(t)$ is random, then $X(e^{j\Omega})$ is *also* random, because

$$x(t, \xi_i) \Leftrightarrow X(e^{j\Omega}, \xi_i) \qquad \forall\, i$$

Both are simply realizations of a stochastic process over the ensemble generated by i. Also, and more importantly, $X(e^{j\Omega})$ for stationary processes almost never exists, because any non-zero realization $x(t, \xi_i)$ is not absolutely summable in the ordinary sense. (These integrals can be modified (see [3] for details).) Even if the Z-transform is applied instead of the DtFT, it can be shown that convergence problems will occur in the inverse transform. So we must somehow modify the constraints on our signal to ensure convergence of the DtFT.

Consider a finite-duration realization defined by

$$x_N(t) := \begin{cases} x_N(t, \xi_i) & |t| \leq N < \infty \\ 0 & |t| > N \end{cases} \tag{4.3-2}$$

Note that $x_N(t)$ will be absolutely summable (N finite) if $x(t)$ has a finite mean-squared value. In fact, $x_N(t)$ will also have finite energy; therefore, it will be

---

[†]We shall return to more mathematically precise notation for this development, that is, $X(\Omega) = X(e^{j\Omega})$.

Fourier transformable. The average power of $x_N(t)$ over the interval $(-N, N)$ is

$$\text{Average power} = \frac{1}{2N+1} \sum_{t=-N}^{N} x^2(t) = \frac{1}{2N+1} \sum_{t=-\infty}^{\infty} x_N^2(t) \quad (4.3\text{–}3)$$

but, by Parseval's theorem for discrete signals, we have

$$\sum_{t=-\infty}^{\infty} x_N^2(t) = \frac{1}{2\pi} \int_{2\pi} |X_N(e^{j\Omega})|^2 d\Omega \quad (4.3\text{–}4)$$

where $|X_N(e^{j\Omega})|^2 = X_N(e^{j\Omega})X_N^*(e^{j\Omega})$ and $*$ is the conjugate. Substituting Eq. (4.3–4) into Eq. (4.3–3), we have

$$\text{Average power} = \int_{2\pi} \left(\frac{|X_N(e^{j\Omega})|^2}{2N+1} \right) \frac{d\Omega}{2\pi} \quad (4.3\text{–}5)$$

The quantity in parentheses represents the average power per unit bandwidth and is called the *power spectral density* of $x_N(t)$. Since x_N is a realization of a stochastic process, we must average over the ensemble of realizations; therefore, we define the *power spectral density* (PSD) of $x_N(t)$ as

$$S_{x_N x_N}(e^{j\Omega}) = E\left\{ \frac{|X_N(e^{j\Omega})|^2}{2N+1} \right\} \quad (4.3\text{–}6)$$

but since $x_N \to x$ as $N \to \infty$, we have

$$S_{xx}(e^{j\Omega}) = \lim_{N\to\infty} E\left\{ \frac{|X_N(e^{j\Omega})|^2}{2N+1} \right\} \quad (4.3\text{–}7)$$

Now let us develop the relationship between the PSD and covariance. We can write Eq. (4.3–7) as

$$S_{xx}(e^{j\Omega}) = \lim_{N\to\infty} \frac{1}{2N+1} E\{X_N(e^{j\Omega})X_N^*(e^{j\Omega})\}$$

$$= \lim_{N\to\infty} \frac{1}{2N+1} E\left\{ \left(\sum_{t=-N}^{N} x(t)e^{-j\Omega t} \right) \left(\sum_{m=-N}^{N} x(m)e^{+j\Omega m} \right) \right\}$$

or $\quad S_{xx}(e^{j\Omega}) = \lim\limits_{N\to\infty} \dfrac{1}{2N+1} E\left\{ \sum\limits_{t=-N}^{N} \sum\limits_{m=-N}^{N} x(t)x(m)e^{-j\Omega(t-m)} \right\}$

Moving the expectation operation inside the summation gives the relation

$$S_{xx}(e^{j\Omega}) = \lim_{N\to\infty} \frac{1}{2N+1} \sum_{t=-N}^{N} \sum_{m=-N}^{N} R_{xx}(t, m)e^{-j\Omega(t-m)} \quad (4.3\text{–}8)$$

Introducing a change of variable k as

$$k = t - m \Rightarrow R_{xx}(m + k, m)$$

and substituting into Eq. (4.3–8), we obtain

$$S_{xx}(e^{j\Omega}) = \lim_{N \to \infty} \frac{1}{2N+1} \sum_{k=-N-m}^{N-m} \sum_{m=-N}^{N} R_{xx}(m+k,m) e^{-j\Omega k}$$

or

$$S_{xx}(e^{j\Omega}) = \sum_{k=-\infty}^{\infty} \left\{ \lim_{N \to \infty} \frac{1}{2N+1} \sum_{m=-N}^{N} R_{xx}(m+k,m) \right\} e^{-j\Omega k}$$

That is,

$$S_{xx}(e^{j\Omega}) = \sum_{k=-\infty}^{\infty} E_T\{R_{xx}(m+k,m)\} e^{-j\Omega k} \tag{4.3–9}$$

which is valid for nonstationary processes as well. Note the implied definition of time average, E_T, as given in Eq. (4.2–23). If we further assume that the process is wide-sense stationary, then $R_{xx}(m+k,m) = R_{xx}(k)$ is no longer a function of time. Therefore,

$$S_{xx}(e^{j\Omega}) = \sum_{k=-\infty}^{\infty} R_{xx}(k) e^{-j\Omega k} \tag{4.3–10}$$

evolves as the well-known *Wiener-Khintchine* relation, with the corresponding covariance given by IDtFT

$$R_{xx}(k) = \frac{1}{2\pi} \int_{2\pi} S_{xx}(e^{j\Omega}) e^{j\Omega k} d\Omega \tag{4.3–11}$$

Note also that if we recall that the DtFT is just the Z-transform of $x(t)$ evaluated on the unit circle, that is,

$$S_{xx}(\Omega) = S_{xx}(z)\big|_{z=e^{j\Omega}}$$

then we obtain the equivalent pair

$$S_{xx}(z) = Z[R_{xx}(k)] = \sum_{k=-\infty}^{\infty} R_{xx}(k) z^{-k}$$

and

$$R_{xx}(k) = \frac{1}{2\pi j} \oint S_{xx}(z) z^{k-1} dz \tag{4.3–12}$$

Example 4.3–1. Suppose we have a discrete signal given by

$$x(t) = A \sin(\Omega_0 t + \phi)$$

where ϕ is uniformly random with $\phi \sim U(0, 2\pi)$.

We are asked to analyze the spectral content of this random signal, implying that we must determine the corresponding PSD. We can determine the power spectrum by applying the Wiener-Khintchine theorem to the covariance function which must be determined. First, we must determine the mean:

$$m_x(t) = E\{A \sin(\Omega_0 t + \phi)\} = A \sin \Omega_0 t E\{\cos \phi\} + A \cos \Omega_0 t E\{\sin \phi\} = 0$$

where $E\{\cos \phi\} = \int_0^{2\pi} \cos \phi \ \text{prob} (\phi) d\phi = (1/2\pi) \int_0^{2\pi} \cos \phi d\phi = 0$ and, similarly,

$E\{\sin\phi\} = 0$. Thus, for a zero-mean process, the covariance or correlation is given by

$$R_{xx}(k) = E\{x(t)x(t+k)\} = E\{A\sin(\Omega_0 t + \phi)A\sin(\Omega_0(t+k) + \phi)\}$$

If we let $y = \Omega_0 t + \phi$, and $z = \Omega_0(t+k) + \phi$, then using the trigonometric identity

$$\sin y \sin z = \frac{1}{2}\cos(y-z) - \frac{1}{2}\cos(y+z)$$

we obtain

$$R_{xx}(k) = \frac{A^2}{2}E\{\cos\Omega_0 k - \cos(2(\Omega_0 t + \phi) + \Omega_0 k)\}$$

$$= \frac{A^2}{2}\cos\Omega_0 k - \frac{A^2}{2}E\{\cos(2(\Omega_0 t + \phi) + \Omega_0 k)\}$$

Using the fact that ϕ is uniform and calculating the expected value shows that the last term is zero. We then have

$$R_{xx}(k) = \frac{A^2}{2}\cos\Omega_0|k|$$

From the theorem, we have

$$S_{xx}(\Omega) = \text{DtFT}\,[R_{xx}(k)] = \text{DtFT}\left[\frac{A^2}{2}\cos\Omega_0 k\right] = \pi A^2[\delta(\Omega - \Omega_0) + \delta(\Omega + \Omega_0)]$$

So, we see that a sinusoidal signal with random phase can be characterized by a cosinusoidal covariance and impulsive power spectrum at the specified frequency, Ω_0. The functions are shown in Fig. 4.3–1.

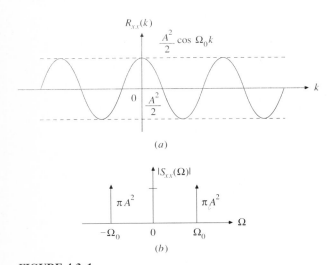

(a)

(b)

FIGURE 4.3–1
Random sinusoidal signal (a) Auto-covariance (b) Power spectrum

We define stochastic processes in terms of their covariance and spectral densities. For example, a purely random or *white-noise* sequence, such as $e(t)$, is a sequence in which all the $e(t)$ are mutually independent; that is, knowing $e(t)$ in no way can be used to predict $e(t + 1)$. The *power spectral density* of white noise is a constant:

$$S_{ee}(\Omega) = R_{ee}$$

And the corresponding covariance is given by

$$R_{ee}(k) = R_{ee}\,\delta(k)$$

where R_{ee} is the variance of the noise.

In fact, white noise of random signals is analogous to the unit impulse of deterministic signals; that is,

$$R_{ee}\,\delta(k) \Leftrightarrow A\,\delta(t)$$

and

$$S_{ee}(\Omega) = R_{ee} \Leftrightarrow H(\Omega) = A$$

As we shall see, white noise is the random counterpart for random systems of the unit impulse excitation for the analysis of LTI systems. We summarize other useful covariance-spectral density pairs in Table 4.3–1.

Note also that random or white sequences are also Markov sequences, since they are uncorrelated:

$$p\big(e(t) \mid e(t-1), \cdots, e(1)\big) = p\big(e(t)\big)$$

Sequences of this type have historically been called white because of their analogy to white light, which possesses all frequencies (constant power spectrum).

If each of the values for $e(t)$ is also gaussian-distributed, then the sequence is called *white-gaussian*. White-gaussian noise is easily simulated on a digital computer by applying the central limit theorem:

$$y(t) = \sum_{i=1}^{N} \frac{x(i) - m_x(i)}{\sqrt{NR_{xx}(i)}} \quad \text{and} \quad y \sim N(0, 1)$$

TABLE 4.3–1
Covariance-spectral density transformation pairs

Discrete process	Covariance	Power spectrum
White noise	$R_{xx}\delta(k)$	R_{xx}
Bandlimited white noise	$\dfrac{R_{xx}\Omega_B}{\pi}\dfrac{\sin k\Omega_B}{k\Omega_B}$	$R_{xx} \quad \lvert\Omega\rvert \le \Omega_B$
Bias (constant)	$B_x^2 \,\forall\, k$	$2\pi B_x^2\delta(\Omega)$
Exponential	$a^{\lvert k\rvert},\ a < 1$	$\dfrac{R_{xx}}{(1 - ae^{-j\Omega})(1 - ae^{j\Omega})}$
Sinusoidal	$\dfrac{A^2}{2}\cos\Omega_0\lvert k\rvert$	$\pi A^2[\delta(\Omega - \Omega_0) + \delta(\Omega + \Omega_0)]$
Triangular	$1 - \dfrac{\lvert k\rvert}{T_B} \quad \lvert k\rvert \le T_B$	$\dfrac{(1/T_B^2)\sin^2(T_B + 1)\Omega/2}{\sin^2(\Omega/2)}$

The random-number generator on the computer generates uniformly distributed random numbers between 0 and 1. Thus, the sequence has a mean of 1/2 and variance 1/12. If we apply the theorem, then we choose $N = 12$, and

$$y(t) = \sum_{i=1}^{12} \frac{x(i) - \frac{1}{2}}{\sqrt{12(\frac{1}{12})}} = \sum_{i=1}^{12} x(i) - 6 \qquad t = 1, 2, \dots$$

Thus, $y \sim N(0, 1)$. If one desires to generate a sequence, say $v(t) \sim N(m_v, R_{vv})$, then

$$v(t) = y(t) \sqrt{R_{vv}} + m_v$$

Example 4.3–2. A white-gaussian sequence, $N(1, 0.01)$ of 256 samples, is generated on a computer using SIG [1] and shown in Fig. 4.3–2. Notice the 2-sigma confidence interval about the sequence (95% of the samples should lie within $\pm 2\sqrt{R_{yy}}$), here only 4.8% of the values for $\{y(t)\}$ exceed the bound, or "we are 95% confident that the true sequence mean lies within $[1 - 0.2, 1 + 0.2]$." To check the whiteness of the sequence, we perform a statistical hypothesis test (see [4]), using the estimated covariance shown in the figure. The 95% confidence interval or the *whiteness test* is given by

$$I = \left[R_{yy}(k) - \frac{1.96 R_{yy}(0)}{\sqrt{N}}, R_{yy}(k) + \frac{1.96 R_{yy}(0)}{\sqrt{N}} \right]$$

But ± 0.00123 is the width [that is, $\pm 1.96(0.009)/\sqrt{256}$], so we see that only 3.1% of the $R_{yy}(k)$ lie outside the limit; therefore, the sequence is "statistically" white. Note that the covariance appears as a unit pulse and that the power spectrum appears flat (constant).

Next, let us examine some properties of the discrete covariance function for stationary processes and see how the properties of the PSD evolve. Recall that the covariance function is given by

$$R_{xx}(k) = R_{xx}(t, t + k) = E\{x(t)x(t + k)\} - m_x^2$$

where E is the ensemble-expectation operator. Covariance functions satisify the following important properties, which we state without proof (see [2] for details):

1. *Maximum value.* The auto-covariance function has a maximum at zero lag, that is,

$$R_{xx}(0) \geq |R_{xx}(k)|$$

Also, the cross-covariance function is bounded by the corresponding maxima,

$$|R_{xy}(k)| \leq \frac{1}{2}(R_{xx}(0) + R_{yy}(0))$$

2. *Symmetry conditions.* The auto-covariance is an even function, given by

$$R_{xx}(k) = R_{xx}(-k) = E\{x(t)x(t + (-k))\}$$

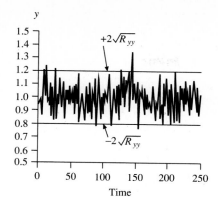

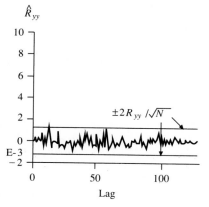

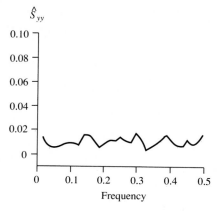

FIGURE 4.3–2
Simulated white noise, auto-covariance,
and power spectrum

105

while the cross-covariance satisfies

$$R_{xy}(k) = R_{yx}(-k)$$

3. *Mean-square value.* The auto-covariance function is the mean square of the process at zero lag,

$$R_{xx}(0) = E\{x^2(t)\}$$

while the cross-covariance mean square is bounded by

$$|R_{xy}(k)|^2 \leq R_{xx}(0)R_{yy}(0)$$

4. *DC component.* The auto-covariance function has a DC component, if x is non-zero mean:

$$R_{xx}(0) = (\text{DC})^2 + R_{xx}(k)$$

5. *Periodic component.* The auto-covariance has a periodic component if x has a periodic component.

Utilizing these properties of covariance functions, it can be shown that the *average* power in the signal $x(t)$ is given by

$$R_{xx}(0) = E\{x^2(t)\} = \frac{1}{2\pi} \int_{2\pi} S_{xx}(e^{j\Omega}) d\Omega$$

which follows directly from the Wiener-Khintchine conditions with $k = 0$. It also follows that the PSD is a real, even, non-negative function. By definition, it follows immediately that

$$S_{xx}(e^{j\Omega}) = E\{X(\Omega)X^*(\Omega)\} = E\{|X^2(\Omega)|\}$$

Therefore, the auto PSD is always real. The PSD is *even*, since

$$S_{xx}(e^{j\Omega}) = \text{DtFT}[R_{xx}(k)] = \text{DtFT}[R_{xx}(-k)] = S_{xx}(e^{j\Omega})$$

The cross PSD satisfies a similar property,

$$S_{xy}(e^{j\Omega}) = S_{yx}(e^{-j\Omega}) = S_{yx}^*(e^{j\Omega})$$

The proof that the auto PSD is non-negative on the unit-circle, that is, $S_{xx}(e^{j\Omega}) \geq 0$, is quite involved; therefore, we refer the interested reader to [5] for details. It is important to recognize that these properties are essential to analyze the information available in a discrete random signal. For instance, if we are trying to determine phase information about a particular measured signal, we immediately recognize that such information is lost in both the auto-covariance and corresponding spectrum (see Example 4.3-1).

With these properties in mind, we can now re-examine the expression for the PSD and decompose it further as[†]

[†]In the subsequent sections, we will perform operations on the PSD using the Z-transform notation, which follows from the fact that the DtFT is merely the Z-transform evaluated on the unit circle. That is, $S(\Omega) = S(z)|_{z=e^{j\Omega}}$.

$$S(z) = \sum_{k=-\infty}^{\infty} R(k)z^{-k} = \sum_{k=0}^{\infty} R(k)z^{-k} + \sum_{k=-\infty}^{0} R(k)z^{-k} - R(0)$$

Letting $i = -k$ in the second sum, and using the property that $R(k)$ is even, we obtain

$$S(z) = \underbrace{\sum_{k=0}^{\infty} R(k)z^{-k}}_{S^+(z)} + \underbrace{\sum_{i=0}^{\infty} R(i)z^{i}}_{S^-(z)} - R(0)$$

This is the *sum decomposition* of the PSD, which is defined by

$$S(z) := S^+(z) + S^-(z) - R(0)$$

where $S^+(z) = Z_I\{R(k)\}$ is the one-sided Z-transform and $S^-(z) = S^+(z^{-1})$. Note that here $S^{\pm}(z)$ employs the $\pm$ to represent positive and negative time, relative to inverse transforms. Also, it can be shown that $S^+(z)$ only has poles inside the unit circle, while $S^-(z)$ only has poles outside the unit circle (see [5] for details). Thus, the sum decomposition can be used to calculate the power spectrum, using one-sided transforms of the covariance function. The inverse process is more complicated, since the PSD must be decomposed into sums (usually by partial fractions): one only having poles inside the unit circle, the other only having poles outside. The corresponding covariance is then determined using inverse transform methods.

The sum decomposition for the power spectrum is analogous in deterministic systems to calculating the two-sided Z-transform (or Laplace transform) given the one-sided transform, since the PSD is two-sided. That is,

$$X_{II}(z) = X_I(z) + X_I(z^{-1}) - x(0)$$

where as before the subscripts "II" and "I" denote the two-sided and one-sided transforms respectively.

The sum decomposition is an important mechanism which can be utilized to simulate random signals, as we shall see in the next section. Consider how it can be used to determine the PSD of a random signal with given covariance in the following example.

Example 4.3–3. Suppose we are given the auto-covariance of a zero-mean process as

$$R(k) = e^{-a|k|} = \begin{cases} e^{-ak} & k > 0 \\ 1 & k = 0 \\ e^{+ak} & k < 0 \end{cases}$$

and we would like to calculate the corresponding PSD. Taking one-sided tranforms, we have

$$S^+(z) = Z_I[R(k)] = Z_I[e^{-ak}] = \frac{z}{z - e^{-a}}$$

$$S^-(z) = S^+(z^{-1}) = \frac{z^{-1}}{z^{-1} - e^{-a}}$$

$$R(0) = 1$$

So the sum decomposition gives

$$S(z) = S^+(z) + S^-(z) - R(0)$$

$$= \left(\frac{z}{z - e^{-a}}\right) + \left(\frac{z^{-1}}{z^{-1} - e^{-a}}\right) - 1$$

Multiplying and combining terms, we obtain

$$S(z) = \frac{1 - e^{-2a}}{(z - e^{-a})(z^{-1} - e^{-a})} = \frac{1 - e^{-2a}}{(1 + e^{-2a}) - e^{-a}(z + z^{-1})}$$

Let us now reverse the problem at hand. Suppose we are given $S(z)$ above; how can we calculate $R(k)$? As discussed previously, we can use the Z-transform inversion technique to obtain the correct result. Using the inversion integral approach (see Sec. 2.1), we have

$$R(k) = \text{Res}\left[S(z)z^{k-1}; z = e^{-a}\right]$$

$$= (z - e^{-a})\left[\frac{(1 - e^{-2a})z^{k-1}}{(z - e^{-a})(z^{-1} - e^{-a})}\right]_{z=e^{-a}} \quad \text{for } k \geq 0$$

or $\qquad R(k) = \dfrac{e^a(1 - e^{-2a})e^{-ak}}{e^a - e^{-a}} = e^{-ak} \qquad k \geq 0$

For values of $k < 0$, we use the fact that the covariance is even, $R(k) = R(-k)$, which gives the desired result.

We summarize the important properties of the covariance and power spectrum in Table 4.3–2.

In practice, the auto-covariance and PSD find most application in the analysis of random signals yielding information about spectral content, periodicities, etc., while the cross-covariance and spectrum are used to estimate the properties of two distinct processes (e.g., the input and output of a system). The following example illustrates how the cross-covariance and spectrum can be used to estimate the time delay of a signal.

Example 4.3–4. Suppose we have a random signal $x(t)$ that is transmitted through some medium (e.g., an acoustical wave in sonar or an electromagnetic wave in radar) and received some time later by a receiver at the same location. We would like to determine the delay time τ_d, an integer multiple of the sampling interval between the transmitted and received signals, where $\tau_d = d/v$, d is the distance, and v the velocity. The received signal, assuming only an attenuation A of the medium, is characterized by

$$r(t) = Ax(t - \tau_d) + n(t)$$

TABLE 4.3–2
Properties of the covariance and power spectrum

Covariance	Power spectrum
Average power: $R_{xx}(0) = E\{x^2(t)\}$	$R_{xx}(0) = \dfrac{1}{2\pi}\displaystyle\int_{2\pi} S_{xx}(z)z^{-1}dz$
Symmetry: $R_{xx}(k) = R_{xx}(-k)$ (even) $R_{xy}(k) = R_{yx}(-k)$	$S_{xx}(z) = S_{xx}(z^{-1})$ (even) $S_{xy}(z) = S^*_{yx}(z)$
Maximum: $R_{xx}(0) \geq \lvert R_{xx}(k)\rvert$ $\frac{1}{2}R_{xx}(0) + \frac{1}{2}R_{yy}(0) \geq \lvert R_{xy}(k)\rvert$ $R_{xx}(0)R_{yy}(0) \geq \lvert R_{xy}(k)\rvert^2$	$S_{xx}(e^{j\Omega}) \geq 0$
Real:	$S_{xx}(z) = E\{\lvert X(z)\rvert^2\}$ is real $S_{xy}(z) = E\{X(z)Y^*(z)\}$ is complex
Sum decomposition:	$S_{xx}(z) = S^+_{xx}(z) + S^-_{xx}(z) - R_{xx}(0)$ $S^+_{xx}(z) = Z_{\mathrm{I}}[R_{xx}(k)]$ $S^-_{xx}(z) = S^+_{xx}(z^{-1})$

where n is white noise. The cross-covariance of the zero-mean transmitted and received signals is given by

$$R_{xr}(k) = E\{x(t)r(t+k)\} = AE\{x(t)x(t-\tau_d+k)\} + E\{x(t)n(t+k)\}$$

or $\quad R_{xr}(k) = AR_{xx}(k-\tau_d)$

since n is white and uncorrelated with x. From the *maximum* property of the auto-covariance function, we see that the cross-covariance function will achieve a maximum value when $k = \tau_d$; that is,

$$R_{xr}(\tau_d) = AR_{xx}(0) \quad \text{for } k = \tau_d$$

If we take the DtFT of the received signal, we have

$$R(\Omega) = AX(\Omega)e^{-j\Omega\tau_d} + N(\Omega)$$

The corresponding cross-spectrum is given by

$$S_{xr}(\Omega) = E\{X(\Omega)R^*(\Omega)\} = AE\{X(\Omega)X^*(\Omega)\}e^{j\Omega\tau_d} + E\{X(\Omega)N^*(\Omega)\}$$

or $\quad S_{xr}(\Omega) = AS_{xx}(\Omega)e^{j\Omega\tau_d}$

Thus, the time delay results in a peak in the cross-covariance function and is characterized by a linear phase component, which can be estimated from the slope of the cross-spectral phase function. We use SIG [1] to simulate a sinusoidal transmitted signal of 1 Hz. The received signal was attenuated by 1/2, delayed by 2.5 sec, and contaminated with white measurement noise with a variance of 0.01. The transmitted signal and spectrum is shown in Fig. 4.3–3(a), while the corresponding received signal and magnitude cross-spectrum is shown in (b). Note the similarity between both magnitude spectra. The corresponding cross-covariance

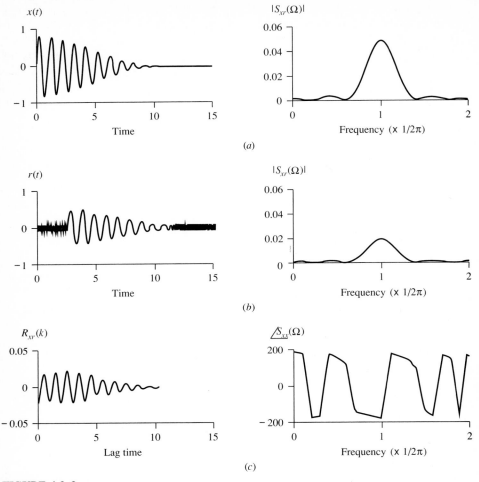

FIGURE 4.3–3
Time delay simulation (*a*) Transmitted signal and spectrum (*b*) Received signal and cross-spectrum magnitude (*c*) Cross-covariance function and phase spectrum

and spectrum are shown in (*c*). From the peak of the estimated cross-covariance function or the corresponding slope of the cross-spectrum, we estimate the delay at approximately $\hat{\tau}_d = 2.5$ sec and the attenuation as the ratio $A = R_{xr}(\tau_d)/R_{xx}(0) = 1/2$. Therefore, we estimate

$$\hat{r}(t) = \frac{1}{2}\sin 2\pi(1)(t - 2.5)$$

So we see in this case that the properties of covariance and spectra can be used not only to analyze the information available in random signal data, but also to estimate various signal characteristics. We also note in passing that this is precisely the principle behind *matched filtering*, used extensively in sonar and radar receivers to detect the presence of a target.

Another important function which is very useful in analyzing the quality of measured data is the *coherence function*, defined by

$$\gamma_{x_i x_j}(\Omega) := \frac{S_{x_i x_j}(\Omega)}{|S_{x_i x_i}(\Omega)S_{x_j x_j}(\Omega)|^{1/2}} \qquad (4.3\text{--}13)$$

The coherence function has the property that it is bounded by

$$0 \le \gamma_{x_i x_j}(\Omega) \le 1$$

This will be discussed in the next section in more detail.

4.4 LINEAR SYSTEMS WITH RANDOM INPUTS

When random inputs are applied to linear systems, covariance and power-spectrum techniques must be applied, replacing the signal and Fourier spectrum in deterministic signal theory. In this section, we develop the relationship between systems theory and random signals. From linear systems theory, we have the convolution or equivalent frequency relation

$$y(t) = h(t) * x(t) = \sum_{k=0}^{\infty} h(k)x(t - k) \qquad (4.4\text{--}1)$$

or, taking discrete-time Fourier transforms, we have

$$Y(\Omega) = H(\Omega)X(\Omega) \qquad (4.4\text{--}2)$$

If we assume that x is a random signal, then (as we have seen in the previous section) we must resort to spectral representations of random processes. Exciting a causal linear system with a zero-mean random signal, we obtain the output covariance at lag k as

$$R_{yy}(k) = E\{y(t + k)y(t)\} = E\left\{\left(\sum_{i=0}^{\infty} h(i)x(t + k - i)y(t)\right)\right\}$$

$$= \sum_{i=0}^{\infty} h(i)E\{x(t + k - i)y(t)\}$$

$$R_{yy}(k) = \sum_{i=0}^{\infty} h(i)R_{xy}(k - i) = h(k) * R_{xy}(k) \qquad (4.4\text{--}3)$$

Suppose we calculate the output power spectrum based on this result:

$$S_{yy}(z) = \sum_{k=-\infty}^{\infty} R_{yy}(k)z^{-k} = \sum_{k=-\infty}^{\infty} \left(\sum_{i=0}^{\infty} h(i)R_{xy}(k-i) \right) z^{-k}$$

Multiplying by $z^i z^{-i}$ and interchanging the order of summation, we obtain

$$S_{yy}(z) = \left(\sum_{i=0}^{\infty} h(i)z^{-i} \right) \left(\sum_{k=-\infty}^{\infty} R_{xy}(k-i)z^{-(k-i)} \right)$$

If we let $m = k - i$, then we have

$$S_{yy}(z) = \sum_{i=0}^{\infty} h(i)z^{-i} \sum_{m=-\infty}^{\infty} R_{xy}(m)z^{-m} = H(z)S_{xy}(z) \qquad (4.4\text{--}4)$$

Similar results can be obtained for auto- and cross-covariances and corresponding spectra. We summarize the results in Table 4.4–1.

Example 4.4–1. Suppose we have a digital filter given by the difference equation

$$y(t) = ay(t-1) + x(t)$$

We would like to analyze the impulse response and corresponding spectrum of this linear system under two conditions: with $x(t)$ being deterministic, and with $x(t)$ being random. First, we assume that x is deterministic, then taking the Z-transform, we have

$$H(z) = \frac{Y(z)}{X(z)} = \frac{1}{1 - az^{-1}} \qquad \text{for } |z| > a, \ a < 1$$

and taking the inverse Z-transform, we have the corresponding impulse response,

$$h(t) = a^t$$

Next we assume x is random, uniform with variance σ_{xx}^2, and using $H(z)$ from

TABLE 4.4–1
Random linear system relations

Covariance	Spectrum
$R_{yy}(k) = h(k) * h(-k) * R_{xx}(k)$	$S_{yy}(z) = H(z)H(z^{-1})S_{xx}(z)$
$R_{yy}(k) = h(k) * R_{xy}(k)$	$S_{yy}(z) = H(z)S_{xy}(z)$
$R_{yx}(k) = h(k) * R_{xx}(k)$	$S_{yx}(z) = H(z)S_{xx}(z)$

where

$R_{yy}(k) = \dfrac{1}{2\pi j} \oint S_{yy}(z)z^{k-1}dz$	$S_{yy}(z) = \sum_{k=-\infty}^{\infty} R_{yy}(k)z^{-k}$
$R_{xy}(k) = \dfrac{1}{2\pi j} \oint S_{xy}(z)z^{k-1}dz$	$S_{xy}(z) = \sum_{k=-\infty}^{\infty} R_{xy}(k)z^{-k}$

the deterministic solution, and the spectral relation from Table 4.4–1, we obtain

$$S_{yy}(z) = H(z)H(z^{-1})S_{xx}(z) = \left(\frac{1}{1 - az^{-1}}\right)\left(\frac{1}{1 - az}\right)\sigma_{xx}^2$$

The covariance can be calculated directly from the PSD using the inversion-integral method (Sec. 2.1):

$$R_{yy}(k) = \text{Res}\,[S_{yy}(z)z^{k-1}; z = a] = \left(\frac{\sigma_{xx}^2}{1 - a^2}\right)a^k \quad k \geq 0$$

or, using the impulse-response and convolution relations of Table 4.4–1,

$$R_{yy}(k) = h(k) * h(-k) * R_{xx}(k) = \sigma_{xx}^2 \sum_{k=0}^{\infty} h(k)h(t + k)$$

Since $R_{xx}(k) = \sigma_{xx}^2\delta(k)$, we obtain for $k \geq 0$

$$R_{yy}(k) = \sigma_{xx}^2 \sum_{t=0}^{\infty} a^t(a^{t+k}) = \sigma_{xx}^2 a^k \sum_{t=0}^{\infty} (a^2)^t = \frac{\sigma_{xx}^2 a^k}{1 - a^2}$$

using the geometric series.

We can now state an important result that is fundamental for the modeling of stochastic processes. Suppose we have a linear time-invariant system that is asymptotically stable with a rational transfer function. If we excite this system with white noise, we get the so-called *spectral factorization theorem*, which states that, given a stable, rational system, a rational *H exists* such that

$$S_{yy}(\Omega) = H(\Omega)H^*(\Omega) \tag{4.4–5}$$

where the poles and zeros of *H* lie within the unit circle. If we can represent the spectral density in factored form, then all stationary processes can be thought of as the output of a dynamic system with white-noise input. Synthesizing stochastic processes with given spectra, then, requires only the generation of white-noise sequences. We summarize this discussion with the *representation theorem*:†

> Given a rational spectral density $S_{yy}(\Omega)$, there exists an asymptotically stable linear system which, when excited by white noise, produces a stationary output y with this spectrum.

The simulation of stochastic processes with given statistics requires the construction of the power spectrum from the sum decomposition, followed by a spectral factorization. This leads to the *spectrum simulation procedure*:

---

†This is really a restricted variant of the famous Wold decomposition for stationary stochastic processes (see [6] for details).

1. Calculate $S_{yy}(z)$ from the sum decomposition
2. Perform the spectral factorization to obtain $H(z)$
3. Generate a white-noise sequence of variance R_{xx}
4. Excite the system $H(z)$ with the sequence

The most difficult part of this procedure is performing the spectral factorization of Step 2. For simple systems, the factorization can be performed by equating coefficients of the known $S_{yy}(z)$, obtained in Step 1, with the unknown coefficients of the spectral factor and solving the resulting nonlinear algebraic equations:

$$S_{yy}(z) = \begin{cases} S_{yy}^+(z) + S_{yy}^-(z) - R_{yy}(0) & = \dfrac{N_s(z, z^{-1})}{D_s(z, z^{-1})} \\[3mm] H(z)H(z^{-1})R_{xx} & = \dfrac{N_H(z)\,N_H(z^{-1})}{D_H(z)\,D_H(z^{-1})}R_{xx} \end{cases}$$

$$N_s(z, z^{-1}) = N_H(z)N_H(z^{-1})$$

and
$$D_s(z, z^{-1}) = D_H(z)D_H(z^{-1}) \qquad (4.4\text{--}6)$$

For higher-order systems, more efficient iterative techniques exist, including multichannel systems (see [7,8,9]).

Example 4.4–2. Suppose we would like to generate a random sequence $\{y(t)\}$ with auto-covariance

$$R_{yy}(k) = a^{|k|} \qquad \text{for } 0 < a < 1$$

From the definition of spectral density, we obtain

$$S_{yy}(z) = S_{yy}^+(z) + S_{yy}^-(z) - R_{yy}(0)$$

$$S_{yy}(z) = \frac{1}{(1 - az^{-1})} + \frac{1}{(1 - az)} - 1$$

$$S_{yy}(z) = \frac{(1 - a^2)}{(1 - az^{-1})(1 - az)} = H(z)H(z^{-1})R_{xx}$$

Since S_{yy} is already in factored form, we identify H by inspection:

$$N_s(z, z^{-1}) = N_H(z)N_H(z^{-1}) = 1$$

and
$$D_s(z, z^{-1}) = 1 - a(z + z^{-1}) + a^2 = D_H(z)D_H(z^{-1})$$
$$= (1 - az^{-1})(1 - az)$$

Using the representation theorem, we identify the white-noise sequence

$$S_{xx}(z) = R_{xx} = (1 - a^2) \qquad \text{and} \qquad H(z) = \frac{1}{(1 - az^{-1})}$$

or
$$y(t) = h(t) * x(t) = \sum_{k=0}^{\infty} a^{|t-k|} x(k)$$

4.5 PARAMETRIC REPRESENTATIONS OF STOCHASTIC PROCESSES: ARMAX MODELS

In the previous section we showed the relationship between linear systems and spectral shaping, that is, generating random signals with specified statistics. To generate such a sequence, we now choose to use models which are equivalent— the input-output or state-space models. Each model set has its own advantages: the input-output models are easy to use, whereas the state-space models are easily generalized.

The *input-output* or *transfer function* model is familiar to engineers and scientists, because it is usually presented in the frequency domain with Laplace transforms. In the discrete-time case, it is called the *pulse transfer function* model and is given by

$$H(z) = \frac{B(z)}{A(z)} \qquad (4.5\text{--}1)$$

where A and B are polynomials in z^{-1}.

$$A(z) = 1 + a_1 z^{-1} + \cdots + a_{N_a} z^{-N_a}$$
$$B(z) = b_0 + b_1 z^{-1} + \cdots + b_{N_b} z^{-N_b}$$

If we consider the equivalent time-domain representation, then we have a *difference equation* relating the output sequence $\{y(t)\}$ to the input sequence $\{u(t)\}$.† Here we use the backward-shift operator q with the property that $q^{-k} y(t) = y(t-k)$.

$$A(q^{-1})y(t) = B(q^{-1})u(t)$$

or
$$y(t) + a_1 y(t-1) + \cdots + a_{N_a} y(t - N_a)$$
$$= b_0 u(t) + \cdots + b_{N_b} u(t - N_b) \qquad (4.5\text{--}2)$$

When the system is excited by random inputs, the models are given by the *autoregressive moving average model with exogenous inputs* (ARMAX)‡

$$\underbrace{A(q^{-1})y(t)}_{\text{AR}} = \underbrace{B(q^{-1})u(t)}_{\text{X}} + \underbrace{C(q^{-1})e(t)}_{\text{MA}}$$

---

†Here we change from the common signal-processing convention of using $x(t)$ for the deterministic excitation to using $u(t)$.

‡The ARMAX model can be interpreted in terms of the Wold decomposition of stationary time series, which states that a time series can be decomposed into a predictable or deterministic component $u(t)$ and a nondeterministic or random component $e(t)$[10].

where A, B, and C, are polynomials, $\{e(t)\}$ is a white-noise source, and

$$C(q^{-1}) = 1 + c_1 q^{-1} + \cdots + c_{N_c} q^{-N_c}$$

The ARMAX model, usually abbreviated by ARMAX(N_a, N_b, N_c), represents the general form for popular time-series and digital-filter models:

1. Pulse-transfer function or infinite-impulse response (IIR) model: $C(\,\cdot\,) = 0$, or ARMAX($N_a, N_b, 0$), that is,

$$A(q^{-1})y(t) = B(q^{-1})u(t)$$

2. Finite-impulse response (FIR) model: $A(\,\cdot\,) = 1$, $C(\,\cdot\,) = 0$, or ARMAX $(1, N_b, 0)$, that is,

$$y(t) = B(q^{-1})u(t)$$

3. Autoregressive (AR) model: $B(\,\cdot\,) = 0$, $C(\,\cdot\,) = 1$, or ARMAX($N_a, 0, 1$), that is,

$$A(q^{-1})y(t) = e(t)$$

4. Moving-average (MA) model: $A(\,\cdot\,) = 1$, $B(\,\cdot\,) = 0$, or ARMAX($1, 0, N_c$), that is,

$$y(t) = C(q^{-1})e(t)$$

5. Autoregressive-moving average (ARMA) model: $B(\,\cdot\,) = 0$, or ARMAX $(N_a, 0, N_c)$, that is,

$$A(q^{-1})y(t) = C(q^{-1})e(t)$$

6. Autoregressive model with exogeneous input (ARX): $C(\,\cdot\,) = 1$, or ARMAX $(N_a, N_b, 1)$, that is,

$$A(q^{-1})y(t) = B(q^{-1})u(t) + e(t)$$

The ARMAX model is shown in Fig. 4.5–1. ARMAX models can easily be used for signal-processing purposes, since they are basically a digital filter with known deterministic $(u(t))$ and random $(e(t))$ excitations. Consider the following example of spectral shaping using the ARMAX model.

Example 4.5–1. Let $a = -0.5$ in the previous example, and $\{e(t)\}$ be $N(0, 1)$. We are asked to generate the output sequence $y(t)$. This can be accomplished using ARMAX($1, 0, 0$), an AR model; that is,

$$H(z) = \frac{Y(z)}{E(z)} = \frac{1}{1 + 0.5z^{-1}}$$

Cross-multiplying and substituting the backward shift operation for z, we obtain the difference equation

$$y(t) = -0.5y(t - 1) + e(t)$$

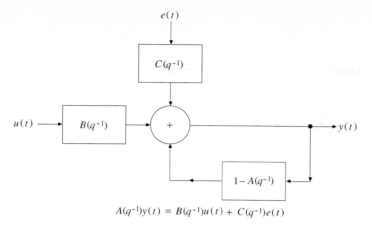

$$A(q^{-1})y(t) = B(q^{-1})u(t) + C(q^{-1})e(t)$$

where u, y, and e are input, output, and noise sequences; A, B, and C are polynomials of degree N_a, N_b, and N_c in the backward shift operator q; and $e(t)$ is white.

FIGURE 4.5–1
ARMAX input-output model

We use SIG [1] to simulate these sequences; the results are shown in Fig. 4.5–2. Note the estimated covariance and spectrum for this process as well.

ARMAX can be quite useful for simulation purposes. In fact, as we shall see subsequently, this model provides the basis for the parametric or modern approach to signal processing.

Since the ARMAX model is used to characterize a stochastic process, we are interested in its statistical properties. The mean value of the output is easily determined by

$$A(q^{-1})E\{y(t)\} = B(q^{-1})E\{u(t)\} + C(q^{-1})E\{e(t)\}$$

or

$$A(q^{-1})m_y(t) = B(q^{-1})u(t) + C(q^{-1})m_e(t) \qquad (4.5\text{–}4)$$

Because the first term in the A-polynomial is unity, we can write the *mean propagation recursion* for the ARMAX model as

$$m_y(t) = \left(1 - A(q^{-1})\right)m_y(t) + B(q^{-1})u(t) + C(q^{-1})m_e(t)$$

$$m_y(t) = -\sum_{i=1}^{N_a} a_i m_y(t-i) + \sum_{i=0}^{N_b} b_i u(t-i) + \sum_{i=0}^{N_c} c_i m_e(t-i)$$

We note that the mean of the ARMAX model is propagated using a recursive digital filter requiring N_a, N_b, N_c past input and output values.

The corresponding variance of the ARMAX model is more complex. First, we note that the mean must be removed; that is,

$$y(t) - m_y(t) = \left[(1 - A(q^{-1}))y(t) + B(q^{-1})u(t) + C(q^{-1})e(t)\right]$$

$$-\left[(1 - A(q^{-1}))m_y(t) + B(q^{-1})u(t) + C(q^{-1})m_e(t)\right]$$

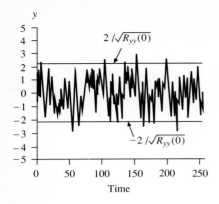

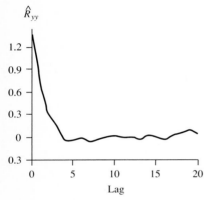

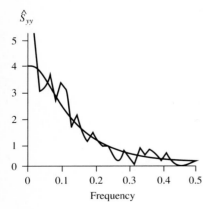

FIGURE 4.5–2
ARMAX(1,0,0) simulation for gaussian excitation

or $\quad y(t) - m_y(t) = \left(1 - A(q^{-1})\right)\left(y(t) - m_y(t)\right) + C(q^{-1})\left(e(t) - m_e(t)\right)$

or $\qquad\qquad A(q^{-1})\left(y(t) - m_y(t)\right) = C(q^{-1})\left(e(t) - m_e(t)\right)$ (4.5–5)

$y - m_y$ is characterized by an ARMAX($N_a, 0, N_b$) model or, equivalently, an ARMA model.

The covariance of the ARMAX model can be calculated using the fact that it is essentially an IIR system:

$$\frac{Y(z)}{E(z)} = H(z) = \sum_{i=0}^{\infty} h(i)z^{-i} \qquad (4.5\text{–}6)$$

Using this fact and the commutativity of the convolution operator, and assuming the mean has been removed, we have

$$R_{yy}(k) = E\{y(t)y(t+k)\} = E\left\{ \sum_{i=0}^{\infty} h(i)e(t-i) \sum_{j=0}^{\infty} h(j+k)e(t-j+k) \right\}$$

or

$$R_{yy}(k) = \sum_{i=0}^{\infty} h(i)h(i+k)E\{e(t-i)e(t-i+k)\}$$

$$+ \sum\sum_{i\neq j} h(i)h(j+k)E\{e(t-i)e(t-j+k)\}$$

The whiteness of $\{e(t)\}$ gives

$$R_{ee}(k) = \begin{cases} R_{ee} & k = 0 \\ 0 & \text{elsewhere} \end{cases}$$

Therefore, applying this whiteness property, we have the covariance of the ARMAX model given by

$$R_{yy}(k) = R_{ee} \sum_{i=0}^{\infty} h(i)h(i+k) \qquad \text{for } k \geq 0 \qquad (4.5\text{–}7)$$

with corresponding variance

$$R_{yy}(0) = R_{ee} \sum_{i=0}^{\infty} h^2(i) \qquad (4.5\text{–}8)$$

We note that one property of a stationary signal is that its impulse response is bounded which implies from Eq. (4.5–7) that the variance is bounded ([11]). Clearly, since the variance is characterized by an ARMA model $\left(\text{ARMAX}(N_a, 0, N_c)\right)$, then we have

$$A(z)H(z) = C(z)U(z), \qquad U(z) = \sqrt{R_{ee}}$$

or, taking the inverse Z-transform,

$$h(t) = (1 - A(q^{-1}))h(t) + C(q^{-1})\delta(t)$$

or
$$h(t) = -\sum_{i=1}^{N_a} a_i h(t-i) + \sum_{i=0}^{N_c} c_i \delta(t), \qquad c_0 = 1 \qquad (4.5\text{--}9)$$

where $\delta(t)$ is an impulse of weight $\sqrt{R_{ee}}$. This recursion coupled with Eq. (4.5–7) provides a method for calculating the variance of an ARMAX model. We summarize these results in Table 4.5–1 and the following example.

Example 4.5–2. Suppose we are given the AR model of the previous example, with $A(z) = 1 + 0.5z^{-1}$ and $R_{ee} = 1$; we would like to calculate the variance. From Eq. (4.5–7), we have

$$
\begin{aligned}
h(t) &= -0.5h(t-1) + \delta(t), \qquad \text{or} \\
h(0) &= 1 \\
h(1) &= -0.5h(0) = -0.5 \\
h(2) &= -0.5h(1) = 0.25 \\
&\vdots \qquad \vdots \\
h(t) &= (-0.5)^t
\end{aligned}
$$

and $\quad R_{yy}(0) = \displaystyle\sum_{i=0}^{\infty} h^2(i) = [1^2 - 0.5^2 + .25^2 - .125^2 + .0875^2 - \ldots] \to 1.333$

TABLE 4.5–1
ARMAX representation

Output propagation

$$y(t) = \left(1 - A(q^{-1})\right)y(t) + B(q^{-1})u(t) + C(q^{-1})e(t)$$

Mean propagation

$$m_y(t) = \left(1 - A(q^{-1})\right)m_y(t) + B(q^{-1})u(t) + C(q^{-1})m_e(t)$$

Impulse propagation

$$h(t) = \left(1 - A(q^{-1})\right)h(t) + C(q^{-1})\delta(t)$$

Variance/covariance propagation

$$R_{yy}(k) = R_{ee}\sum_{i=0}^{\infty} h(i)h(i+k) \qquad k \geq 0$$

---

where
y = the output or measurement sequence
u = the input sequence
e = the process (white) noise sequence with variance R_{ee}
h = the impulse-response sequence
δ = the impulse input of amplitude $\sqrt{R_{ee}}$
m_y = the mean output or measurement sequence
m_e = the mean-process noise sequence
R_{yy} = the stationary-output covariance at lag k
A = the N_ath order system-characteristic (poles) polynomial
B = the N_bth order input (zeros) polynomial
C = the N_cth order noise (zeros) polynomial

The covariance is calculated from Eq. (4.5–7) using SIG [1] and is shown in Fig. 4.5–3, along with the corresponding power spectrum and impulse response.

Before we leave this discussion let us consider a more complex example to illustrate the use of the ARMAX model.

Example 4.5–3. Suppose we would like to simulate an ARMAX(2,1,1) model with the following difference equation,

$$(1 + \tfrac{3}{4}q^{-1} + \tfrac{1}{8}q^{-2})y(t) = (1 + \tfrac{1}{8}q^{-1})u(t) + (1 + \tfrac{1}{16}q^{-1})e(t)$$

where $u(t) = \sin 2\pi(0.025)t$ and $e \sim N(1, 0.01)$. Then the corresponding mean-propagation equation is

$$(1 + \tfrac{3}{4}q^{-1} + \tfrac{1}{8}q^{-2})m_y(t) = (1 + \tfrac{1}{8}q^{-1})u(t) + (1 + \tfrac{1}{16}q^{-1})m_e(t)$$

for $m_e(t) = 1 \; \forall \, t$. The impulse propagation model is

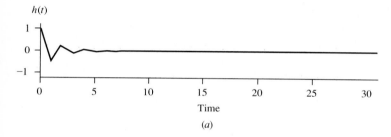

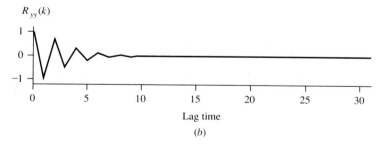

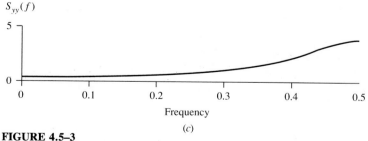

FIGURE 4.5–3
Autoregressive example (*a*) Impulse response (*b*) Simulated covariance (*c*) Simulated power spectrum

$$(1 + \tfrac{3}{4}q^{-1} + \tfrac{1}{8}q^{-2})h(t) = (1 + \tfrac{1}{16}q^{-1})\sqrt{R_{ee}}\,\delta(t) \qquad \text{for } \sqrt{R_{ee}} = 0.1$$

and the variance/covariance propagation model is

$$R_{yy}(k) = 0.01 \sum_{i=0}^{\infty} h(i)h(i+k) \qquad k \geq 0$$

Using SIG [1], we perform the simulation with $\Delta T = 0.1$ sec; the results are shown in Fig. 4.5–4. In the figure, we see the simulated process, mean (sinusoidal), and corresponding ARMA (mean removed) signals as developed in Eqs. (4.5–3 to 4.5–5). The statistical relations are also shown with the corresponding impulse-response and variance-propagation models of Eqs. (4.5–7,9) and estimated power spectrum.

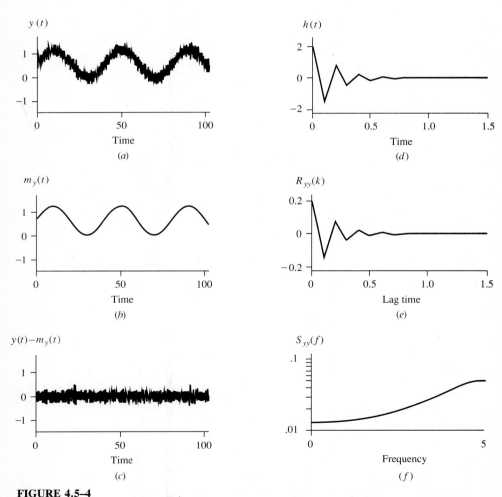

FIGURE 4.5–4
ARMAX(2,1,1) simulation (*a*) Simulated signal (*b*) Mean propagation (*c*) ARMA (mean removed) signal (*d*) Impulse response propagation (*e*) Covariance propagation (*f*) PSD

It should also be noted that for certain special cases of the ARMAX model, it is particularly simple to calculate the mean and covariance. For instance, the MA model $(\text{ARMAX}(1,0,N_c))$ has mean

$$m_y(t) = E\{C(q^{-1})e(t)\} = C(q^{-1})m_e(t) \qquad (4.5\text{--}10)$$

and covariance (directly from Eq. (4.5–7) with $h \to c$)

$$R_{yy}(k) = R_{ee} \sum_{i=0}^{N_c} c_i c_{i+k} \qquad \text{for } k \geq 0 \qquad (4.5\text{--}11)$$

We summarize these results for the MA model in Table 4.5–2.

Another special case of interest is the AR $(\text{ARMAX}(N_a, 0, 1))$ model with mean

$$m_y(t) = \left(1 - A(q^{-1})\right)m_y(t) + m_e(t) \qquad (4.5\text{--}12)$$

and *covariance*, which is easily derived by direct substitution,

$$R_{yy}(k) = E\{y(t)y(t+k)\} = \left(1 - A(q^{-1})\right)R_{yy}(k) = -\sum_{i=1}^{N_a} a_i R_{yy}(k-i) \quad \text{for } k > 0$$

$$(4.5\text{--}13)$$

In fact, the covariance AR model of Eq. (4.5–13) is essentially a recursive (all-pole) digital filter, which can be propagated by exciting it with the variance $R_{yy}(0)$ as initial condition. In this case, the variance is given by

$$R_{yy}(0) = E\{y^2(t)\} = E\left\{\left(-\sum_{i=1}^{N_a} a_i y(t-i) + e(t)\right)y(t)\right\} = -\sum_{i=1}^{N_a} a_i R_{yy}(i) + R_{ee}$$

$$(4.5\text{--}14)$$

TABLE 4.5–2
MA representation

Output propagation

$$y(t) = C(q^{-1})e(t)$$

Mean propagation

$$m_y(t) = C(q^{-1})m_e(t)$$

Variance/covariance propagation

$$R_{yy}(k) = R_{ee} \sum_{i=0}^{N_c} c_i c_{i+k} \qquad k \geq 0$$

where
y = the output or measurement sequence
e = the process (white) noise sequence with variance R_{ee}
m_y = the mean output or measurement noise sequence
m_e = the mean process noise sequence
R_{yy} = the stationary output covariance at lag k
C = the N_cth order noise (zeros) polynomial

So, combining Eqs. (4.5–13,14), we have the *covariance propagation* equations for the AR model given by

$$R_{yy}(k) = \begin{cases} -\sum_{i=1}^{N_a} a_i R_{yy}(i) + R_{ee} & k = 0 \\ -\sum_{i=1}^{N_a} a_i R_{yy}(k-i) & k > 0 \end{cases} \qquad (4.5\text{–}15)$$

We summarize these results for the AR model in Table 4.5–3.

Example 4.5–5. Consider the AR model of the previous examples; we would like to determine the corresponding mean and variance, using the recursions of Eqs. (4.5–12,15) with $A(q^{-1}) = 1 + 0.5q^{-1}$. The mean is

$$m_y(t) = -0.5m_y(t-1)$$

and the covariance is

$$R_{yy}(k) = \begin{cases} -0.5R_{yy}(1) + 1 & k = 0 \\ -0.5R_{yy}(k-1) & k > 0 \end{cases}$$

The variance is obtained directly from these recursions, since

$$R_{yy}(1) = -0.5R_{yy}(0)$$

and therefore $R_{yy}(0) = -0.5R_{yy}(1) + 1$

Substituting for $R_{yy}(1)$ we obtain

$$R_{yy}(0) = -0.5(-0.5R_{yy}(0)) + 1$$

or $R_{yy}(0) = 1.333$

TABLE 4.5–3
AR representation

Output propagation

$$y(t) = (1 - A(q^{-1}))\, y(t) + e(t)$$

Mean propagation

$$m_y(t) = (1 - A(q^{-1}))\, m_y(t) + m_e(t)$$

Variance/covariance propagation

$$R_{yy}(k) = (1 - A(q^{-1})) R_{yy}(k) + R_{ee}\delta(k) \qquad k \geq 0$$

---

where y = the output or measurement sequence
e = the process (white) noise sequence with variance R_{ee}
m_y = the mean output or measurement sequence
m_e = the mean process noise sequence
R_{yy} = the stationary output covariance at lag k
A = the N_ath order system characteristic (poles) polynomial
δ = the Kronecker delta function or unit impulse

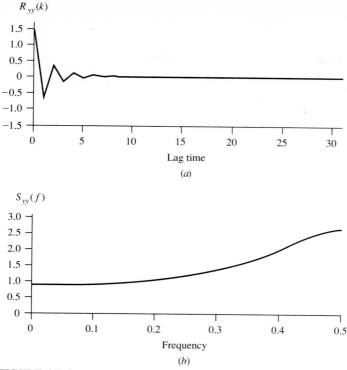

FIGURE 4.5–5
Autoregressive example (*a*) Simulated covariance (*b*) Simulated power spectrum

as before. Using SIG [1], we simulated the covariance propagation equations; the results are depicted in Fig. 4.5–5. The corresponding PSD is also shown.

4.6 PARAMETRIC REPRESENTATIONS OF STOCHASTIC PROCESSES: LATTICE MODELS

The design of digital filters has evolved many ways to realize and implement a filter transfer function. Some implementations use a minimum number of elements, while others may use a minimum number of multipliers. The lattice filter (sometimes called the ladder filter) does not possess minimizing properties, but it offers other advantages in the implementation of digital filters (e.g., good roundoff). Lattice filters find their roots in analog filter designs and in the design of digital filters as well [11,12,13]. More recently, lattice filters have evolved as equivalent representations of stochastic processes. Estimation filters based on the lattice design perform well because of their orthogonality properties [14]. In any case, the scalar *lattice* model is defined by the following fundamental propagation equation:

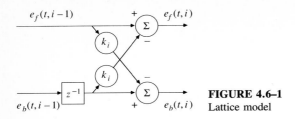

FIGURE 4.6–1
Lattice model

$$e_f(t, i) = e_f(t, i - 1) - k_i e_b(t - 1, i - 1)$$
$$e_b(t, i) = e_b(t - 1, i - 1) - k_i e_f(t, i - 1) \tag{4.6-1}$$

where $e_f(t, i)$ and $e_b(t, i)$ are the respective forward and backward signals of the ith lattice section (or stage) at time t, and k_i is the *reflection coefficient* of the ith section. The lattice model is depicted in Fig. 4.6–1. Here we see that the model is made up of a series of connected sections, with each section having identical structure, except that the values of the reflection coefficients change.

The lattice model derives its physical origins from the fact that it mathematically represents a *wave model* or *wave* propagating through a layered medium. For instance, lattice models are employed to characterize seismic waves propagating through a layered earth in the exploration of oil [11], an acoustic wave propagating through a tube in speech synthesis [15], or an electromagnetic wave propagating in a transmission line. In each of these cases, the reflection coefficient, which has magnitude between ± 1, indicates the percentage of the wave transmitted and the percentage reflected at a layer boundary (or lattice stage).

Structurally, the lattice model possesses some very attractive properties because of its sectional or stagewise symmetry. Consider a filter of length N, consisting of N lattice sections arranged in a feed-forward configuration, as shown in Fig. 4.6–2. We first note that the lattice model operation for the ith section can be written in matrix form, using the backward shift operator, as

$$\begin{bmatrix} e_f(t, i) \\ e_b(t, i) \end{bmatrix} = \begin{bmatrix} 1 & -k_i q^{-1} \\ -k_i & q^{-1} \end{bmatrix} \begin{bmatrix} e_f(t, i - 1) \\ e_b(t, i - 1) \end{bmatrix} \tag{4.6-2}$$

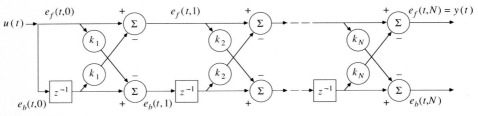

FIGURE 4.6–2
Feed-forward lattice (all-zero) structure

Taking the Z-transform of Eq. (4.6–2) gives

$$\begin{bmatrix} E_f(z,i) \\ E_b(z,i) \end{bmatrix} = \begin{bmatrix} 1 & -k_i z^{-1} \\ -k_i & z^{-1} \end{bmatrix} \begin{bmatrix} E_f(z,i-1) \\ E_b(z,i-1) \end{bmatrix}$$

or, more compactly,

$$\underline{E}(z,i) = L(z,i)\underline{E}(z,i-1) \qquad (4.6\text{–}3)$$

where $L(z,i)$ can be thought of as a multichannel transfer function matrix[†] between the ith and $(i-1)$-th section of the lattice. In fact, inverting $L(z,i)$ enables us to determine the inverse relation between the ith and the $(i-1)$-th stage as

$$\underline{E}(z,i-1) = L^{-1}(z,i)\underline{E}(z,i) \qquad (4.6\text{–}4)$$

where
$$L^{-1}(z,i) = \frac{1}{1-k_i^2} \begin{bmatrix} 1 & k_i \\ k_i z & z \end{bmatrix}$$

The lattice transfer function is useful for predicting the response of the lattice at various stages. The transfer function for the N-stage lattice can easily be derived using Eq. (4.6–3) and proceeding backwards; that is,

$$\begin{aligned} \underline{E}(z,N) &= L(z,N)\underline{E}(z,N-1) \\ &= L(z,N)[L(z,N-1)\underline{E}(z,N-2)] \qquad (4.6\text{–}5) \\ &= L(z,N)L(z,N-1)\cdots L(z,1)\underline{E}(z,0) \end{aligned}$$

which gives the stage-wise transfer function. If we start with $(i=1)$ in Eq. (4.6–4), we can similarly derive relations for the inverse transfer function. Thus, the overall *stage transfer function* of an N-stage lattice and its inverse are given by

$$L(z) := L(z,N)L(z,N-2)\cdots L(z,1) \qquad (4.6\text{–}6)$$
$$L^{-1}(z) := L^{-1}(z,1)L^{-1}(z,2)\cdots L^{-1}(z,N) \qquad (4.6\text{–}7)$$

These stage transfer functions must be related to our standard input-output models of the previous section. Before we explore these relations, consider the following example.

Example 4.6–1. We have a two-stage lattice $\{k_1, k_2\}$ and would like to determine the overall transfer function $L(z)$, as well as the inverse transfer function $L^{-1}(z)$. From Eq. (4.6–6) we see that

$$L(z) = L(z,2)L(z,1) = \begin{bmatrix} 1 & -k_2 z^{-1} \\ -k_2 & z^{-1} \end{bmatrix} \begin{bmatrix} 1 & -k_1 z^{-1} \\ -k_1 & z^{-1} \end{bmatrix}$$

[†]We note that the lattice filter is a two-port network [12], and that $L(z,i)$ is the so-called transfer matrix from input-to-output.

or
$$L(z) = \begin{bmatrix} 1 + k_1 k_2 z^{-1} & -k_1 z^{-1} - k_2 z^{-2} \\ -k_2 - k_1 z^{-1} & k_1 k_2 z^{-1} + z^{-2} \end{bmatrix}$$

The inverse transfer function is given by

$$L^{-1}(z) = L^{-1}(z,1)L^{-1}(z,2) = \frac{1}{(1-k_1^2)(1-k_2^2)} \begin{bmatrix} 1 + k_1 k_2 z & k_2 + k_1 z \\ k_1 z + k_2 z^2 & k_1 k_2 z + z^2 \end{bmatrix}$$

The overall lattice transfer function is related to the standard IIR and FIR transfer functions of the previous section. If we examine the FIR, MA or equivalently all-zero forms of the ARMAX model, that is, ARMAX$(1, N_b, 0)$ or ARMAX$(1, 0, N_c)$, then we can develop the required relations. The FIR model is characterized by the input-output difference equation[†]

$$y(t) = B(q^{-1})u(t)$$

and, equivalently,
$$H(z) = \frac{Y(z)}{U(z)} = B(z) \qquad (4.6\text{–}8)$$

If we use a feed-forward lattice structure, that is,

$$e_f(t,0) = e_b(t,0) := u(t) \qquad \text{and} \qquad e_f(t,N) := y(t)$$

which is equivalent to exciting both input ports and measuring the output, then Eq. (4.6–5) becomes

$$\underline{E}(z,N) = L(z,N) \cdots L(z,1)\underline{E}(z,0) = L(z)\begin{bmatrix} U(z) \\ U(z) \end{bmatrix} = L(z)\begin{bmatrix} 1 \\ 1 \end{bmatrix}U(z)$$

or, dividing both sides by $U(z)$, we obtain

$$\begin{bmatrix} Y(z)/U(z) \\ Y^R(z)/U(z) \end{bmatrix} = \begin{bmatrix} H(z) \\ H^R(z) \end{bmatrix} = \begin{bmatrix} B(z) \\ B^R(z) \end{bmatrix} = L(z)\begin{bmatrix} 1 \\ 1 \end{bmatrix} = \begin{bmatrix} L_{11}(z) + L_{12}(z) \\ L_{21}(z) + L_{22}(z) \end{bmatrix} \qquad (4.6\text{–}9)$$

where $H^R(z)$ is the Nth order *reverse polynomial* [13] given by

$$H^R(z) = z^{-N}H(z^{-1})$$

So we see that if we equate the coefficients $\{b_i\}$ of the all-zero transfer function $H(z)$ to the sum of the row elements of $L(z)$ resulting from the iteration of Eq. (4.6–6), then we obtain the N-stage recursion (usually called the Levinson recursion)

$$\begin{aligned} b(1,1) &= -k_1 \\ b(i,i) &= -k_i & \text{for } i = 2,\ldots,N_b \\ b(j,i) &= b(j,i-1) - k_i b(i-j,i-1) & \text{for } j = 1,\ldots,i-1 \end{aligned} \qquad (4.6\text{–}10)$$

where
$$H(z,i) = 1 + \sum_{j=1}^{N_b} b(j,i)z^{-j}$$

[†]We could have selected the stochastic model as well.

and
$$H(z, N_b) = 1 + b(1, N_b)z^{-1} + \ldots + b(N_b, N_b)z^{-N_b}$$

The ARMAX polynomial is equivalent to the final stage. The following example illustrates the operation of the recursion.

Example 4.6–2. We are given a three-stage feed-forward lattice structure and we would like to convert it to an all-zero model. Using the recursion of Eq. (4.6–10) with the following parameters $N_b = 3$ and $\{k_1, k_2, k_3\}$ then we have

$i = 1$
$b(1, 1) = -k_1$

$\quad H(z, 1) = 1 + b(1, 1)z^{-1}$
$\quad\quad\quad\quad = 1 - k_1 z^{-1}$

$i = 2$
$b(1, 2) = b(1, 1) - k_2 b(1, 1)$
$\quad\quad\quad = -k_1(1 - k_2)$
$b(2, 2) = -k_2$

$\quad H(z, 2) = 1 + b(1, 2)z^{-1} + b(2, 2)z^{-2}$
$\quad\quad\quad\quad = 1 - k_1(1 - k_2)z^{-1} - k_2 z^{-2}$

$i = 3$
$b(1, 3) = b(1, 2) - k_3 b(2, 2)$
$\quad\quad\quad = -k_1(1 - k_2) + k_2 k_3$
$b(2, 3) = b(2, 2) - k_3 b(1, 2)$
$\quad\quad\quad = -k_2 + k_1 k_3(1 - k_2)$
$b(3, 3) = -k_3$

$\quad H(z, 3) = 1 + b(1, 2)z^{-1} + b(2, 2)z^{-2}$
$\quad\quad\quad\quad\quad + b(3, 3)z^{-3}$
$\quad\quad\quad\quad = 1 - [k_1(1 - k_2) - k_2 k_3]z^{-1}$
$\quad\quad\quad\quad\quad - [k_2 - k_1 k_3(1 - k_2)]z^{-2}$
$\quad\quad\quad\quad\quad - k_3 z^{-3}$

The resulting all-zero filter is simply found by inspection of the final stage coefficients,

$$\{b_i\} = \{b(i, N_b)\} \qquad \text{with } b_0 = 1$$

The all-zero ARMAX model can also be transformed to the lattice form by using the inverse operation $L^{-1}(z)$ of Eqs. (4.6–4,5), which leads to the following recursion[†]

$$k_i = -b(i, i) \qquad\qquad \text{for } i = N_b, \ldots, 1$$

$$b(j, i - 1) = \frac{b(j, i) + k_i b(i - j, i)}{1 - k_i^2} \qquad \text{for } j = 1, \ldots, i - 1 \qquad (4.6\text{–}11)$$

where we again use the coefficient relations

$$\{b_i\} = \{b(i, N_b)\}$$

Example 4.6–3. Suppose we are given the following ARMAX(1,2,0) model, with $\{1, b_1, b_2\}$, and we would like to find the corresponding reflection coefficients $\{k_1, k_2\}$. Using the recursions of Eq. (4.6–11) with $N_b = 2$, we have

---

[†]It should be noted that this procedure is equivalent to the Jury or Schur-Cohn stability test for the roots of the polynomial lying inside the unit circle. In fact, the necessary and sufficient condition for stability is that $|k_i| < 1$.

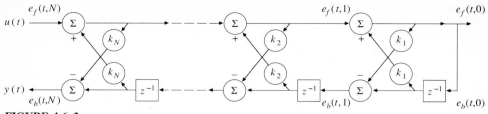

FIGURE 4.6–3
Feedback lattice (all-pole) structure.

$$i = 2 \quad k_2 = -b(2,2) = -b_2$$

$$b(1,1) = \frac{b(1,2) + k_2 b(1,2)}{1 - k_2^2} = \frac{b_1(1 + k_2)}{1 - k_2^2}$$

$$= \frac{b_1(1 + b_2)}{1 - b_2^2} = \frac{b_1}{1 - b_2}$$

$$k_1 = -b(1,1)$$

which completes the recursion.

The inverse of the feed-forward lattice is the feedback lattice shown in Fig. 4.6–3. The feedback lattice is an all-pole transfer function; thus, if we apply an input signal to a feed-forward lattice, and the result is applied to a feedback lattice with identical coefficients, then the original signal will result. From Mason's rule of circuit theory, we know that the reflection coefficients parameterize *both* the feedback as well as feed-forward lattices with appropriate changes in signal flow. Thus, the conversion from the reflection coefficients to tapped delay line coefficients is identical for the all-zero or all-pole ARMAX models, where the signal flow specifies which transfer function is implied. We summarize the recursions in Table 4.6–1. The final useful transformation of the ARMAX model is the pole-

TABLE 4.6–1
Lattice recursions

Lattice to all-pole or all-zero model

$$b(1,1) = -k_1$$
$$b(i,i) = -k_i \qquad\qquad i = 2,\ldots,N_b$$
$$b(j,i) = b(j,i-1) - k_i b(i-j,i-1) \qquad j = 1,\ldots,i-1$$

All-pole or all-zero to lattice model

$$k_i = -b(i,i) \qquad\qquad i = N_b,\ldots,1$$
$$b(j,i-1) = \frac{b(j,i) + k_i b(i-j,i)}{1 - k_i^2} \qquad j = 1,\ldots,i-1$$

where

$$H(z,k) = \sum_{i=0}^{k} b(i,k)z^{-k} \qquad \text{and} \qquad \{b_i\} = \{b(i,N_b)\}$$

zero form, converting from ARMAX($N_a, N_b, 0$) or ARMAX($N_a, 0, N_c$) to lattice models. The symmetric two-multiplier rational lattice, pole-zero transfer function is shown in Fig. 4.6–4 where the tap coefficients are given by $\{g_i\}$, and the pole-zero transfer function is given by

$$H(z) = H(z, N) = \frac{\sum_{j=0}^{N} b(j, N)z^{-j}}{1 + \sum_{j=1}^{N} a(j, N)z^{-j}} = \frac{\sum_{j=0}^{N} b_j z^{-j}}{1 + \sum_{j=1}^{N} a_j z^{-1}} = \frac{B(z)}{A(z)} \quad (4.6\text{–}12)$$

Using procedures similar to the all-zero case (see [13] for details), we have the recursion

$$k_i = -a(i, i) \qquad \text{for } i = N, \ldots, 1$$

$$g_i = b(i, i)$$

$$a(j, i - 1) = \frac{a(j, i) + k_i a(i - j, i)}{1 - k_i^2} \qquad \text{for } j = 1, \ldots, i - 1$$

$$b(j, i - 1) = b(j, i) - g_i a(i - j, i)$$

$$b(0, i - 1) = b(0, i) + g_i k_i$$

$$g_0 = b(0, 0) \qquad (4.6\text{–}13)$$

Although these coefficients are related in a nonlinear manner, this recursion is invertible, so that the rational lattice structure can be converted uniquely to a direct-form, pole-zero filter, and vice versa.

Example 4.6–4. Consider a second-order transfer function with coefficients $\{b_0, b_1, b_2, a_1, a_2\}$. We would like to calculate the corresponding lattice coefficients $\{k_1, k_2, g_0, g_1, g_2\}$, using the recursion of Eq. (4.6–13) with $N = 2$.

$$i = 2 \qquad \qquad k_2 = -a(2, 2) = -a_2$$

$$g_2 = b(2, 2) = b_2$$

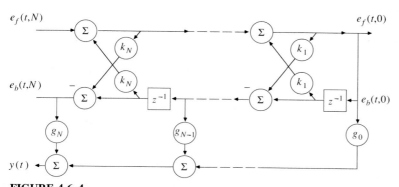

FIGURE 4.6–4
Rational lattice (pole-zero) structure

TABLE 4.6–2
Rational transfer function to lattice transformation

$$k_i = -a(i,i) \qquad\qquad i = N, \ldots, 1$$
$$g_i = b(i,i)$$
$$a(j, i-1) = \frac{a(j,i) + k_i a(i-j,i)}{1 - k_i^2} \qquad j = 1, \ldots, i-1$$
$$b(j, i-1) = b(j,i) - g_i a(i-j,i)$$
$$b(0, i-1) = b(0,i) + g_i k_i$$
$$g_0 = b(0,0)$$

where
$$H(z, N) = \frac{\sum_{j=0}^{N} b(j,N) z^{-j}}{1 + \sum_{j=1}^{N} a(j,N) z^{-j}}$$

and
$$\{b_j\} = \{b(j,N)\}$$

$$j = 1 \qquad a(1,1) = \frac{a(1,2) + k_2 a(1,2)}{1 - k_2^2} = \frac{a_1(1 + k_2)}{1 - k_2^2} = \frac{a_1}{1 - k_2} = \frac{a_1}{1 + a_2}$$

$$b(1,1) = b(1,2) - g_2 a(1,2) = b_1 - b_2 a_1$$

$$b(0,1) = b(0,2) + g_2 k_2 = b_0 - b_2 a_2$$

$$i = 1 \qquad k_1 = -a(1,1)$$

$$g_1 = b(1,1)$$

$$b(0,0) = b(0,1) + g_1 k_1 = b_0 - b_1\left(\frac{a_1}{1 + a_2}\right) - b_2\left(a_2 + \frac{a_1^2}{1 + a_2}\right) = g_0$$

We summarize the recursion in Table 4.6–2.

So, we see that the lattice structure can be transformed to all-pole (IIR, AR), all-zero (FIR, MA), and pole-zero (IIR, ARX, ARMA) models, and vice versa. It should also be noted that in the stochastic case where we have various versions of the ARMAX model, the signal variance can be calculated directly from the white-noise variance, R_{ee}, with knowledge of the reflection coefficients ([16], see Sec. 7.5):

$$R_{yy}(0) = \text{var}\{y(t)\} = \frac{R_{ee}}{\prod_{i=1}^{N}(1 - k_i^2)} \qquad (4.6\text{–}14)$$

Compare this result with the recursions developed in the previous section for the variance of an ARMAX model.

4.7 PARAMETRIC REPRESENTATIONS OF STOCHASTIC PROCESSES: STATE-SPACE MODELS

In addition to the ARMAX and lattice models, we introduce the state-space model as an alternative representation of stochastic processes. As we stated previously, state-space models are easily generalized to multi-channel, non-stationary, and non-linear processes. The state-space models are very popular for model-based signal processing, primarily because most physical phenomena modeled by mathematical relations occur naturally in state-space form (see [17] for details). The state-space model or internal representation of a discrete-time process is given by the *system* or *process* model as

$$x(t) = A(t-1)x(t-1) + B(t-1)u(t-1) \qquad (4.7\text{--}1)$$

with the corresponding *output* or *measurement* model as

$$y(t) = C(t)x(t) \qquad (4.7\text{--}2)$$

where x, u, and y are the respective N_x-state, N_u-input, N_y-output vectors, and A, B, and C are the $(N_x \times N_x)$-system, $(N_x \times N_u)$-input, and $(N_y \times N_x)$-output matrices.

The state-space representation for linear, time-invariant, discrete systems is characterized by constant system, input, and output matrices; that is,

$$A(t) = A, \qquad B(t) = B, \qquad \text{and} \qquad C(t) = C$$

Time-invariant state-space systems can also be represented in *input-output* or *transfer function* form using Z-transforms to give

$$H(z) = C(zI - A)^{-1}B \qquad (4.7\text{--}3)$$

The solution to the state-difference equations can easily be derived by induction [18] and is given by the relations

$$x(t) = \Phi(t,0)x(0) + \sum_{k=0}^{t-1} \Phi(t, k+1)B(k)u(k) \qquad \text{for } t > k \qquad (4.7\text{--}4)$$

where Φ is the *state-transition* matrix. For time-varying systems, it can be shown that the state-transition matrix[†] satisfies

$$\Phi(t,k) = A(t-1) \cdot A(t-2) \cdots A(k)$$

while for time-invariant systems the state-transition matrix is given by

$$\Phi(t,k) = A^{t-k} \qquad \text{for } t > k$$

If the excitation is a random signal, then the state is also random. Restricting the input to be deterministic $u(t-1)$ and zero-mean, white, random gaussian $w(t-1)$, the *Gauss-Markov* model evolves as

$$x(t) = A(t-1)x(t-1) + B(t-1)u(t-1) + W(t-1)w(t-1) \quad (4.7\text{--}5)$$

where $w \sim N(0, R_{ww})$ and $x(0) \sim N(\hat{x}(0), P(0))$.

The solution to the Gauss-Markov equations can be obtained by induction to give

---

[†]Recall that for a sampled-data system the state-transition matrix is $\Phi(t,k) = e^{A_c(t-k)}$, where A_c is the continuous-time system matrix.

$$x(t) = \Phi(t,k)x(k) + \sum_{i=k}^{t-1}\Phi(t,i+1)B(i)u(i) + \sum_{i=k}^{t-1}\Phi(t,i+1)W(i)w(i)$$

$$(4.7\text{--}6)$$

which is Markov depending only on the previous state, and since $x(t)$ is merely a linear transformation of gaussian processes, it is also gaussian. Thus, we can easily represent a Gauss-Markov process using the space-state models. Since the process is assumed to be Gaussian, it is completely characterized by its first two moments. When the measurement model is also included, then we have

$$y(t) = C(t)x(t) + v(t) \qquad (4.7\text{--}7)$$

where $v \sim N(0, R_{vv})$. The model is shown in Fig. 4.7–1.

Since the Gauss-Markov model of Eq. (4.7–5) is gaussian, it is completely specified statistically by its mean and variance. Therefore, if we take the expectation of Eqs. (4.7–5) and (4.7–7) respectively, we obtain the *state mean vector* m_x as

$$m_x(t) = A(t-1)m_x(t-1) + B(t-1)u(t-1) \qquad (4.7\text{--}8)$$

and the *measurement mean vector* m_y as

$$m_y(t) = C(t)m_x(t) \qquad (4.7\text{--}9)$$

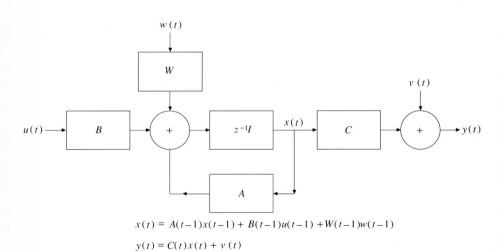

$$x(t) = A(t-1)x(t-1) + B(t-1)u(t-1) + W(t-1)w(t-1)$$

$$y(t) = C(t)x(t) + v(t)$$

where x, u, and y are the N_x-state, N_u-input, and N_y-output vectors; A, B, C, and W are appropriately dimensioned matrices; and w and v are zero-mean, white, gaussian noise sequences with respective covariances $R_{ww}(t)$ and $R_{vv}(t)$; $x(0)$ is gaussian with mean $\hat{x}(0)$ and covariance $P(0)$.

FIGURE 4.7–1
Gauss-Markov model of a discrete process

The *state variance* $P(t) := \text{var } \{x(t)\}$ † is given by the discrete Lyapunov equation:

$$P(t) = A(t-1)P(t-1)A'(t-1) + W(t-1)R_{ww}(t-1)W'(t-1) \quad (4.7\text{--}10)$$

and the measurement variance $R_{yy}(t) := \text{var } \{y(t)\}$ is

$$R_{yy}(t) = C(t)P(t)C'(t) + R_{vv}(t) \quad (4.7\text{--}11)$$

Similarly, it can be shown that the *state covariance* propagates according to the following equations:

$$P(t, k) = \Phi(t, k)P(k) \qquad \text{for } t \geq k$$

$$P(t, k) = P(t)\Phi'(t, k) \qquad \text{for } t \leq k \qquad (4.7\text{--}12)$$

We summarize the Gauss-Markov and corresponding statistical models in Table 4.7–1.

If we restrict the Gauss-Markov model to the stationary case, then

$$A(t) = A, \qquad B(t) = B, \qquad C(t) = C, \qquad W(t) = W,$$

$$R_{ww}(t) = R_{ww}, \qquad \text{and} \qquad R_{vv}(t) = R_{vv}$$

---

†We use the shorthand notation, $P(t) := P_{xx}(t, t) = \text{cov } \{x(t), x(t)\} = \text{var } \{x(t)\}$, throughout this text.

TABLE 4.7–1
Gauss-Markov representation

State propagation

$$x(t) = A(t-1)x(t-1) + B(t-1)u(t-1) + W(t-1)w(t-1)$$

State mean propagation

$$m_x(t) = A(t-1)m_x(t-1) + B(t-1)u(t-1)$$

State variance/covariance propagation

$$P(t) = A(t-1)P(t-1)A'(t-1) + W(t-1)R_{ww}(t-1)W'(t-1)$$

$$P(t, k) = \begin{cases} \Phi(t, k)P(k) & t \geq k \\ P(t)\Phi'(t, k) & t \leq k \end{cases}$$

Measurement propagation

$$y(t) = C(t)x(t) + v(t)$$

Measurement mean propagation

$$m_y(t) = C(t)m_x(t)$$

Measurement variance/covariance propagation

$$R_{yy}(t) = C(t)P(t)C'(t) + R_{vv}(t)$$

$$R_{yy}(t, k) = C(t)P(t, k)C'(t) + R_{vv}(t, k)$$

and the variance equations become

$$P(t) = AP(t-1)A' + WR_{ww}W'$$

and
$$R_{yy}(t) = CP(t)C' + R_{vv}$$

(4.7–13)

At steady-state $(t \to \infty)$, we have

$$P(t) = P(t-1) = \cdots = P_{ss} := P$$

and, therefore, the measurement covariance relations become

$$R_{yy}(0) = CPC' + R_{vv} \qquad \text{for lag } k = 0 \qquad (4.7–14)$$

and it is possible to show that

$$R_{yy}(k) = CA^{|k|}PC' \qquad \text{for } k \neq 0 \qquad (4.7–15)$$

The measurement power spectrum is easily obtained by taking the Z-transform of Eq. (4.7–7) for the time-invariant case,

$$S_{yy}(z) = CS_{xx}(z)C' + S_{vv}(z) \qquad (4.7–16)$$

where $S_{xx}(z) = T(z)S_{ww}(z)T'(z^{-1})$ for $T(z) = (zI - A)^{-1}W$

and $S_{ww}(z) = R_{ww}$ and $S_{vv}(z) = R_{vv}$

Thus, using $H(z) = CT(z)$, the spectrum is given by

$$S_{yy}(z) = H(z)R_{ww}H'(z^{-1}) + R_{vv} \qquad (4.7–17)$$

So we see that going to the Gauss-Markov state-space model enables us to obtain a much more general representation than with the ARMAX or lattice models. In fact, we are able to handle the multi-channel and non-stationary statistical cases easily. Generalizations are also possible with the so-called vector ARMA models, but the forms become quite complicated and require some knowledge of multivariable systems theory and canonical forms (see [18] for details). Before we leave this subject, let us reapproach the input-output example given previously with Gauss-Markov models. The following example indicates the equivalence between input-output and state-space models.

Example 4.7–1. Suppose we are given the ARMA relations of Example 4.5–1,

$$y(t) = -ay(t-1) + e(t-1)$$

The corresponding state-space representation is obtained as

$$x(t) = -ax(t-1) + w(t-1) \qquad \text{and} \qquad y(t) = x(t)$$

Using these models, we obtain the variance equation

$$P(t) = a^2P(t-1) + R_{ww}$$

However, $P(t) = P$ for all t, so

$$P = \frac{R_{ww}}{1 - a^2}$$

Therefore,

$$R_{yy}(k) = CA^{|k|}PC' = \frac{a^{|k|}R_{ww}}{1 - a^2} \quad \text{and} \quad R_{yy}(0) = CPC' + R_{vv} = \frac{R_{ww}}{1 - a^2}$$

Choosing $R_{ww} = 1 - a^2$ gives $R_{yy}(k) = a^{|k|}$ and

$$S_{yy}(z) = H(z)R_{ee}H'(z^{-1}) + R_{vv} = \frac{1}{1 - az^{-1}}R_{ww}\frac{1}{1 - az}$$

as before. So we conclude that for stationary processes these models are equivalent. If we let $a = -0.75, x(0) \sim N(1, 2.3), w \sim N(0, 1)$, and $v \sim N(0, 4)$, then the corresponding Gauss-Markov model is given by

$$x(t) = 0.75x(t - 1) + w(t - 1) \quad \text{and} \quad y(t) = x(t) + v(t)$$

The corresponding statistics are given by the mean relations

$$m_x(t) = 0.75m_x(t - 1) \quad m_x(0) = 1$$
$$m_y(t) = m_x(t) \quad m_y(0) = m_x(0)$$

and the variance equations are

$$P(t) = 0.5625P(t - 1) + 1 \quad \text{and} \quad R_{yy}(t) = P(t) + 4$$

We apply the simulator available in SSPACK (Appendix E or [19]) to develop a 100-point simulation. The results are shown in Fig. 4.7–2(a) through (c). In (a) and (b) we see the mean and simulated states and corresponding confidence interval about the mean,

$$[m_x(t) \pm 1.96\sqrt{P(t)}]$$

and

$$P = \frac{R_{ww}}{1 - a^2} = 2.286$$

We expect 95 percent of the samples to lie within $(m_x \rightarrow 0) \pm 3.03$. From the figure, we see that only 2 of the 100 samples exceed the bound, indicating a statistically acceptable simulation. We observe similar results for the simulated measurements. The steady-state variance is given by

$$R_{yy} = P + R_{vv} = 2.286 + 4 = 6.286$$

Therefore, we expect 95 percent of the measurement samples to lie within $(m_y \rightarrow 0)$ ± 5.01 at steady-state.

4.8 EQUIVALENCE OF LINEAR MODELS: ARMAX, LATTICE TRANSFORMATIONS TO STATE–SPACE

In this section, we show the equivalence between the ARMAX, lattice and state-space models (for scalar processes). That is, we show how to obtain by inspection the state-space model given the ARMAX or lattice models. We choose a particular coordinate system in the state-space (observer canonical form) and obtain a relationship between entries of the state-space system and coefficients

m_x, m_y

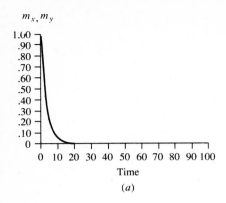

Time

(a)

x

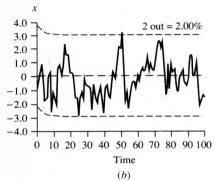

Time

(b)

y

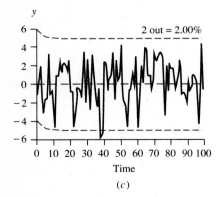

Time

(c)

FIGURE 4.7–2
Gauss-Markov simulation of first-order process (a) State and measurement means (b) State propagation (c) Measurement propagation

of the ARMAX model. An example shows how these models can be applied to realize a stochastic process. First, we consider the ARMAX to state-space tranformation.

Recall from Eq. (4.5–3) that the general difference equation form of the ARMAX model is given by

$$y(t) = -\sum_{i=1}^{N_a} a_i y(t-i) + \sum_{i=0}^{N_b} b_i u(t-i) + \sum_{i=0}^{N_c} c_i e(t-i) \quad (4.8\text{–}1)$$

where $N_a \geq N_b$ and N_c and $\{e(t)\}$ is a zero-mean white sequence with spectrum given by R_{ee}.

It is straightforward but tedious to show that the ARMAX model can be represented in *observer canonical form* (see [5]):

$$x(t) = A_0 x(t-1) + B_0 u(t-1) + W_0 e(t-1)$$
$$y(t) = C'_0 x(t) + b_0 u(t) + c_0 e(t) \tag{4.8--2}$$

where x, u, e, and y are the N_a-state vector, scalar input, noise, and output with

$$A_0 := \left[\begin{array}{c|c} 0 & -a_{N_a} \\ \hline & \vdots \\ I_{N_a - 1} & -a_1 \end{array} \right] \qquad B_0 := \left[\begin{array}{c} -a_{N_a} b_0 \\ \vdots \\ -a_{N_b + 1} b_0 \\ \hline b_{N_b} - a_{N_b} b_0 \\ \vdots \\ b_1 - a_1 b_0 \end{array} \right]$$

$$W_0 := \left[\begin{array}{c} -a_{N_a} c_0 \\ \vdots \\ -a_{N_c + 1} c_0 \\ \hline c_{N_c} - a_{N_c} c_0 \\ \vdots \\ c_1 - a_1 c_0 \end{array} \right] \qquad C'_0 := [0 \cdots 0 \; 1]$$

Noting this structure, we see that each of the matrix or vector elements $\{A_{i,N_a}, B_i, W_i, C_i\}$ $i = 1, \ldots, N_a$ can be determined from the relations

$$A_{i,N_a} = -a_i \qquad i = 1, \ldots, N_a$$
$$B_i = b_i - a_i b_0$$
$$W_i = c_i - a_i c_0$$
$$C_i = \delta(N_a - i) \tag{4.8--3}$$

where
$$b_i = 0 \qquad \text{for } i > N_b$$
$$c_i = 0 \qquad \text{for } i > N_c$$
$$\delta(i - j) \text{ is the Kronecker delta}$$

The following example illustrates these relations.

Example 4.8--1. Let $N_a = 3$, $N_b = 2$, and $N_c = 1$; then the corresponding ARMAX model is

$$y(t) = -a_1 y(t-1) - a_2 y(t-2) - a_3 y(t-3) + b_0 u(t)$$
$$+ b_1 u(t-1) + b_2 u(t-2) + c_0 e(t) + c_1 e(t-1)$$

Using the observer canonical form of Eq. (4.8–2), we have

$$
x(t) = \left[\begin{array}{cc|c}
0 & 0 & -a_3 \\
\hline
1 & 0 & -a_2 \\
0 & 1 & -a_1
\end{array}\right] x(t-1) + \left[\begin{array}{c}
-a_3 b_0 \\
\hline
b_2 - a_2 b_0 \\
b_1 - a_1 b_0
\end{array}\right] u(t-1)
$$

$$
+ \left[\begin{array}{c}
-a_3 c_0 \\
-a_2 c_0 \\
\hline
c_1 - a_1 c_0
\end{array}\right] e(t-1)
$$

$$
y(t) = [0\ 0\ 1]x(t) + b_0 u(t) + c_0 e(t)
$$

If we assume that $\{e(t)\}$ is gaussian, that is, $e \sim N(0, R_{ee})$, then a particular Gauss-Markov model evolves from Eq. (4.8–2), called the *innovations model* (see [20]):

$$
x(t) = Ax(t-1) + Bu(t-1) + We(t-1)
$$

$$
y(t) = C'x(t) + b_0 u(t) + c_0 e(t) \tag{4.8–4}
$$

where

$$
R^*_{ee} := \mathrm{cov}\left(\left[\begin{array}{c} We(t) \\ c_0 e(t) \end{array}\right], \left[\begin{array}{c} We(k) \\ c_0 e(k) \end{array}\right]\right) = \left[\begin{array}{c|c}
WR_{ee}W' & WR_{ee}c_0 \\
\hline
c_0 R_{ee}W' & c_0 R_{ee}c_0
\end{array}\right] \delta(t-k) \tag{4.8–5}
$$

It is important to note that the innovations model has implications in Wiener-Kalman filtering, because R^*_{ee} in Eq. (4.8–5) can be represented in *factored* form:

$$
R^*_{ee} := \left[\begin{array}{c} W\sqrt{R_{ee}} \\ c_0\sqrt{R_{ee}} \end{array}\right] \left[\begin{array}{cc} \sqrt{R_{ee}}\,W' & \sqrt{R_{ee}}\,c_0 \end{array}\right] \delta(t-k) \tag{4.8–6}
$$

Comparing the innovations model to the Gauss-Markov model of Eq. (4.7–5), we see that they are equivalent, except that Eq. (4.8–4) corresponds to the case when w and v are correlated. That is, the *standard* Gauss-Markov model for *correlated process and measurement noise* is given by

$$
x(t) = Ax(t-1) + Bu(t-1) + w^*(t-1)
$$

$$
y(t) = Cx(t) + v^*(t) \tag{4.8–7}
$$

where $R^*(t, k) := R^* \delta(t-k)$ and

$$
R^* := \left[\begin{array}{c|c}
R_{w^*w^*} & R_{w^*v^*} \\
\hline
R_{v^*w^*} & R_{v^*v^*}
\end{array}\right] = \left[\begin{array}{c|c}
WR_{ww}W' & WR_{wv} \\
\hline
R_{vw}W' & R_{vv}
\end{array}\right]
$$

Next we consider lattice transformations to state-space. Recall from Eq. (4.6–1) that the general lattice recursion is given by

$$
e_f(t, i) = e_f(t, i-1) - k_i e_b(t-1, i-1)
$$

$$
e_b(t, i) = e_b(t-i, i-1) - k_i e_f(t, i-1) \tag{4.8–8}
$$

It is possible to show that an N-stage lattice model can be represented in *state-space feed-forward lattice form* (see [20]):

$$x(t) = A_{ff}x(t - 1) + B_{ff}u(t - 1)$$
$$y(t) = C'_{ff}x(t) + u(t) \qquad (4.8\text{--}9)$$

where

$$x'(t) = [e_b(t - 1, 0) \cdots e_b(t - 1, N - 1)]$$
$$y(t) = e_f(t, N)$$

and

$$A_{ff} = \begin{bmatrix} 0 & & \cdots & & 0 \\ 1 & & & & \\ k_1 k_2 & 1 & & & \vdots \\ \vdots & \vdots & & \ddots & \\ k_1 k_{N-1} & k_2 k_{N-1} & \cdots & k_{N-2} k_{N-1} & 1 & 0 \end{bmatrix}, \quad B_{ff} = \begin{bmatrix} 1 \\ -k_1 \\ \vdots \\ -k_{N-1} \end{bmatrix}$$

$$C'_{ff} = [-k_1 \cdots -k_N]$$

This MA model of the feed-forward lattice structure leads to the following relations:

$$A_{ij} = \begin{cases} 1 & i = j + 1 \\ k_{i-j}k_i & i > j \\ 0 & i \le j \end{cases}$$

$$B_{i+1} = -k_i, \qquad i = 1, \ldots, N - 1 \text{ and } B_1 = 1$$
$$C_i = -k_i, \qquad i = 1, \ldots, N \qquad (4.8\text{--}10)$$

The corresponding *state-space feedback lattice form* is obtained from the feed-forward lattice form, using the matrix inversion lemma [21], as

$$A_{fb} = A_{ff} - B_{ff}C'_{ff}$$
$$B_{fb} = B_{ff}$$
$$C'_{fb} = -C'_{ff} \qquad (4.8\text{--}11)$$

where

$$A_{fb} = \begin{bmatrix} k_1 & k_2 & \cdots & k_N \\ 1 - k_1^2 & -k_1 k_2 & \cdots & -k_1 k_N \\ & & \ddots & \vdots \\ 0 & \cdots & 1 - k_{N-1}^2 & -k_{N-1} k_N \end{bmatrix}$$

Finally, the *state-space rational lattice form* is simply a combination of the feed-forward and feedback lattice results,

$$A_R = A_{fb}$$
$$B_R = B_{fb}$$

and

$$C'_R = [-g_1 \cdots -g_N] \qquad (4.8\text{--}12)$$

TABLE 4.8–1
State-space/ARMAX equivalence relations

ARMAX to state-space

$$A_{iN_a} = -a_i \qquad i = 1, \ldots, N_a$$
$$B_i = b_i - a_i b_0$$
$$W_i = c_i - a_i c_0$$
$$C_i = \delta(N_a - i)$$

where
$$b_i = 0 \qquad i > N_b$$
$$c_i = 0 \qquad i > N_c$$
$$\delta(i - j) = \text{Kronecker delta}$$

and
$$A_0 = \begin{bmatrix} 0 & | & A_{iN_a} \\ ----- & | & \vdots \\ I & | & A_{N_a N_a} \end{bmatrix}, \quad B_0 = \begin{bmatrix} B_1 \\ \vdots \\ B_{N_a} \end{bmatrix}, \quad W_0 = \begin{bmatrix} W_1 \\ \vdots \\ W_{N_a} \end{bmatrix}, \quad C_0' = [0 \cdots 1]$$

where values for $\{g_i\}$ are the coefficients of the rational lattice recursion of Eq. (4.6–13). Let us consider an example to illustrate these forms.

Example 4.8–2. Assume we have a three-stage lattice with $\{k_1, k_2, k_3\}$, and we would like to construct the corresponding feed-forward and feedback lattice state-space forms. Starting with the feed-forward form and Eq. (4.8–10), we have

$$A_{\text{ff}} = \begin{bmatrix} 0 & 0 & 0 \\ 1 & 0 & 0 \\ k_1 k_2 & 1 & 0 \end{bmatrix}, \qquad B_{\text{ff}} = \begin{bmatrix} 1 \\ -k_1 \\ -k_2 \end{bmatrix}, \qquad C_{\text{ff}}' = [-k_1 \quad -k_2 \quad -k_3]$$

The feedback lattice form is determined by inspection from Eq. (4.8–11) as

$$A_{\text{fb}} = \begin{bmatrix} k_1 & k_2 & k_3 \\ 1 - k_1^2 & -k_1 k_2 & -k_1 k_3 \\ 0 & 1 - k_2^2 & -k_3^2 \end{bmatrix}, \qquad B_{\text{fb}} = \begin{bmatrix} 1 \\ -k_1 \\ -k_2 \end{bmatrix}, \qquad C_{\text{fb}}' = [k_1 \quad k_2 \quad k_3]$$

If we further assume that we have a rational lattice with coefficients $\{g_1, g_2, g_3\}$, then we obtain the rational lattice form as

$$A_R = A_{\text{fb}}, \qquad B_R = B_{\text{fb}}, \qquad C_R' = [-g_1 \ -g_2 \ -g_3]$$

We summarize the relations of ARMAX and lattice models to state-space forms in Tables 4.8–1 and 4.8–2.

4.9 SUMMARY

In this chapter, we discussed the evolution of stochastic processes as models of phenomenological events, ranging from coin-flipping to a linear dynamic system. We also discussed special processes (e.g., Gaussian, Markov) and properties of stochastic processes. Assuming stationary processes, we developed spectral representations and the concept of simulating a stationary process with given

TABLE 4.8–2
State-space/lattice equivalence relations

Feed-forward lattice to state-space

$$A_{ij} = \begin{cases} k_{i-1}k_i & i > j \\ 1 & i = j + 1 \\ 0 & i \le j \end{cases}$$

$$B_{i+1} = -k_i, \quad B_1 = 1 \qquad\qquad i = 1, \dots, N-1$$
$$C_i = -k_i \qquad\qquad\qquad\qquad i = i, \dots, N$$

where

$$A_{\text{ff}} = \begin{bmatrix} 0 & \cdots & & 0 \\ 1 & & & 0 \\ k_1 k_2 & \ddots & & \vdots \\ k_1 k_{N-1} & \cdots & 1 & 0 \end{bmatrix}, \quad B_{\text{ff}} = \begin{bmatrix} 1 \\ -k_1 \\ \vdots \\ -k_{N-1} \end{bmatrix}, \quad C_{\text{ff}}' = [-k_1 \cdots -k_N]$$

Feedback lattice to state-space

$$A_{\text{fb}} = \begin{bmatrix} k_1 & k_2 & \cdots & k_N \\ 1 - k_1^2 & -k_1 k_2 & \cdots & -k_1 k_N \\ \vdots & & \ddots & \vdots \\ 0 & \cdots & 1 - k_{N-1}^2 & \vdots & -k_{N-1} k_N \end{bmatrix}, \quad B_{\text{fb}} = B_{\text{ff}}, \quad C_{\text{fb}}' = -C_{\text{ff}}'$$

Rational lattice to state-space

$$A_R = A_{\text{fb}}, \quad B_R = B_{\text{fb}}, \quad C_R' = [-g_1 \cdots -g_N]$$

where values for $\{g_i\}$ are given in Table 4.6–2.

covariance by driving a linear system with white noise. We introduced basic models which will be employed throughout the text, including the autoregressive moving average model with exogeneous inputs (ARMAX), and the lattice and the Gauss-Markov or state-space stochastic models. We showed the equivalence of these models and discussed how to transform one to the other. We also introduced the innovations model as a special Gauss-Markov state-space model and showed how it simplifies the simulation of correlated noise sources.

REFERENCES

1. D. Lager and S. Azevedo, "SIG—A General Purpose Signal Processing Code," *Proc. IEEE,* 1987.
2. A. Papoulis, *Probability, Random Variables and Stochastic Processes* (New York: McGraw-Hill, 1965).
3. A. Jazwinski, *Stochastic Processes and Filtering Theory* (New York: Academic Press, 1970).
4. R. Hogg and A. Craig, *Introduction to Mathematical Statistics* (New York: Macmillan, 1970).
5. S. Tretter, *Introduction to Discrete-Time Signal Processing* (New York: Wiley, 1976).
6. K. Astrom, *Introduction to Stochastic Control Theory* (New York: Academic Press, 1970).
7. F. Bauer, "A Direct Iterative Process for the Hurwitz Decomposition of a Polynomial," *Arch. Elect. Ubertragung,* Vol. 9, 1955.
8. T. Kailath and J. Rissanen, "Partial Realization of Random Systems," *Automatica,* Vol. 8, 1972.

9. J. Rissanen,"Algorithm for the Triangular Decomposition of Block Hankel and Toeplitz Matrices with Application to Factorizing Positive Matrix Polynomials," *Mathematics of Comput.*, Vol. 17, 1973.

10. G. Goodwin and K. Sin, *Adaptive Filtering, Prediction and Control* (Englewood Cliffs, N. J.: Prentice-Hall, 1984).

11. E. Robinson and M. Silvia, *Digital Foundations of Time Series Analysis*, Vol. 1 (San Francisco: Holden-Day, 1979).

12. S. Mitra and R. Sherwood, "Digital Ladder Networks," *IEEE Trans. Audio Electroacoust.*, Vol. AU-21, 1973.

13. A. Gray and J. Markel, "Digital Lattice and Ladder Filter Synthesis," *IEEE Trans. Audio Electroacoust.* , Vol. AU-21, 1975.

14. B. Friedlander, "Lattice Filters for Adaptive Processing," *Proc. IEEE*, Vol. 70, 1982.

15. J. Markel and A. Gray, *Linear Prediction of Speech* (New York: Springer-Verlag, 1976).

16. S. Orfanidis, *Optimum Signal Processing* (New York: Macmillan, 1985).

17. J. Candy, *Signal Processing: The Model-Based Approach* (New York: McGraw-Hill, 1986).

18. T. Kailath, *Linear Systems* (Englewood Cliffs, N. J.: Prentice-Hall, 1980).

19. S. Azevedo, J. Candy, and D. Lager, "SSPACK—An Interactive, Multi-channel, Model-based Signal Processing Package," *Proc. IEEE Confr. Circuits Systems*, 1986.

20. D. Lee, *Canonical Ladder Form Realizations and Fast Estimation Algorithms*, Ph.D. Dissertation, Stanford, 1980.

21. T. Kailath, *Lectures on Kalman and Wiener Filtering Theory* (New York: Springer-Verlag, 1981).

SIG NOTES

SIG can be used to simulate deterministic as well as random signals and systems. Besides processing a full complement of signal simulation commands consisting of the deterministic CHIRP (changing sinusoids), CONSTANT (step function), RAMP, PULSETRAIN, and the random UNIFORMRANDOM, PSEUDORANDOM (binary sequence), and GAUSSIANRANDOM, it also contains linear system commands which can be used to simulate correlated signals, as discussed in Sec. 4.4. Filtered noise (bandpass, low-pass, etc.), using the signal simulation commands in conjunction with the digital filtering commands as well as the CONVOLUTION command, can be used. The spectral simulation procedure of Sec. 4.4 can be implemented in the time domain, using the LSSABONLY and LSSABCD $(\text{ARMAX}(N_a, N_b, N_c, N_d)$ model$)$. The ARMAX model and associated statistics of Table 4.5–1, with the output and mean are simulated using LSSABCD, the impulse simulated with LSSIMPULSE $(C \rightarrow B)$ command, and the covariance propagation is simulated using the scaled CONVOLUTION command. SIG does *not* simulate the Gauss-Markov (state-space) models; SSPACK can be used to simulate these signals and systems (see Chapter 9 and Appendix E).

EXERCISES

4.1. Suppose the stochastic process $\{y(t)\}$ is generated by

$$y(t) = a \exp(-t) + ct, \qquad a, b \text{ random}$$

(a) What is the mean of the process?
(b) What is the corresponding covariance?
(c) Is the process stationary, if $E\{a\} = E\{b\} = 0$, and $E\{ab\} = 0$?

4.2. Derive the following properties of conditional expectations:
(a) $E_x\{X \mid Y\} = E\{X\}$ if X and Y are independent
(b) $E\{X\} = E_y\{E\{X \mid Y\}\}$
(c) $E_x\{g(Y) X\} = E_y\{g(Y) E\{X \mid Y\}\}$
(d) $E_{xy}\{g(Y) X\} = E_y\{g(Y) E\{X \mid Y\}\}$

(e) $E_x\{c \mid Y\} = c$

(f) $E_x\{g(Y) \mid Y\} = g(Y)$

(g) $E_{xy}\{cX + dY \mid Z\} = cE\{X \mid Z\} + dE\{Y \mid Z\}$

4.3. Suppose x, y, and z are gaussian random variables, with corresponding means m_x, m_y, m_z and variance R_{xx}, R_{yy}, R_{zz}. Show that

(a) if $y = ax + b$, a, b constants, then show $y \sim N(am_x + b, a^2R_x)$

(b) if x and y are uncorrelated, then show they are independent

(c) if $x(i)$ is gaussian with mean $m(i)$ and variance $R_{xx}(i)$, then for

$$y = \sum_i K_i x(i), \qquad \text{show } y \sim N\left(\sum_i K_i m(i), \sum_i K_i^2 R_{xx}(i)\right)$$

(d) if x and y are jointly (conditionally) gaussian, then show

$$E\{x \mid y\} = m_x + R_{xy}R_{yy}^{-1}(y + m_y), \text{ and}$$

$$R_{x|y} = R_{xx} + R_{xy}R_{yy}^{-1}R_{yx}$$

(e) the random variable $x = E\{x \mid y\}$ is orthogonal to y

(f) if y and z are independent, then show

$$E\{x \mid y, z\} = E\{x \mid y\} + E\{x \mid z\} - m_x$$

(g) if y and z are not independent, then show

$$E\{x \mid y, z\} = E\{x \mid y, e\} = E\{x \mid y\} + E\{x \mid e\} - m_x$$

for $e = z - E\{x \mid y\}$.

4.4. Assume $y(t)$ is a zero-mean, ergodic process with covariance $R_{yy}(k)$, and calculate the corresponding power spectra, $S_{yy}(z)$ if

(a) $R_{yy}(k) = Ca^{|k|}$

(b) $R_{yy}(k) = C \cos(\omega|k|)$, $|k| < \pi/2$

(c) $R_{yy}(k) = C \exp(-a^{|k|})$

4.5. Verify the covariance-spectral density pairs of Table 4.3–1 for these discrete processes:

(a) Bandlimited white noise

(b) Triangular

4.6. Let the impulse response of a linear system with random input $u(t)$ be given by $h(t)$; then show that

(a) $R_{yy}(k) = \sum_{m=0}^{\infty} \sum_{i=0}^{\infty} h(m)h(i)R_{uu}(k + i - m)$ and $S_{yy}(z) = H(z)H(z^{-1})S_{uu}(z)$

(b) $R_{yy}(k) = \sum_{m=0}^{\infty} h(m)R_{yu}(k - m)$ and $S_{yy}(z) = H(z)S_{yu}(z)$

(c) $R_{uy}(k) = \sum_{m=0}^{\infty} h(m)R_{uu}(k - m)$ and $S_{uy}(z) = H(z)S_{uu}(z)$

4.7. Derive the *sum decomposition* relation,

$$S_{yy}(z) = S_{yy}^+(z) + S_{yy}^-(z) - R_{yy}(0)$$

4.8. Develop a computer program to simulate the ARMA process

$$y(t) = -ay(t - 1) + e(t)$$

where $a = 0.75$, $e \sim N(0, 0.1)$ for 100 data points.

(a) Calculate the analytic covariance $R_{yy}(k)$.

(b) Determine an expression to "recursively" calculate $R_{yy}(k)$.

(c) Plot the simulated results and construct the $\pm 2\sqrt{R_{yy}(0)}$ bounds.

(d) Do 95% of the samples fall within these bounds?

4.9. Develop the digital filter to simulate a sequence, $y(t)$, with covariance $R_{yy}(k) = 4e^{-3|k|}$.

4.10. Suppose we are given a linear system characterized by transfer function

$$H(z) = \frac{1 - \frac{1}{2}z^{-1}}{1 - \frac{1}{3}z^{-1}}$$

which is excited by discrete exponentially correlated noise

$$R_{xx}(k) = (1/2)^{|k|}$$

(a) Determine the output PSD, $S_{yy}(z)$.
(b) Determine the output covariance, $R_{yy}(k)$.
(c) Determine the cross-spectrum, $S_{yx}(z)$.
(d) Determine the cross-covariance, $R_{yx}(k)$.

4.11. Suppose we are given a causal LTI system characterized by its impulse response, $h(t)$. If this system is excited by zero-mean, unit-variance white noise, then
(a) determine the output variance, $R_{yy}(0)$;
(b) determine the covariance, $R_{yy}(k)$ for $k > 0$;
(c) suppose the system transfer function is given by

$$H(z) = \frac{1 + b_0 z^{-1}}{1 + a_1 z^{-1} + a_2 z^{-2}}$$

and find a method to recursively calculate $h(t)$ and therefore $R_{yy}(0)$.

4.12. Given the covariance function

$$R_{yy}(k) = e^{-1/2|k|} \cos \pi|k|$$

find the digital filter which, when driven by unit-variance white noise, produces a sequence $\{y(t)\}$ with these statistics.

4.13. Suppose we have a process characterized by difference equation

$$y(t) = x(t) + \frac{1}{2}x(t - 1) + \frac{1}{3}x(t - 2)$$

(a) Determine a recursion for the output covariance, $R_{yy}(k)$.
(b) If $x(t)$ is white, with variance σ_{xx}^2, determine $R_{yy}(k)$.
(c) Determine the output PSD, $S_{yy}(z)$.

4.14. Suppose we are given a linear system characterized by the difference equation

$$y(t) - \frac{1}{5}y(t - 1) = \frac{1}{\sqrt{3}}x(t)$$

and the system is excited by (i) white Gaussian noise, $x \sim N(0, 3)$, and (ii) exponentially correlated noise, $R_{ee}(k) = (1/2)^{|k|}$. In both cases find
(a) output PSD, $S_{yy}(z)$;
(b) output covariance, $R_{yy}(k)$;
(c) cross-spectrum, $S_{ye}(z)$;
(d) cross covariance, $R_{ye}(k)$.

4.15. Suppose we have a MA process (two-point averager)

$$y(t) = \frac{e(t) + e(t - 1)}{2}$$

(*a*) Develop an expression for $S_{yy}(z)$ when e is white with variance R_{ee}.

(*b*) Let $z = \exp\{j\Omega\}$ and sketch the "response" of $S_{yy}(e^{j\Omega})$.

(*c*) Calculate an expression for the covariance, $R_{yy}(k)$, in closed and recursive form.

4.16. Suppose we are given a zero-mean process with covariance

$$R_{yy}(k) = 10\exp(-0.5|k|)$$

(*a*) Determine the digital filter which, when driven by white noise, will yield a sequence with the above covariance.

(*b*) Develop a computer program to generate $y(t)$ for 100 points.

(*c*) Plot the results and determine if 95 percent of the samples fall within $\pm 2\sqrt{R_{yy}(0)}$.

4.17. Suppose we have the following transfer functions

(*a*) $H_1(z) = 1 - \frac{1}{8}z^{-1}$

(*b*) $H_2(z) = \dfrac{1}{1 - \frac{3}{4}z^{-1} + \frac{1}{8}z^{-2}}$

(*c*) $H_3(z) = \dfrac{1 - \frac{1}{8}z^{-1}}{1 - \frac{3}{4}z^{-1} + \frac{1}{8}z^{-2}}$

Find the corresponding lattice model for each case, that is, all-zero, all-pole, and pole-zero.

4.18. Suppose we are given the factored power spectrum $S_{yy}(z) = H(z)H(z^{-1})$, with

$$H(z) = \frac{1 + \beta_1 z^{-1} + \beta_2 z^{-2}}{1 + \alpha_1 z^{-1} + \alpha_2 z^{-2}}$$

(*a*) Develop the ARMAX model for the process.

(*b*) Develop the corresponding Gauss-Markov model for *both* the standard and innovations representation of the process.

4.19. Suppose we are given the following Gauss-Markov model:

$$x(t) = \frac{1}{3}x(t-1) + \frac{1}{2}w(t-1)$$

$$y(t) = 5x(t) + v(t)$$

$$w \sim N(0, 3) \qquad v \sim N(0, 2)$$

(*a*) Calculate the state power spectrum, $S_{xx}(z)$

(*b*) Calculate the measurement power spectrum, $S_{yy}(z)$.

(*c*) Calculate the state covariance recursion, $P(t)$.

(*d*) Calculate the steady-state covariance, $P(t) = \cdots = P = P_{ss}$.

(*e*) Calculate the output covariance recursion, $R_{yy}(t)$.

(*f*) Calculate the steady-state output covariance, R_{yy}.

4.20. Suppose we are given the Gauss-Markov process characterized by the state equations

$$x(t) = 0.97x(t-1) + u(t-1) + w(t-1)$$

if $u(t)$ is a step of amplitude 0.03, $w \sim N(0, 10^{-4})$, and $x(0) \sim N(2.5, 10^{-12})$.

(*a*) Calculate the covariance of x, i.e., $P(t) = \text{cov } x(t)$.

(*b*) Since the process is stationary, we know that

$$P(t + k) = P(t + k - 1) = \cdots = P(0) = P$$

What is the steady state covariance, P, of this process?

(c) Develop a computer program to simulate this process.

(d) Plot the process $x(t)$ with the corresponding confidence limits $\pm 2\sqrt{P(t)}$ for 100 data points. Do 95 percent of the samples lie within the bounds?

4.21. Suppose the process in Problem 4.20 is measured using an instrument with uncertainty $v \sim N(0, 4)$, such that

$$y(t) = 2x(t) + v(t)$$

(a) Calculate the output covariance $R_{yy}(k)$.

(b) Develop a computer program to simulate the output, using the results of Problem 4.20.

(c) Plot the process $y(t)$ with the corresponding confidence limits $\pm 2\sqrt{R_{yy}(0)}$ for 100 data points. Do 95 percent of the samples lie within the bounds?

4.22. Suppose we are given the ARMAX model

$$y(t) = -0.5y(t-1) - 0.7y(t-2) + u(t) + 0.3u(t-1) + e(t) + 0.2e(t-1) + 0.4e(t-2)$$

(a) What is the corresponding innovations model in state-space form for $e \sim N(0, 10)$?

(b) Calculate the corresponding covariance matrix R_{ee}^*.

CHAPTER
5

RANDOM SIGNAL ANALYSIS: CLASSICAL APPROACH

In this chapter we are concerned with the analysis and extraction of information from a random signal. As in the deterministic case, we must develop techniques to estimate the spectral representations of the signals under investigation. First, we review basic concepts in estimation; then, we see how they can be applied to estimate the covariance from raw data. Next, we investigate some of the classical nonparametric methods of spectral estimation. Finally, we introduce a complementary function requiring spectral estimates which can also be used in the analysis of random signals: the coherence function.

5.1 ESTIMATION CONCEPTS

In the previous chapter we saw that a discrete random signal can be completely characterized by the probabilistic concept of a stochastic process. The filtering of random signals is referred to as *estimation*, and the particular algorithm is called a *signal estimator* or simply an *estimator*. The process of estimation is concerned with the design of a rule or algorithm, the estimator, to extract useful signal information from random data.

149

There are many different estimators and algorithms. Suppose we are given two estimators and are asked to evaluate their performance to decide which one is superior. We must have a reasonable measure to make this decision. Thus, we must develop techniques to investigate various properties of estimators, as well as a means to decide how well they perform. Sometimes we are interested in estimators that are not optimal for various reasons, such as simplicity or implementation ease and we would like to judge how well they perform compared to the optimum. In this section we investigate desirable properties of estimators and criteria to evaluate their performance, and look at some popular schemes.

The need for an agreed-upon rule(s) to measure estimator performance is necessary. The two primary statistical measures employed are the estimator mean (accuracy) and the variance (precision). These measures lead to desirable estimator properties; that is, they lead to estimators that are accurate or unbiased and precise. More formally, an *unbiased estimator* is one whose expected value is identical to the parameter being estimated. Suppose we wish to estimate Θ; the estimator $\hat{\Theta}$ is *unconditionally unbiased* if

$$E\{\hat{\Theta}\} = E\{\Theta\} \qquad \text{for all } \Theta \tag{5.1–1}$$

If Θ is a function of the measurements Y, then it is equivalent to have

$$E_{\hat{\Theta}Y}\{\hat{\Theta}(Y)\} = E\{\Theta\} \qquad \text{for all } \Theta$$

If the estimator is conditioned on Θ, then we desire a *conditionally unbiased* estimator

$$E\{\hat{\Theta} \mid \Theta = \theta\} = E\{\Theta\} \tag{5.1–2}$$

A biased estimator, then, is given by

$$E\{\hat{\Theta}\} = E\{\Theta\} + E\{B(\Theta)\}$$

where $B(\cdot)$ is the bias, which may be known or unknown. Known biases are easily removed.

An estimator is *consistent* if the estimate improves as the number of measurements increases; or, equivalently, it is said to *converge in probability*, if

$$\lim_{t \to \infty} P([\Theta - \hat{\Theta}(Y_t)] = 0) = 1$$

It is said to be *mean-squared convergent* if

$$\lim_{t \to \infty} E\{[\Theta - \hat{\Theta}(Y_t)][\Theta - \hat{\Theta}(Y_t)]'\} = 0$$

and it is *asymptotically efficient* if, for any other estimator $\tilde{\Theta}^*$, the corresponding estimation error ($\tilde{\Theta}$) variance is

$$\text{var}\,(\tilde{\Theta}) < \text{var}\,(\tilde{\Theta}^*)$$

Finally, an estimator is called *sufficient* if it possesses all the information contained in the set of measurements regarding the parameter to be estimated (see [1] for more details). These properties are desirable in an estimator and are checked in its evaluation.

The quality of an estimator is usually measured in terms of its *estimation error*,

$$\tilde{\Theta} = \Theta - \hat{\Theta}$$

A common measure of estimator quality is called the *Cramer-Rao* (CRB) *lower bound*. The CRB bound offers a means of assessing estimator quality prior to processing the measured data. We restrict discussion of the CRB bound to the case of unbiased estimates $\hat{\Theta}$ of a "non-random" parameter Θ. The bound is easily extended to more complex cases for biased estimates as well as the case of random parameters [2,3]. The Cramer-Rao bound[†] for any unbiased estimate $\hat{\Theta}$ of Θ based on the measurement is given by

$$R_{\tilde{\Theta}|\Theta} = \text{cov}\,(\Theta - \hat{\Theta}(Y)\,|\,\Theta = \theta) \geq \mathcal{I}^{-1} \qquad (5.1\text{--}3)$$

where $\mathcal{I}$ is the $N \times N$ *information matrix*, given by

$$\mathcal{I} := -E_y\{\nabla_\Theta(\nabla_\Theta \ln p(Y\,|\,\Theta))'\}$$

with the *gradient vector* $\nabla_\Theta \in R^n$ defined by

$$\nabla_\Theta := \left[\frac{\partial}{\partial\Theta_1} \cdots \frac{\partial}{\partial\Theta_n}\right]'$$

Any estimator satisfying the CRB bound with equality is called *efficient*. The bound is easily calculated, using the *chain rule* from vector calculus [4], as

$$\nabla_\Theta(a'b) = (\nabla_\Theta a')b + (\nabla_\Theta b')a \qquad a, b \in R^n \qquad (5.1\text{--}4)$$

where a and b are functions of Θ. The following example illustrates the calculation of the CRB bound.

Example 5.1–1. Suppose we would like to estimate a nonrandom but unknown parameter Θ from a measurement y contaminated by additive gaussian noise; that is,

$$y = \Theta + v$$

where $v \sim N(0, R_{vv})$ and Θ is unknown. Thus, we have

$$E\{Y\,|\,\Theta\} = E\{\Theta + v\,|\,\Theta\} = \Theta$$

and $\quad \text{var}\,(Y\,|\,\Theta) = E\{(y - E\{Y\,|\,\Theta\})^2\,|\,\Theta\} = E\{v^2\,|\,\Theta\} = R_{vv}$

which gives $\quad P(Y\,|\,\Theta) \sim N(\Theta, R_{vv})$

and therefore $\quad \ln P(Y\,|\,\Theta) = -\frac{1}{2}\ln 2\pi R_{vv} - \frac{1}{2}\frac{(y - \Theta)^2}{R_{vv}}$

Differentiating according to Eq. (5.1–3) and taking the expectation, we obtain

$$\mathcal{I} = -E\left\{\frac{\partial^2}{\partial\Theta^2}\ln P(Y\,|\,\Theta)\right\} = -E\left\{-\frac{\partial}{\partial\Theta}\frac{(y - \Theta)}{R_{vv}}(-1)\right\} = \frac{1}{R_{vv}}$$

[†]We choose the matrix-vector version, since parameter estimators are typically vector estimates.

and, therefore, the CRB bound is given by

$$R_{\tilde{\Theta}|\Theta} \geq R_{vv}$$

The utility of the CRB bound is twofold: (1) it enables us to measure estimator quality, because it indicates the "best" (minimum error covariance) that any estimator can achieve, and (2) it allows us to decide whether or not the estimator is efficient, a desirable statistical property. The properties of an estimator can be calculated prior to estimation (in some cases) and these properties can be used to answer the question "how well does this estimator perform."

5.2 MINIMUM-VARIANCE/ MAXIMUM-LIKELIHOOD ESTIMATION

In this section we briefly review the concepts of minimum variance and maximum likelihood estimation. The general techniques developed will be used in subsequent sections.

The development of an estimation procedure consists of the selection of a criterion function, to be minimized or maximized, and the development and implementation of the corresponding algorithm. The *minimum (error) variance estimator* evolves from the minimization of the error variance, or mean-squared error criterion,

$$J(\Theta) = E_{\Theta}\{[\Theta - \hat{\Theta}(Y)]'[\Theta - \hat{\Theta}(Y)] \mid Y\} \qquad (5.2\text{--}1)$$

where Θ is the true random n-vector
              Y is the measured random p-vector (data)
              $\hat{\Theta}$ is the estimate of Θ, given Y

If we minimize $J(\Theta)$ using the chain rule of Eq. (5.1–4), we obtain

$$\nabla_{\Theta}J(\Theta) = -2[E\{\Theta \mid Y\} - \hat{\Theta}(Y)]$$

Setting this equation to zero and solving gives

$$\hat{\Theta}_{MV} = \hat{\Theta}(Y) = E\{\Theta \mid Y\} \qquad (5.2\text{--}2)$$

Thus, the minimum-variance estimator is the *conditional mean*. It can easily be shown that this estimator is linear, unconditionally and conditionally unbiased, and possesses the following orthogonality properties (see [4] for details). If the *estimation error* is given by

$$\tilde{\Theta} = \Theta - \hat{\Theta}(Y) \qquad (5.2\text{--}3)$$

then the error is uncorrelated or *orthogonal* to all past Y; that is,

$$E_{\Theta Y}\{Y\tilde{\Theta}'\} = 0$$

or in fact $E_{\Theta}\{Y\tilde{\Theta}' \mid Y\} = 0 \qquad (5.2\text{--}4)$

The well-known minimum-variance estimator results in the linear case [5]. For

$$y = C\Theta + v$$

where $y, v \in R^N, \Theta \in R^n, C \in R^{N \times n}$, and v is white with R_{vv}, then the mean-squared error criterion

$$J(\Theta) = E\{\tilde{\Theta}'\tilde{\Theta}\}$$

is minimized to determine the estimate. The minimization results in the orthogonality condition of Eq. (5.2–4), which we write as

$$E\{y\tilde{\Theta}'\} = E\{y\Theta'\} - E\{y\hat{\Theta}'_{MV}\} = 0 \tag{5.2–5}$$

for $\hat{\Theta}_{MV} = K_{MV}y$, a linear function of the data vector. Substituting for y and $\hat{\Theta}$, we obtain

$$K_{MV} = R_{\Theta\Theta}C'(CR_{\Theta\Theta}C' + R_{vv})^{-1} \tag{5.2–6}$$

The corresponding quality is obtained as

$$R_{\tilde{\Theta}\tilde{\Theta}} = (R_{\Theta\Theta}^{-1} + C'R_{vv}^{-1}C)^{-1} \tag{5.2–7}$$

So we see that the minimum-variance estimator for the linear case is

Criterion: $J = E\{\tilde{\Theta}'\tilde{\Theta}\}$
Algorithm: $\hat{\Theta}_{MV} = K_{MV}y$
Quality: $R_{\tilde{\Theta}\tilde{\Theta}}$

It is also interesting to note that the fundamental Wiener result is easily obtained from the orthogonality condition of Eq. (5.2–5):

$$E\{y\tilde{\Theta}'\} = E\{y\Theta'\} - E\{yy'\}K'_{MV} = R_{y\Theta} - R_{yy}K'_{MV} = 0 \tag{5.2–8}$$

This is called the *discrete Wiener-Hopf equation*. Solving for K_{MV}, we obtain the Wiener solution for a linear (batch) estimation scheme,

$$K_{MV} = R_{\Theta y}R_{yy}^{-1} \tag{5.2–9}$$

We also mention in passing that *least-squares estimation* is similar to that of minimum variance, except that no statistical information is assumed to be known about the process; that is, the least-squares estimator minimizes the squared-error criterion

$$J = \tilde{y}'\tilde{y} \tag{5.2–10}$$

for $\tilde{y} = y - \hat{y}_{LS}$.

In contrast to the least-squares approach requiring *no* statistical information, we introduce two popular estimators: the "most probable" or maximum a posteriori (MAP) estimator, and the maximum-likelihood (ML) estimator, which is a special case of MAP.

Suppose we are trying to estimate a random parameter, say Θ, from data $Y = y$. Then the associated conditional mass $p(\Theta \mid Y = y)$ is called the *a posteriori* mass, because the estimate is conditioned "after (*post*) the measurements" have been acquired. Estimators based on the a posteriori mass are usually called bayesian because they are constructed from Bayes' rule, since $p(\Theta \mid Y)$ is difficult to obtain directly. That is,

$$p(\Theta \mid Y) = p(Y \mid \Theta)\frac{p(\Theta)}{p(Y)} \qquad (5.2\text{--}11)$$

where $p(\Theta)$ is called the *a priori* (before measurement) mass. To solve the estimation problem, the first step requires the determination of the a posteriori mass. A logical solution to this problem leads us to find the "most probable" value of $p(\Theta \mid Y)$ that is its maximum. The maximum a posteriori (MAP) estimate is the value that yields the maximum value of the a posteriori mass. The optimization is carried out in the usual manner by

$$\nabla_\Theta p(\Theta \mid Y)\big|_{\Theta = \hat{\Theta}_{\text{MAP}}} = 0$$

For convenience, we consider the $\ln p(\Theta \mid Y)$ instead. Since the logarithm is a monotonic function, the maximum of $p(\Theta \mid Y)$ and $\ln p(\Theta \mid Y)$ occur at the same value of Θ. Therefore, the MAP *equation* is

$$\nabla_\Theta \ln p(\Theta \mid Y)\big|_{\Theta = \hat{\Theta}_{\text{MAP}}} = 0 \qquad (5.2\text{--}12)$$

If we apply Bayes' rule to Eq. (5.2–12), then

$$\ln p(\Theta \mid Y) = \ln p(Y \mid \Theta) + \ln p(\Theta) - \ln p(Y) \qquad (5.2\text{--}13)$$

Since $p(Y)$ is not a function of the parameter Θ, the MAP equation can be written

$$\nabla_\Theta \ln p(\Theta \mid Y)\big|_{\Theta = \hat{\Theta}_{\text{MAP}}} = \nabla_\Theta (\ln p(Y \mid \Theta) + \ln p(\Theta))\big|_{\Theta = \hat{\Theta}_{\text{MAP}}} = 0 \quad (5.2\text{--}14)$$

Of course, the second derivative must be checked to ensure that a global maximum is obtained. It can be shown that for a variety of circumstances (cost function convex, symmetrical) the minimum-variance estimate of the previous section is in fact identical to the MAP estimate [5].

Suppose we have no a priori knowledge of the parameter Θ (i.e., we do not know $p(\Theta)$, or we model Θ as a nonrandom parameter); in that case, we obtain the maximum-likelihood estimate as a special case of the MAP estimate, if we take the MAP equation of Eq. (5.2–9) and ignore the a priori mass $p(\Theta)$ (assume it is unknown, or Θ is nonrandom). Using the same arguments as before, we prefer the $\ln p(\Theta \mid Y)$, instead of $p(\Theta \mid Y)$. Thus, we obtain the maximum-likelihood estimate by solving the *log-likelihood equation* and checking for the existence of a maximum; that is,

$$\nabla_\Theta \ln p(\Theta \mid Y)\big|_{\Theta = \hat{\Theta}_{\text{ML}}} = 0$$

and, of course, we check that $\nabla_\Theta (\nabla_\Theta \ln p(\Theta \mid Y)') < 0$. Again applying Bayes' rule, as in Eq. (5.2–13), and ignoring $p(\Theta)$, we have

$$\nabla_\Theta \ln p(\Theta \mid Y) = \nabla_\Theta \ln p(Y \mid \Theta)\big|_{\Theta = \hat{\Theta}_{\text{ML}}}$$

The maximum-likelihood (ML) estimator has some very interesting properties, which we list without proof (see [6] for details):

1. ML estimates are *consistent*.
2. ML estimates are *asymptotically efficient*.
3. ML estimates are *asymptotically gaussian* with $N(\Theta, R_{\tilde{\Theta}\tilde{\Theta}})$.

These properties are asymptotic and therefore imply that a large amount of data must be available for processing.

Example 5.2–2. Consider estimating an unknown constant from a noisy measurement, as in the previous example. Further, assume that the noise is an independent gaussian random variable, such that $v \sim N(0, R_{vv})$; the measurement model is given by

$$y = \Theta + v$$

First, we asssume no "a priori" knowledge of Θ except that it is an unknown, nonrandom constant. Thus we require the maximum-likelihood estimate, since no prior information is assumed about $p(\Theta)$. The associated conditional mass is

$$p(Y \mid \Theta) = \frac{1}{\sqrt{2\pi R_{vv}}} e^{-(1/2)(y-\Theta)^2/R_{vv}}$$

The maximum-likelihood estimate of Θ is found by solving the log-likelihood equation:

$$\nabla_{\Theta} \ln p(Y \mid \Theta)\big|_{\Theta = \hat{\Theta}_{ML}} = 0$$

or
$$\frac{\partial}{\partial \Theta} \ln p(Y \mid \Theta) = \frac{\partial}{\partial \Theta}\left\{ \frac{-1}{2} \ln 2\pi R_{vv} - \frac{1}{2R_{vv}}(y - \Theta)^2 \right\}$$

$$= \frac{1}{R_{vv}}(y - \Theta)$$

Setting this expression to zero and solving for Θ, we obtain

$$\hat{\Theta}_{ML} = y$$

That is, the best estimate of Θ in a maximum likelihood sense is the raw data y. The corresponding error variance is easily calculated as

$$R_{\tilde{\Theta}|\Theta} = R_{vv}$$

Next we model Θ as a gaussian random variable, that is, $\Theta \sim N(\overline{\Theta}, R_{\Theta\Theta})$, and we require the maximum a posteriori estimate. The MAP equation is

$$\nabla_{\Theta}(\ln p(Y \mid \Theta) + \ln p(\Theta))$$

$$= \frac{\partial}{\partial \theta}\left\{ -\frac{1}{2} \ln 2\pi R_{vv} - \frac{1}{2R_{vv}}(y - \Theta)^2 - \frac{1}{2} \ln 2\pi R_{\Theta\Theta} - \frac{\frac{1}{2}(\Theta - \overline{\Theta})^2}{R_{\Theta\Theta}} \right\}$$

or $\qquad \nabla_\Theta(\ln p(Y \mid \Theta) + \ln p(\Theta)) = \dfrac{1}{R_{vv}}(y - \Theta) - \dfrac{1}{R_{\Theta\Theta}}(\Theta - \overline{\Theta})$

Setting this expression to zero and solving for $\Theta = \hat{\Theta}_{\text{MAP}}$, we obtain

$$\hat{\Theta}_{\text{MAP}} = \frac{y + (R_{vv}/R_{\Theta\Theta})\overline{\Theta}}{1 + (R_{vv}/R_{\Theta\Theta})}$$

It can be shown that the corresponding error variance is

$$R_{\tilde{\Theta}\mid\Theta} = \frac{R_{vv}}{1 + R_{vv}/R_{\Theta\Theta}}$$

Examining the results of this example, we see that when the parameter variance is large ($R_{\Theta\Theta} \gg R_{vv}$), the MAP and ML estimates perform equivalently. However, when the variance is small, the MAP estimator performs better, because the corresponding error variance is smaller.

The main point to note is that the MAP estimate provides a means to incorporate a priori information, while the ML estimate does not. Therefore, for some problems MAP is the efficient estimator. In this example, if Θ were actually gaussian, then the ML solution, which models Θ as an unknown parameter, is not an efficient estimator, whereas the MAP solution incorporates this information by using $p(\Theta)$.

5.3   COVARIANCE ESTIMATION

With a means to assess the performance of estimators and the various estimation schemes available, let us consider estimating the covariance from random data. Assume that we have a stationary random process $\{x(t)\}$, and for convenience assume $m_x(t) = 0$ for all t. The covariance is then given by

$$R_{xx}(k) = E\{x(t)x(t + k)\} \tag{5.3–1}$$

There are two basic approaches to estimating covariances: parametric and nonparametric. The *parametric approach* assumes that the true covariance function has a particular analytic form, for instance,

$$R(k) = Ae^{-a|k|} \qquad |a| < 1$$

which is completely specified by the parameters A and a. The *nonparametric approach* attempts to provide a numerical estimator for each sample of $\{R(k)\}$, $k = 0, \ldots, N-1$ for some choice of N. We shall begin with nonparametric covariance estimates, which will lead to the corresponding classical spectral estimates.

One of the most popular nonparametric estimates of the covariance evolves from the sample average used to estimate the variance,

$$\hat{R}_{xx}(0) = \frac{1}{N}\sum_{t=0}^{N-1} x^2(t) \tag{5.3–2}$$

which is motivated by the fact that the average of independent realizations of $x^2(t)$ approaches the expectation $E\{x^2(t)\}$. Note that this estimator is unbiased, since

$$E\{\hat{R}_{xx}(0)\} = E\left\{\frac{1}{N}\sum_{t=0}^{N-1}x^2(t)\right\} = \frac{1}{N}\sum_{t=0}^{N-1}E\{x^2(t)\} = \frac{1}{N}\sum_{t=0}^{N-1}R_{xx}(0) = R_{xx}(0)$$

and is consistent:

$$\lim_{N\to\infty} E\left\{\left|\hat{R}_{xx}(0) - R_{xx}(0)\right|^2\right\} = 0$$

Similarly, if we estimate the covariance at other lags, we have

$$\hat{R}_{xx}(k) = \frac{1}{N}\sum_{t=0}^{N-|k|-1} x(t)x(t + |k|) \tag{5.3-3}$$

which is usually called the *sample covariance* estimator, since it is based on the average of samples. Again, if we calculate the mean of this estimator, we obtain[†]

$$E\{\hat{R}_{xx}(k)\} = \frac{1}{N}\sum_{t=0}^{N-|k|-1} R_{xx}(k) = \left(\frac{N-|k|}{N}\right)R_{xx}(k) = \left(1 - \frac{|k|}{N}\right)R_{xx}(k) \tag{5.3-4}$$

which is clearly biased, since it does not equal $R_{xx}(k)$. However, as N gets larger, we see that this estimator is *asymptotically unbiased*; that is,

$$\lim_{N\to\infty} E\{\hat{R}_{xx}(k)\} = \lim_{N\to\infty}\left(1 - \frac{|k|}{N}\right)R_{xx}(k) = R_{xx}(k)$$

Next, we investigate the variance of the sample covariance estimator:

$$\mathrm{var}\,\{\hat{R}_{xx}(k)\} = E\{\hat{R}_{xx}^2(k)\} - E^2\{\hat{R}_{xx}(k)\} = E\{\hat{R}_{xx}^2(k)\} - \left(1 - \frac{|k|}{N}\right)^2 R_{xx}^2(k)$$

$$= \frac{1}{N^2}\sum_{m=0}^{N-|k|-1}\sum_{i=0}^{N-|k|-1} E\{x(m)x(m + |k|)x(i)x(i + |k|)\} - \left(1 - \frac{|k|}{N}\right)^2 R_{xx}^2(k) \tag{5.3-5}$$

If we assume that the values for x are gaussian, then from the fourth-moment property of gaussian processes (see Sec. 4.2), we have

---

[†]Note that the term $(1/N)$ can be selected to be $1/(N - |k| - 1)$, and $\hat{R}_{xx}(k)$ would then be unbiased. However, it has been shown that the mean-squared error for this estimate is larger than Eq. (5.3-4) [7].

$$E\{x(m)x(m + k)x(i)x(i + k)\} = E\{x(m)x(m + k)\}E\{x(i)x(i + k)\}$$

$$+ E\{x(m)x(i)\}E\{x(m + k)x(i + k)\} + E\{x(m)x(i + k)\}E\{x(m + k)x(i)\}$$

$$= R_{xx}^2(k) + R_{xx}^2(m - i) + R_{xx}(m - i + k)R_{xx}(m - i - k)$$

Substituting into Eq. (5.3–5) and subtracting the squared mean, we obtain

$$\text{var}\,\{\hat{R}_{xx}(k)\} = \frac{1}{N^2} \sum_{m=0}^{N-|k|-1} \sum_{i=0}^{N-|k|-1} R_{xx}^2(m - i) + R_{xx}(m - i + k)R_{xx}(m - i - k)$$

(5.3–6)

If we let $m - i = j$, then Eq. (5.3–6) becomes

$$\text{var}\,\{\hat{R}_{xx}(k)\} = \frac{1}{N} \sum_{j=-(N-|k|-1)}^{N-|k|-1} \left(1 - \frac{|k| + |j|}{N}\right) \left[R_{xx}^2(j) + R_{xx}(j + k)R_{xx}(j - k)\right]$$

(5.3–7)

It can be argued that for large N,

$$\text{var}\,\{\hat{R}_{xx}(k)\} \approx \frac{1}{N} \sum_{j=-\infty}^{\infty} R_{xx}^2(j) + R_{xx}(j + k)R_{xx}(j - k)$$

which implies that the variance is proportional to $1/N$ [8]. Thus, we see that the sample covariance estimator is asymptotically unbiased and has a variance which decreases as $1/N$ for large N:

$$\lim_{N \to \infty} E\{\hat{R}_{xx}(k)\} = \lim_{N \to \infty} \left(1 - \frac{|k|}{N}\right) R_{xx}(k) \to R_{xx}(k)$$

and

$$\lim_{N \to \infty} \text{var}\,\{\hat{R}_{xx}(k)\} \propto \frac{1}{N} R_{xx}^2(k)$$

Thus, when $N \gg k$, $\hat{R}_{xx}(k) \to R_{xx}(k)$.

The sample covariance estimator is equivalent to performing a convolution without the negative reflection. Recall that convolution is accomplished graphically by taking a function $x(k)$, performing a negative reflection about the origin $x(-k)$, shifting it by t, $x(t - k)$, and calculating the common area. Covariance is performed in a similar manner, with the exception of the negative reflection; that is,

$$y(t) = \sum_{k} h(k)x(t - k) \Leftrightarrow R_{xx}(k) = \frac{1}{N} \sum_{t} x(t)x(t + k)$$

As an example, we calculate the sample covariance of a rectangular pulse graphically.

Example 5.3–1. Consider the rectangular pulse shown in Fig. 5.3–1. Perform a graphical covariance calculation by shifting, multiplying, and adding according to Eq. (5.3–3). From this figure, we see that the evolution of the covariance function is identical to that of graphical convolution after the negative reflection has been performed. The covariance function in this case is triangular.

This example is very important from a practical viewpoint as well. We can consider the pulse as a rectangular window function and can therefore see that its covariance is triangular. If our data is of finite duration, then (as in the deterministic case) we will have a triangularly windowed covariance whose Fourier transform is convolved with the spectrum of the data.

Covariance functions can be calculated using DFT, analogous to convolution of deterministic signals. That is,

$$y(t) = \sum_k h(k)x(t-k) \Leftrightarrow H(\Omega_m)X(\Omega_m)$$

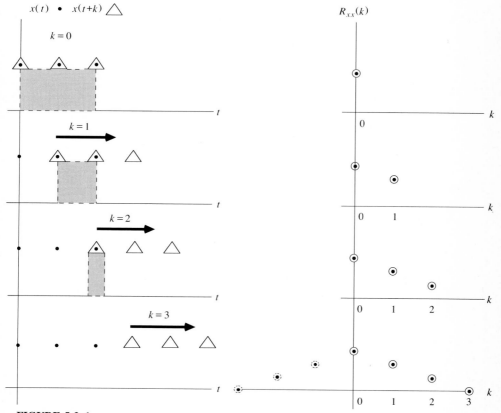

FIGURE 5.3–1
Covariance of a rectangular pulse by graphical techniques

We have
$$\hat{R}_{xx}(k) = \sum_{t} x(t)x(t+k) \Leftrightarrow X(\Omega_m)X^*(\Omega_m) \qquad (5.3\text{--}8)$$

where $\Omega_m = 2\pi m/N$ (recall from this step the DFT). To see this, let us define the finite-duration covariance as

$$\hat{R}_{xx}(k) = \frac{1}{N} \sum_{t=0}^{N-1} x(t)x(t+k) = \text{IDFT}[X(\Omega_m)X^*(\Omega_m)] = \text{IDFT}[|X(\Omega_m)|^2] \qquad (5.3\text{--}9)$$

or, substituting into the IDFT, we have

$$\hat{R}_{xx}(k) = \frac{1}{N} \sum_{m=0}^{N-1} X(\Omega_m)X^*(\Omega_m)e^{jk\Omega_m}$$

$$= \frac{1}{N} \sum_{m=0}^{N-1} \left\{ \sum_{l=0}^{N-1} x(l)e^{-jl\Omega_m} \right\} \left\{ \sum_{t=0}^{N-1} x(t)e^{+jt\Omega_m} \right\} e^{jk\Omega_m}$$

If we reorder the summations, then we obtain

$$\hat{R}_{xx}(k) = \frac{1}{N} \sum_{l=0}^{N-1} \sum_{t=0}^{N-1} x(l)x(t) \left\{ \sum_{m=0}^{N-1} e^{j(2\pi m/N)(k-l+t)} \right\}$$

$$= \frac{1}{N} \sum_{l=0}^{N-1} \sum_{t=0}^{N-1} x(l)x(t)\delta(k-l+t)$$

which occurs from the DFT relations, leading to

$$\hat{R}_{xx}(k) = \frac{1}{N} \sum_{t=0}^{N-1} x(t)x(t+k)$$

which is the desired result.

So we see that the DFT can be used to estimate the covariance function. If the FFT algorithm (see Appendix A) is used in this estimate, then a computational savings can be realized. The computation using lagged sums is proportional to $N \cdot k$ where k is the number of lags, while with the FFT approach the procedure is proportional to $(N + k) \log (N + k)$. Recall that the variance can be reduced by making N large, implying that for large N the FFT approach should be used.

5.4 NONPARAMETRIC METHODS OF SPECTRAL ESTIMATION

With the initial application of Fourier analysis techniques to raw sun-spot data over 200 years ago, the seeds of spectral estimation were sown by Schuster [9]. Fourier analysis for random signals evolved rapidly after the discovery of the Wiener-Khintchine theorem relating the covariance and power spectrum. Finally, with the evolution of the fast Fourier transform (see [10]) and digital computers, all of the essential ingredients were present to establish the classical approach to nonparametric spectral estimation.

Classical spectral estimators typically fall into two categories, direct and indirect, as shown in Fig. 5.4–1. Here we see that the direct methods operate directly on the raw data to transform it to the frequency domain and produce the estimate. Indirect methods first estimate the covariance sequence and then transform it to the frequency domain—an application of the Wiener-Khintchine theorem. In this section, we develop two basic nonparameteric spectral estimation techniques: the correlation method (indirect) and the periodogram method (direct).

The *correlation method*, sometimes called the Blackman-Tukey method, is simply an implementation of the Wiener-Khintchine theorem: the covariance is obtained using the sample covariance estimator of the previous section, and then the PSD is estimated by calculating the discrete Fourier transform. That is,

$$\hat{S}_{xx}(\Omega_m) = \text{DFT } [\hat{R}_{xx}(k)]$$

This technique tends to produce a noisy spectral estimate; however, as will be shown, a smoothed estimate can be obtained by multiplying R_{xx} by a *window function*, W, usually called a *lag window*. The window primarily reduces spectral leakage and therefore improves the estimate. It is also interesting to note that the sample covariance estimator of Eq. (5.3–3) does not guarantee the positivity of the PSD (auto) when estimated directly from the Wiener-Khintchine theorem. However, as we shall see, if the estimator is implemented directly in the Fourier domain by Eq. (5.3–8), then it will preserve this property, since it is the *square* of the Fourier spectrum.

We summarize the *correlation method* of spectral estimation in the following steps (see Fig. 5.4–2):[†]

[†]Note also that if we replace X^* by Y^*, we can estimate the cross correlation $\hat{R}_{xy}(k)$ and corresponding cross spectrum $\hat{S}_{xy}(\Omega_m)$ using this method as well.

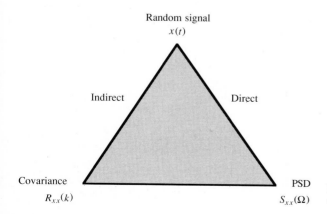

Random signal
$x(t)$

Indirect

Direct

Covariance
$R_{xx}(k)$

PSD
$S_{xx}(\Omega)$

FIGURE 5.4–1
Classical spectral estimation

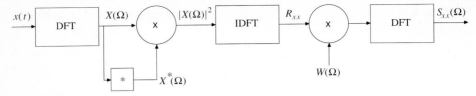

FIGURE 5.4–2
Correlation method block diagram

1. Calculate the DFT of $x(t)$, that is, $X(\Omega_m)$.
2. Multiply $X(\Omega_m)$ by its conjugate to obtain $X(\Omega_m)X^*(\Omega_m)$.
3. Estimate the covariance from the IDFT, $\hat{R}_{xx}(k) = \text{IDFT}\left[|X(\Omega_m)|^2\right]$.
4. Multiply the covariance by the lag window $W(k)$.
5. Estimate the PSD from the DFT of the windowed covariance, that is, $\hat{S}_{xx}(\Omega_m) = \text{DFT}\left[\hat{R}_{xx}(k)W(k)\right]$.

We can investigate the "smoothing" properties of windows by considering the spectral estimator performance for a gaussian white-noise process—the statistical equivalent of a deterministic impulse. In fact, we use this approach to analyze other classical estimators as well, in order to gain insight into their performance. Note well, however, that these results are *only* valid for white gaussian processes. Statistically, spectral estimators of white gaussian processes employing the discrete Fourier transform possess a variance that is independent of the number of observations N, suggesting that they are not consistent in that their distribution does not tend to cluster more closely around the true spectrum as the number of samples increases. To see why this is true, consider the Fourier transform of the process $X(\Omega)$ as

$$X(\Omega) = A(\Omega) + jB(\Omega)$$

with the spectrum given by

$$\hat{S}_{xx}(\Omega) = \frac{1}{N}|X(\Omega)|^2 = \frac{1}{N}\left(A^2(\Omega) + B^2(\Omega)\right)$$

We can study the statistical properties of the spectrum by investigating the random variables A and B above. If we further assume that $\{x(t)\}$ is a zero-mean, white gaussian process with variance R_{xx}, then, using the discrete Fourier representation with A and B defined above, we know that they are also normal from the linearity property of gaussian processes. Thus, the random variables A^2 and B^2 are uncorrelated and independent with each chi-squared distributed, χ_1^2; therefore, it can be shown that

$$\frac{2\hat{S}_{xx}(\Omega_m)}{R_{xx}} \sim \chi_2^2 \qquad \text{and} \qquad \frac{\hat{S}_{xx}(0)}{R_{xx}} \sim \chi_1^2$$

with the properties that

$$E\{\hat{S}_{xx}(\Omega_m)\} = S_{xx}(\Omega_m) \qquad \text{and} \qquad \text{var}\{\hat{S}_{xx}(\Omega_m)\} = S_{xx}^2(\Omega_m)$$

implying that the estimate is *independent* of N, the sample size. Even if x is not normal, from the central limit theorem, $\hat{S}_{xx}(\Omega_m)$ will be nearly χ_2^2, regardless of how x is distributed. It should also be noted that since A and B are uncorrelated and independent at harmonic frequencies $m = 0, \pm 1, \pm 2, \ldots$ then S_{xx} is also *uncorrelated* and *independent at these harmonic frequencies*. Thus, this property explains the erratic behavior of the sample PSD.

The correlation spectral estimates can be statistically improved by using a lag or, equivalently, spectral window. The window function is called a *lag window* in the time or lag domain, $W(k)$, and a *spectral window* in the frequency domain, $W(\Omega_m)$. The lag window, $W(k)$, must satisfy the following conditions:

(i) $W(0) = 1$

(ii) $W(k) = W(-k)$

(iii) $W(k) = 0$ $|k| \geq M, M < N$ where M is the maximum lag value and N is the data length.

These conditions are equivalent to the spectral window, $W(\Omega_m)$, which satisfies the following properties:

(i) $\int_{-\infty}^{\infty} W(\Omega)d\Omega = W(0) = 1$

(ii) $W(\Omega) = W(-\Omega)$

(iii) $W(\Omega)$ is a slit with base width† of order $2/M$

Recall that for large M the spectral window function approaches a delta function,

$$\lim_{M \to \infty} W(\Omega) \to \delta(\Omega)$$

and therefore it follows that

$$\lim_{M \to \infty} E\{\hat{S}_{xx}(\Omega)\} = \lim_{M \to \infty} W(\Omega_m) * S_{xx}(\Omega) \approx S_{xx}(\Omega_m)$$

Thus $\hat{S}_{xx}$ is asymptotically unbiased. However, for finite-data lengths the bias is

$$B(\Omega_m) = E\{\hat{S}_{xx}(\Omega_m)\} - S_{xx}(\Omega_m)$$

The variance of smoothed spectral estimators can be approximated by applying Parseval's theorem, relating the spectral to the lag window energy to obtain

$$\text{var}\,\{\hat{S}_{xx}(\Omega_m)\} \approx \frac{S_{xx}^2(\Omega_m)I}{N\Delta T}$$

where $I = \int_{-\infty}^{\infty} W^2(\beta)d\beta$.

---

†The *base width* of a window is defined as the distance between the origin and the first zeros of the function.

The application of spectral windows improves estimates. For example, consider the Bartlett lag window given by

$$W(k) = 1 - \frac{|k|}{M} \Rightarrow I = \int_{-M}^{M} \left(1 - \frac{|\beta|}{M}\right)^2 d\beta = \left(\frac{2}{3}\right) M$$

The variance then becomes

$$\text{var}\{\hat{S}_{xx}(\Omega_m)\} = S_{xx}^2(\Omega_m)\left(\frac{2M}{3N\Delta T}\right)$$

This shows that the Bartlett window applied to smooth the spectral estimates reduces the variance in the estimate by reducing M—the truncation point of the lag window. However, we recall that in the frequency domain, we would like the spectral window to behave like an impulse function in order to reduce bias (distortion) of the true spectrum. This implies that the base width must be large (slit $\propto 2/M \rightarrow 0$ for impulse), leading us to the fundamental compromise between the bias and variance. Thus, the bias can only be made small by making $W(\Omega_m)$ narrow, but a narrow spectral window results in a large variance. In practice, this compromise is accomplished by making the mean-squared error as small as possible; that is, M is selected so that

$$\Psi(\Omega_m) = \text{var}(\hat{S}_{xx}(\Omega_m)) + B^2(\Omega_m) \rightarrow 0$$

Since the base width of the window is inversely proportional to M, the variance can be decreased by making M small (base width large); however, the bias increases.

The performance of spectral window functions can be evaluated in terms of various quantities. The ability of a smoothing window to reduce the variance of the spectral estimate is called the *variance reduction ratio* and is given by

$$\frac{\text{var}(\hat{S}_{xx}(\Omega_m))}{\text{var}(S_{xx}(\Omega_m))} \approx \frac{S_{xx}^2(\Omega_m)I/N\Delta T}{S_{xx}^2(\Omega_m)} = \frac{I}{N\Delta T} \qquad (5.4\text{--}1)$$

Two other measures of window performance are the degrees of freedom and the bandwidth. It can be shown (see [7]) that the *degrees of freedom* ν are given by

$$\nu = \frac{2N\Delta T}{I} \qquad (5.4\text{--}2)$$

implying that, the larger the number of degrees of freedom (smaller I), the smaller the variance and, therefore, the more reliable the estimate.

The *bandwidth* is defined as

$$b = \frac{1}{I} \qquad (5.4\text{--}3)$$

so that the variance is inversely proportional to the bandwidth; that is,

$$\text{variance} \times \text{bandwidth} = \text{constant}$$

TABLE 5.4–1
Spectral window performance

Description	Variance ratio ($I/N\Delta T$)	Degrees of freedom (ν)	BW (b)
Rectangular	$2\left(\dfrac{M}{N\Delta T}\right)$	$\dfrac{N\Delta T}{M}$	$\dfrac{1}{2}M$
Bartlett	$\dfrac{2}{3}\dfrac{M}{N\Delta T}$	$3\dfrac{N\Delta T}{M}$	$\dfrac{3}{2}M$
Tukey	$\dfrac{3}{4}\dfrac{M}{N\Delta T}$	$\dfrac{8}{3}\dfrac{N\Delta T}{M}$	$\dfrac{4}{3}M$
Parzen	$0.54\dfrac{M}{N\Delta T}$	$3.71\dfrac{N\Delta T}{M}$	$1.86M$
Hanning	$\dfrac{1}{2}\dfrac{M}{N\Delta T}$	$4\dfrac{N\Delta T}{M}$	$\dfrac{2}{M}$
Hamming	$0.503\dfrac{M}{N\Delta T}$	$3.975\dfrac{N\Delta T}{M}$	$\dfrac{1.987}{M}$

The degrees of freedom and the bandwidth are also related by

$$\nu = 2N\Delta Tb \qquad (5.4\text{--}4)$$

We summarize these performance measures for various windows in Table 5.4–1. All of these parameters can be used to design an experiment for a particular level of variance reduction, as in the following example.

Example 5.4–1. Suppose $M = 0.1(N\Delta T)$ (one tenth of the record length). For a rectangular window $I/N\Delta T = 2M/N\Delta T = 0.2$ or 20%, so by truncating the lag window to 10% of the record, the smoothed spectral estimator is reduced to 20% of the variance of the sample spectrum. The improvement using the Bartlett, Tukey, Parzen, and Hanning windows for this case is 6.7%, 7.5%, 5.4%, and 5%, respectively.

With practical window selection and long data records, the correlation method can be effectively utilized to estimate the PSD. The following example of spectral estimation uses the correlation method.

Example 5.4–2. Suppose we generate a 256-point test signal, sampled at 0.001 sec (using SIG [11]) by passing sinusoids at 35, 50, 56, and 90 Hz contaminated with unit variance white noise (SNR[†] $= -20$ dB) through a bandpass Butterworth (third-order) filter, with cutoff frequencies at 30 and 70 Hz. We develop the correlation spectral estimator for various choices of window length and type. The true (dashed lines) and raw spectra are shown in Fig. 5.4–3(a). Note that the spectral lines at 50 and 56 have merged for this realization. Choosing lag window lengths of ($M =$

[†]Here we define $\text{SNR}_{\text{OUT}} := $ signal energy/noise variance.

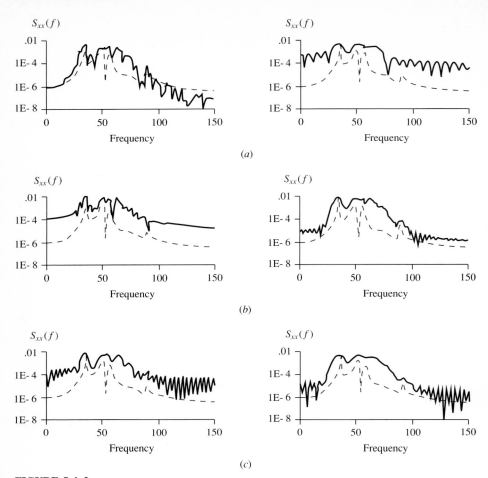

FIGURE 5.4–3
Correlation method spectral estimation for simulated bandpass sinusoids (*a*) Raw/true (dashed) spectra and rectangular windowed ($M = 0.032$) estimate (*b*) Rectangular and Hamming windowed ($M = 0.128$) spectra (*c*) Rectangular and Hamming windowed ($M = 0.064$) spectra

0.128, 0.064, 0.032) the variance ratios for rectangular and Hamming windows are (100%, 50%, 25%) and (25%, $12\frac{1}{2}$%), respectively. The rectangular and Hamming windows for $M = 0.128, 0.064$ are shown in (*b*) and (*c*). Note the "smoothing" effect and leakage reduction caused by the Hamming window, along with the associated loss of "line" resolution, showing the tradeoff between bias and variance. The lines at 35 and 90 Hz are enhanced using the Hamming window, but the lines at 50 and 56 Hz are only resolved at $M = 0.128$ sec.

Next we consider a more direct approach to estimating the PSD. We introduce the concept of a periodogram estimator with statistical properties equivalent to the correlation method, then we show how to improve these estimates by statistical averaging and window smoothing, leading to Welch's method of spectral estimation. The periodogram was devised by statisticians to detect periodicities in noisy data records. The improved method of spectral estimation based on the *periodogram* is defined by

$$P_{xx}(\Omega_m) := \frac{1}{N}|X(\Omega_m)|^2 = \frac{1}{N}\sum_{k=-(N-1)}^{N-1} R_{xx}(k)e^{-jk\Omega_m} \qquad (5.4\text{--}5)$$

where R_{xx} is the biased estimator of Eq. (5.3–3). The bias is found from

$$E\{P_{xx}(\Omega_m)\} = \frac{1}{N}\sum_{k=-(N-1)}^{N-1} E\{R_{xx}(k)\}e^{-jk\Omega_m} = \frac{1}{N}\sum_{k=-(N-1)}^{N-1}\left(1 - \frac{|k|}{N}\right)R_{xx}(k)e^{-jk\Omega_m}$$

$$(5.4\text{--}6)$$

which is biased not just because of $|k|/N$ but also because of the finite limits on the summation. If we identify the term $(1 - |k|/N)$ as a triangular lag window function, $W(k) = 1 - |k|/N$, then

$$E\{P_{xx}(\Omega_m)\} = \text{DFT}\,[R_{xx}(k)W(k)] = W(\Omega_m) * S_{xx}(\Omega_m) \qquad (5.4\text{--}7)$$

So we see that the expected value of the periodogram is the true spectrum $S_{xx}(\Omega_m)$ observed through the spectral window $W(\Omega_m)$. Note that a rectangular data window results in a triangular correlation or Bartlett window, as shown in Example 5.3–1, and

$$W(\Omega_m) = \frac{1}{N}\left[\frac{\sin(\Omega_m N/2)}{\sin(\Omega_m/2)}\right]^2 \qquad (5.4\text{--}8)$$

for the spectral window. Recall that most windows trade off main-lobe width for sidelobe height. For large N, the spectral window W will have a high, narrow main peak along with narrow sidelobes. In this case, we see that

$$E\{P_{xx}(\Omega_m)\} \approx S_{xx}(\Omega_m)$$

and that the periodogram is asymptotically unbiased. In order for the periodogram estimate to be good, it must have a small variance as N increases. Unfortunately, $\text{var}\{P_{xx}(\Omega_m)\}$ is generally not small, even for large N. For example, if x is a white gaussian process, then it can be shown (see [7]) that

$$\lim_{N\to\infty} \text{var}\,\{P_{xx}(\Omega_m)\} = S_{xx}^2(\Omega_m)$$

That is, the variance of the periodogram approaches the square of the true spectrum at each m. In fact, it can be shown in general that [12]

$$\text{var}\,\{P_{xx}(\Omega_m)\} = S_{xx}^2(\Omega_m)\left[1 + \left(\frac{\sin(N\Omega_m)}{N\sin(\Omega_m)}\right)^2\right] \qquad (5.4\text{--}9)$$

which shows that as N increases, the variance is proportional to S_{xx}^2 as before, and therefore P_{xx} is *not* consistent. Similarly, it can be shown that

$$\text{cov}\{P_{xx}(\Omega_m)P_{xx}(\Omega_j)\}$$

$$\approx P_{xx}(\Omega_m)P_{xx}(\Omega_j)\left\{\left(\frac{\sin N/2(\Omega_m + \Omega_j)}{N\sin(\Omega_m + \Omega_j)/2}\right)^2 + \left(\frac{\sin N/2(\Omega_m - \Omega_j)}{N\sin(\Omega_m - \Omega_j)/2}\right)^2\right\} \quad (5.4\text{--}10)$$

Since this relation is evaluated at equally spaced frequency samples, we see that the samples are uncorrelated, giving the periodogram a wildly fluctuating appearance and χ_2^2 distribution as before. Since the samples of P_{xx} are uncorrelated, this suggests that one way of reducing the variance in P_{xx} is to average individual periodograms, which is accomplished by sectioning the original N-point data record into K, L-point sections; that is,

$$\hat{S}_{xx}(\Omega_m) = \frac{1}{K}\sum_{i=1}^{K}\hat{P}_{xx}(\Omega_m, i) \quad (5.4\text{--}11)$$

where $\hat{P}_{xx}(\Omega_m, i)$ is the ith, L-point periodogram. If x is stationary, then

$$E\{\hat{S}_{xx}(\Omega_m)\} = \frac{1}{K}\sum_{i=1}^{K}E\{\hat{P}_{xx}(\Omega_m, i)\} = E\{\hat{P}_{xx}(\Omega_m, i)\} = S_{xx}(\Omega_m)$$

which is unbiased. If we introduce a smoothing window as before, then

$$E\{\hat{S}_{xx}(\Omega_m)\} = \frac{1}{L}S_{xx}(\Omega_m) * W(\Omega_m) \qquad \text{where } L = \frac{N}{K} \quad (5.4\text{--}12)$$

which for impulse window functions gives

$$E\{\hat{S}_{xx}(\Omega_m)\} \propto \frac{K}{N}S_{xx}(\Omega_m)$$

Again assuming independence of the x, we have

$$\text{var}\{\hat{S}_{xx}(\Omega_m)\} = \frac{1}{K}\text{var}\{\hat{P}_{xx}(\Omega_m, i)\} \approx \frac{1}{K}S_{xx}^2(\Omega_m)\left[1 + \left(\frac{\sin K\Omega_m}{K\sin \Omega_m}\right)^2\right] \quad (5.4\text{--}13)$$

So we see that this estimate is consistent, since the variance approaches zero as the number of sections become infinite. We conclude that for the periodogram estimator

$$\text{var} \propto \frac{1}{K} \qquad \text{and} \qquad \text{bias} \propto \frac{K}{N}$$

So we see that for large K, the variance becomes small but the bias increases. Therefore, for a *fixed* record length N, as the number of periodograms increases, variance decreases but the bias increases. This is the basic tradeoff between variance and resolution (bias), which can be used to determine a priori the required record length $N = LK$ for an acceptable variance. If we use a lag window to obtain a smoothed spectral estimate, then the bias is given by Eq. (5.4–12), while the variance can be approximated by

$$\text{var}\{\hat{S}_{xx}(\Omega_m)\} \approx \left(\frac{1}{K\Delta T} \sum_{k=-(K-1)}^{K-1} W^2(k) \right) S_{xx}^2(\Omega_m)$$

assuming the window length is narrow relative to variations of $S_{xx}(\Omega_m)$, yet wide compared to the triangular lag window, for the correlation estimate.

Welch [13] introduced a modification of Bartlett's original procedure. The data is sectioned into K records of length L as before; however, the window is applied directly to the segmented records before periodogram computation. The modified periodograms are then

$$\hat{P}(\Omega_m, i) = \frac{1}{U} |\text{DFT} [x_i(t)W(t)]|^2 \qquad i = 1, \ldots, K$$

where

$$U = \frac{1}{L} \sum_{t=0}^{L-1} W^2(t)$$

and

$$\hat{S}_{xx}(\Omega_m) = \frac{1}{K} \sum_{i=1}^{K} \hat{P}(\Omega_m, i)$$

We summarize the *periodogram method* (Welch's procedure) in these steps:

1. Section the data, $\{x(t)\}$, $t = 1, \ldots, N$ into K sections each of length L, where $K = N/L$, that is,

$$x_i(t) = x(t + L(i - 1)), \qquad i = 1, \ldots, K, \qquad t = 0, \ldots, L - 1$$

2. Window the data to obtain $x_i(t)W_i(t)$

3. Compute K periodograms using the DFT as

$$\hat{P}(\Omega_m, i) = \frac{1}{U} |\text{DFT} [x_i(t)W(t)]|^2 \qquad i = 1, \ldots, K$$

with

$$U = \frac{1}{L} \sum_{t=0}^{L-1} W^2(t)$$

4. Estimate the spectrum using

$$\hat{S}_{xx}(\Omega_m) = \frac{1}{K} \sum_{i=1}^{K} \hat{P}(\Omega_m, i)$$

with var $\{\hat{S}_{xx}(\Omega_m)\} \propto 1/K$ and $B\{\hat{S}_{xx}(\Omega_m)\} \propto K/N$ adjusted for particular windows. The periodogram method is depicted in Fig. 5.4-4.

We note also in closing that since we know that

$$\hat{S}_{xx}(\Omega_m) \sim \chi_\nu^2$$

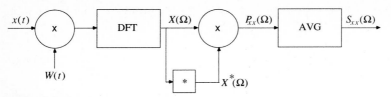

FIGURE 5.4–4
Periodogram method block diagram

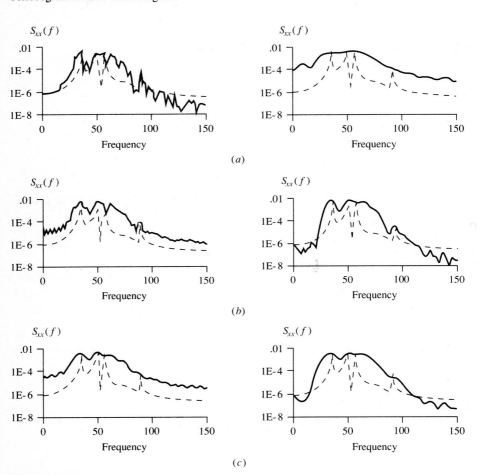

FIGURE 5.4–5
Periodogram method spectral estimation for simulated bandpass sinusoids (*a*) Raw/true (dashed) spectra and rectangular windowed estimate for $K = 8$ averages (*b*) Rectangular and Hamming windowed spectra for $K = 2$ averages (*c*) Rectangular and Hamming windowed spectra for $K = 4$ averages

170

we can construct confidence-interval estimates for the spectrum. Based on an α significance level, we know that

$$\text{Prob}\left\{\chi_\nu\left(\frac{\alpha}{2}\right) < \frac{\nu\hat{S}_{xx}(\Omega_m)}{S_{xx}(\Omega_m)} \le \chi_\nu\left(1 - \frac{\alpha}{2}\right)\right\} = 1 - \alpha$$

leading to the interval

$$\left[\frac{\nu\hat{S}_{xx}(\Omega_m)}{\chi_\nu(1 - \alpha/2)}, \frac{\nu\hat{S}_{xx}(\Omega_m)}{\chi_\nu(\alpha/2)}\right]$$

which is the $100(1 - \alpha)\%$ confidence interval for $S_{xx}(\Omega_m)$. ν can be calculated for a specific spectral window, and χ_ν is obtained from χ^2 tables; that is, Prob$\{\chi_\nu^2 \le \chi_\nu(\alpha/2)\} = \alpha/2$. Since the limits are multiplicative, we can calculate the spectrum using log plots,

$$\left[\log \hat{S}_{xx}(\Omega_m) + \log\left(\frac{\nu}{\chi_\nu(1 - \alpha/2)}\right), \log \hat{S}_{xx}(\Omega_m) + \log\left(\frac{\nu}{\chi_\nu(\alpha/2)}\right)\right]$$

Example 5.4–3. Here we revisit the bandpass sinusoidal simulation and apply the periodogram spectral estimator for various section lengths (averages) and windows. Using SIG [11], we see the results of applying this technique in Fig. 5.4–5. Note the variance reduction caused by the averaging for $K = 2, 4, 8$ as well as the rectangular and Hamming windows. The raw and true (dashed) spectral are shown in (a), along with the averaged ($K = 8$) periodogram estimator. Note the enhancement of the bandpassed response at 30 and 70 Hz. The periodogram spectral estimates for both rectangular and Hamming windowed data with $K = 2$ and 4 are shown in (b) and (c), respectively. Note the tradeoff of variance and bias, as well as spectral resolution caused by windowing. For $K = 2$, note the enhancement of the 35 and 90 Hz spectral lines, along with the decreased resolution of the lines at 50 and 56 Hz.

Even though they are considered classical techniques with limited resolution capability, nonparametric methods of spectral estimation still can provide us with reasonable information if we have data with good signal levels.

5.5 COHERENCE ANALYSIS

In this section we show how a coherence function can be used as a tool to investigate the quality of measured data as well as the linearity of the system under investigation (see [14] for more details).

As before, we define the ordinary *squared-coherence* function by

$$\gamma_{xy}^2(\Omega) := \frac{|S_{xy}(\Omega)|^2}{S_{xx}(\Omega)S_{yy}(\Omega)} \tag{5.5–1}$$

where x is the input and y the measurement or output. It follows immediately from the properties of the PSD (see Table 4.4–1) that the coherence function satisfies

$$0 \le \gamma_{xy}^2(\Omega) \le 1 \qquad (5.5\text{--}2)$$

It also follows that if the system under investigation is linear, then from the spectral properties

$$S_{xy}(\Omega) = H(\Omega)S_{xx}(\Omega) \qquad \text{and} \qquad S_{yy}(\Omega) = |H(\Omega)|^2 S_{xx}(\Omega) \qquad (5.5\text{--}3)$$

and then

$$\gamma_{xy}^2(\Omega) = \frac{|H(\Omega)S_{xx}(\Omega)|^2}{S_{xx}(\Omega)(|H(\Omega)|^2 S_{xx}(\Omega))} = 1 \quad \forall \Omega \qquad (5.5\text{--}4)$$

Another way of examining this function follows directly from the first relation of Eq. (5.5–3):

$$S_{yx}(\Omega) = H^{-1}(\Omega)S_{yy}(\Omega) = S_{xy}^*(\Omega)$$

or

$$H^{-1}(\Omega) = \frac{S_{xy}^*(\Omega)}{S_{yy}(\Omega)} \qquad (5.5\text{--}5)$$

Clearly, from Eq. (5.5–1) we see that

$$\gamma_{xy}^2(\Omega) = \left[\frac{S_{xy}(\Omega)}{S_{xx}(\Omega)}\right]\left[\frac{S_{xy}^*(\Omega)}{S_{yy}(\Omega)}\right] = H(\Omega)H^{-1}(\Omega) \qquad (5.5\text{--}6)$$

and in the ideal case, Eq. (5.5–4) must be satisfied. From this equation we see that the coherence function can be used to determine the frequency band over which data or a system can be considered linear. If it is perfectly linear, then the transfer function multiplied by its inverse leads to a perfectly *coherent* ($\gamma_{xy}^2(\Omega) = 1$) system.

On the other hand, if x and y are uncorrelated, then $S_{xy}(\Omega) = 0$ and, therefore, is completely *incoherent* ($\gamma_{xy}^2(\Omega) = 0$). We summarize these notions as

$$\gamma_{xy}^2(\Omega) = \begin{cases} 0 & \text{incoherent } S_{xy}(\Omega) = 0 \\ 1 & \text{coherent} \end{cases}$$

In practice, the coherence function ranges between 0 and 1 under the following conditions:

(i) measurement noise
(ii) nonlinearities in the system
(iii) biased spectral estimates
(iv) output is due to other inputs

The coherence function can be estimated in a number of ways. We can use any of the spectral estimation techniques of the previous section for the PSD. So, for example, we can use either the correlation or periodogram method to give the estimate

$$\hat{\gamma}_{xy}^2(\Omega) = \frac{|\hat{S}_{xy}(\Omega)|^2}{\hat{S}_{xx}(\Omega)\hat{S}_{yy}(\Omega)} \qquad (5.5\text{--}7)$$

or we can use some of the modern parameter estimation techniques (see the next chapter) to produce transfer function estimates using Eq. (5.5–6), that is,

$$\hat{\gamma}_{xy}^2(\Omega) = \hat{H}(\Omega)\hat{H}^{-1}(\Omega) \tag{5.5–8}$$

Note that in using this relation, we obtain the parameter estimation for the inverse transfer function by reversing the input and output sequences.

To see the utility of the coherence function in assessing the quality of measured data, consider the following example.

Example 5.5–1. We investigate the coherence function estimate for our bandpass sinusoids in noise problem. Using SIG [11], we implement the coherence function estimator using the Welch periodogram estimator. The results are shown in Fig. 5.5–1. In this figure, we show estimates for the noisy sinusoidal input sequence (solid line), with corresponding bandpass output, as well as the noise-free sinusoidal input. We note that the simulated data are coherent at each of the spectral lines (35, 50, 56, 90 Hz), as shown by the dashed lines. Incoherence is caused by the filter roll-off. Now compare the noisy coherence. Here we see that the noise tends to produce "incoherent" notches at random frequencies.

Next consider a case when the measurement is corrupted by random noise.

Example 5.5–2. Suppose we have a linear system contaminated by zero-mean white noise of variance R_{nn}; then

$$y(t) = h(t) * x(t) + n(t)$$

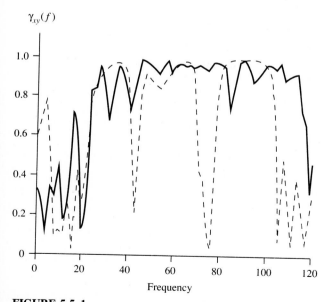

FIGURE 5.5–1
Coherence function estimation (periodogram method) for sinusoidal bandpass simulation: noisy (solid) and deterministic (dashed) spectrum

We would like to analyze the effect of this measurement noise on the quality of our measured data. The output PSD is easily determined from

$$S_{yy}(\Omega) = E\{Y(\Omega)Y^*(\Omega)\} = |H(\Omega)|^2 S_{xx}(\Omega) + S_{nn}(\Omega)$$

and
$$S_{xy}(\Omega) = E\{X(\Omega)Y^*(\Omega)\} = H(\Omega)S_{xx}(\Omega) + S_{xn}(\Omega) = H(\Omega)S_{xx}(\Omega)$$

Substituting into the expression for the coherence function, we obtain

$$\gamma_{xy}^2(\Omega) = \frac{|H(\Omega)|^2 S_{xx}^2(\Omega)}{S_{xx}(\Omega)S_{yy}(\Omega)} = \frac{|H(\Omega)|^2 S_{xx}(\Omega)}{|H(\Omega)|^2 S_{xx}(\Omega) + S_{nn}(\Omega)} = \frac{1}{1 + R_{nn}/|H(\Omega)|^2 S_{xx}(\Omega)}$$

So we see that for small R_{nn}, $\gamma_{xy}^2(\Omega) \to 1$, while large R_{nn} implies that $\gamma_{xy}^2(\Omega) \to 0$. If we are interested in determining that function of the output spectrum linearly due to the input, we can define

$$S_{\hat{y}\hat{y}}(\Omega) = |H(\Omega)|^2 S_{xx}(\Omega) = \left[\frac{S_{xy}(\Omega)}{S_{xx}(\Omega)}\right]^2 S_{xx}(\Omega)$$

$$= \left(\frac{|S_{xy}(\Omega)|^2}{S_{xx}(\Omega)S_{yy}(\Omega)}\right) S_{yy}(\Omega) = \gamma_{xy}^2(\Omega)S_{yy}(\Omega)$$

Here $S_{\hat{y}\hat{y}}(\Omega)$ is called the *coherent output spectrum*. We see from this example that the coherence function can be used to analyze the linearity of a system under investigation. We can also use the coherence function to determine the quality of our measured data by assuring ourselves that it is reasonably close to unity over the frequency range of interest.

5.6   SUMMARY

In this chapter we have discussed the fundamental concepts underlying random signal analysis. Starting with the statistical notions of estimation, we developed desirable properties of an estimator and showed how to compare and evaluate its performance. Next we developed the minimum-variance, maximum a-posteriori, and maximum-likelihood estimators and showed how they are related. Using these ideas, we developed covariance estimators and employed them in the *classical* nonparametric spectral estimation schemes: the correlation and periodogram methods. Finally, we developed the notion of a coherence function and showed how it could be estimated using the classical PSD methods.

REFERENCES

1. R. Hogg and A. Craig, *Introduction to Mathematical Statistics* (New York: Macmillan, 1970).
2. H. Van Trees, *Detection, Estimation and Modulation Theory* (New York: Wiley, 1968).
3. J. Candy, "The Cramer-Rao Bound: A Measure of Estimator Quality," Lawrence Livermore National Laboratory Report, UCID-17660, 1977.
4. J. Candy, *Signal Processing: The Model-Based Approach* (New York: McGraw-Hill, 1986).
5. A. Sage and J. Melsa, *Estimation Theory with Applications to Communications and Control* (New York: McGraw-Hill, 1971).

6. T. Anderson, *An Introduction to Multivariate Statistical Analysis* (New York: Wiley, 1958).
7. G. Jenkins and D. Watts, *Spectral Analysis and Its Applications* (San Francisco: Holden-Day, 1968).
8. M. Schwartz and L. Shaw, *Signal Processing: Discrete Analysis, Detection, and Estimation* (New York: McGraw-Hill, 1975).
9. A. Schuster, "On the Investigation of Hidden Periodicities with Application to a Supposed 26 Day Period of Meteorological Phenomena," *Terrestrial Magnetism*, Vol. 3, 1898.
10. J. Cooley and J. Tukey, "An Algorithm for Machine Calculation of Complex Fourier Series," *Math. Comput.*, 1965.
11. D. Lager and S. Azevedo, "SIG—A General Purpose Signal Processing Code," *Proc. IEEE*, 1987.
12. A. Oppenheim and R. Shafer, *Digital Signal Processing* (Englewood Cliffs, N.J.: Prentice-Hall, 1975).
13. P. Welch, "The Use of Fast Fourier Transforms for the Estimation of Power Spectra: A Method Based on Time Averaging Over Short Modified Periodograms," *IEEE Trans. Audio Electroacoust.*, AU-15, 1967.
14. J. Bendat and A. Piersol, *Engineering Applications of Correlation and Spectral Analysis* (New York: Wiley, 1980).

SIG NOTES

SIG can be used to estimate the covariance (auto and cross) from data using the nonparametric, FFT approach of Sec. 5.3 and the CORRELATE command. Power spectral density estimators are also available (SD***) for random signal analysis. The classical nonparametric methods of Sec. 5.4 are obtained using the SDCORRELATION and SDWELCH commands for the respective correlation and periodogram (Welch's) methods. SIG calculates two coherence function estimators: COHERENCE, using the periodogram approach of Sec. 5.5, and ECOHERENCE, for an ensemble of signals, usually transients.

EXERCISES

5.1. Verify the following properties:

(a) $\nabla_x(a'b) = (\nabla_x a')b + (\nabla_x b')a$, for $a, b \in R^n$, and functions of x

(b) $\nabla_x(b'x) = b$

(c) $\nabla_x(x'C) = C, C \in R^{n \times m}$

(d) $\nabla_x(x') = I$

(e) $\nabla_x(x'x) = 2x$

(f) $\nabla_x(x'Ax) = Ax + A'x$, for A not a function of x

5.2. Suppose we have the following signal in additive gaussian noise:

$$y = x + n \qquad \text{with } n \sim N(0, R_{nn})$$

(a) Find the Cramer-Rao bound if x is assumed to be unknown.

(b) Find the Cramer-Rao bound if $p(x) = xe^{-x^2/R_{nn}}, x \geq 0$.

5.3. We would like to estimate a vector of unknown parameters $p \in R^n$ from a sequence of N noisy measurements given by

$$y = Cp + n$$

where $C \in R^{p \times n}$, $y, n \in R^p$, and v is zero-mean and white with covariance R_{nn}.

(a) Find the minimum-variance estimate $\hat{p}_{MV}$.

(b) Find the corresponding quality cov $(p - \hat{p}_{MV})$.

5.4. Suppose we have N samples $\{x(0) \cdots x(N-1)\}$ of a process $x(t)$, to be estimated by N complex sinusoids of arbitrary frequencies $\{f_0, \cdots, f_{n-1}\}$. Then

$$x(k\Delta T) = \sum_{m=0}^{N-1} a_m \exp(j2\pi f_m k\Delta T) \qquad \text{for } k = 0, \ldots, N-1$$

Find the least-squares estimate $\hat{a}_{LS}$ of $\{a_m\}$.

5.5. Suppose we are given a measurement modeled by

$$y(t) = s + n(t)$$

where s is random and zero-mean, with variance $\sigma_s^2 = 4$, and n is zero-mean and white, with a unit variance. Find the two-weight minimum-variance estimate of s, that is,

$$\hat{s}_{MV} = \sum_{i=1}^{2} w_i y(i)$$

that minimizes $$J = E\{(s - \hat{s})^2\}$$

5.6. Find the maximum likelihood and maximum a posteriori estimate of the parameter x, if

$$p(x) = \alpha e^{-\alpha x} \qquad x \geq 0, \qquad \alpha \geq 0$$

and $$p(y \mid x) = xe^{-xy} \qquad x \geq 0, \qquad y \geq 0$$

5.7. Suppose we are given a measurement system

$$y(t) = x(t) + v(t) \qquad t = 1, \ldots, N$$

where $v(t) \sim N(0, 1)$.

(a) Find the maximum-likelihood estimate of $x(t)$, that is, $\hat{x}_{ML}(t)$, for $t = 1, \ldots, N$.
(b) Find the maximum a posteriori estimate of $x(t)$, that is, $\hat{x}_{MAP}(t)$, if $p(x) = e^{-x}$.

5.8. Suppose we have a simple AM receiver with signal

$$y(t) = \Theta s(t) + v(t) \qquad t = 1, \ldots, N$$

where Θ is a random amplitude, s is the known carrier, and $v \sim N(0, R_{vv})$.

(a) Find the maximum-likelihood estimate $\hat{\Theta}_{ML}$.
(b) Assume $\Theta \sim N(\Theta_0, R_{\Theta\Theta})$. Find the maximum a posteriori estimate $\hat{\Theta}_{MAP}$.
(c) Assume Θ is Rayleigh-distributed (a common assumption). Find $\hat{\Theta}_{MAP}$.

5.9. We would like to estimate a signal from a noisy measurement

$$y = s + v$$

where $v \sim N(0, 3)$ and s is *Rayleigh-distributed*, as

$$p(s) = se^{-s^2/2}$$

with $$p(y \mid s) = \frac{1}{\sqrt{6\pi}} e^{-1/2 \frac{(y-s)^2}{3}}$$

(a) Find the maximum-likelihood estimate.
(b) Find the maximum a posteriori estimate.
(c) Calculate the Cramer-Rao bound (ignoring $p(s)$).

5.10. Suppose we use the covariance estimator

$$\hat{R}_{xx}(k) = \frac{1}{N - k - 1} \sum_{t=0}^{N-k-1} x(t)x(t + k) \qquad \text{for } k \geq 0$$

(a) Is this a biased estimate for $R_{xx}(k) = E\{x(t)x(t + k)\}$?

(b) What is the variance of this estimator?

5.11. Suppose you are required to design classical spectral estimators for a real-time data acquisition system. From other considerations, the sampling rate is fixed to $\Delta T = 0.001$ sec, and the maximum data storage available is 2048 points. Determine the achievable variance reduction for each of the windows in Table 5.4–1 which are 10% and 25% of the data length.

CHAPTER
6

RANDOM SIGNAL ANALYSIS: MODERN APPROACH

In this chapter we develop the modern approach to random signal analysis; that is, we investigate modern parametric spectral estimators. After establishing the basic parametric method, we examine AR (all-pole), MA (all-zero), and ARMA (pole-zero) techniques. Next, we discuss the lattice approach to solving the spectral estimation problem, which is followed by discussion of the minimum-variance distortionless response (MVDR) method. Finally, we consider various harmonic methods of spectral estimation; we develop the Pisarenko harmonic decomposition (PHD) method as well as the Prony methods. For comparison we use the example of sinusoids in bandpass noise from the previous chapter, in order to evaluate the performance of the various techniques on simulated data.

6.1 PARAMETRIC METHODS OF SPECTRAL ESTIMATION

The parametric or modern approach to spectral estimation is based on the assumption that the measured data under investigation evolves from a process that can be represented, or approximately represented, by a selected model. The parametric spectral estimation approach is a three-step procedure: (1) we select the model set, (2) we estimate the model parameters from the data or covariance lags, and (3) we obtain the spectral estimate by using the estimated model parameters. The major implied advantage of these model-based methods is that we can achieve higher frequency resolution since the data is not windowed [1,2]. It can be shown that these methods imply an infinite extension of the autocorrelation sequence (see [3]). We also note that this approach is indirect, since we must first find the appropriate model and then use it to estimate the PSD.

Let us recall the basic spectral relations developed previously (see Sec. 4.8) for linear systems with random inputs. Recall that the measurement of the output spectrum of a process is related to the input spectrum by the factorization relations:

$$S_{yy}(z) = H(z)H^*(z)S_{xx}(z) = |H(z)|^2 S_{xx}(z) \tag{6.1-1}$$

where
$\quad x$ is the input process
$\quad H$ is the linear system transfer function
$\quad y$ is the measurement

Since we are taking the signal-processing viewpoint, then we will assume that *only* the measured data, y, is available and not the excitation, x. In fact, if both x and y are available, then the problem is called a system identification problem (see [4] for details), and we will use the ARMAX model. However, in this section we restrict ourselves to signal representations and select the AR, MA, and ARMA† model sets as representative for spectral estimation.

The modern method of *parametric spectral estimation* is summarized in these steps:

1. Select a representative model set (AR, MA, ARMA, etc.).
2. Estimate the model parameters from the data; that is, given $\{y(t)\}$, find

$$\hat{\Theta} = \left\{ \hat{A}(q^{-1}), \hat{C}(q^{-1}), \hat{R}_{ee} \right\}$$

3. Estimate the PSD using these parameters, that is,

$$S_{yy}(z) \approx \hat{S}_{yy}(z, \hat{\Theta}) = \hat{H}(z)\hat{H}^*(z)S_{ee}(z)$$

where $H(z) = C(z)/A(z)$ and $S_{ee}(z) = R_{ee}$.

The parametric spectral estimation method is depicted in Fig. 6.1–1. Next we develop the various modern methods to estimate the PSD from noisy measurement data.

---

†Recall from Sec. 4.7 that after the mean is subtracted from the ARMAX model, the deterministic input is removed, and an ARMA model results.

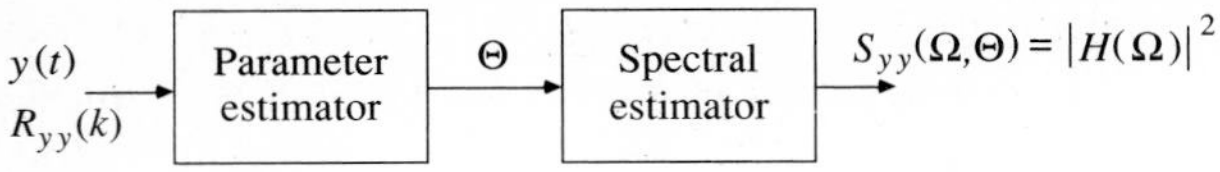

$$S_{yy}(\Omega,\Theta) = |H(\Omega)|^2$$

$y(t)$
$R_{yy}(k)$ → Parameter estimator → Θ → Spectral estimator →

$$
\begin{array}{lll}
\text{AR}: & H(\Omega) = 1/A(\Omega) \\
\text{MA}: & H(\Omega) = C(\Omega) \\
\text{ARMA}: & H(\Omega) = C(\Omega)/A(\Omega)
\end{array}
$$

FIGURE 6.1–1
Parametric spectral estimation method

6.2   AUTOREGRESSIVE (ALL-POLE) SPECTRAL ESTIMATION

The autoregressive (AR) or all-pole model is characterized by the difference equation

$$A(q^{-1})y(t) = e(t) \qquad (6.2\text{--}1)$$

where y is the measured data
e is a zero-mean, white-noise sequence with variance R_{ee}
A is an N_ath order polynomial in backward-shift operator q^{-1}

Taking Z-transforms of Eq. (6.2–1), we have

$$H_{AR}(z) = \frac{Y(z)}{E(z)} = \frac{1}{A(z)} \qquad (6.2\text{--}2)$$

where $$A(z) = 1 + a_1 z^{-1} + \cdots + a_{N_a} z^{-N_a}$$

If we substitute H_{AR} for H in Eq. (6.1–1), we obtain

$$S_{AR}(z) := S_{yy}(z) = H_{AR}(z)H_{AR}^*(z)S_{ee}(z) = |H_{AR}(z)|^2 S_{ee}(z) \qquad (6.2\text{--}3)$$

or $$S_{AR}(z) = \left(\frac{1}{A(z)}\right)\left(\frac{1}{A(z^{-1})}\right) R_{ee} = \frac{R_{ee}\Delta T}{|A(z)|^2}$$

which is the desired representation of the all-pole spectral density. Note that the spectral estimation procedure requires that we first estimate the AR model parameters ($\{\hat{a}_i\}$, $\hat{R}_{ee}$) from the data and then form the estimate,[†]

$$\hat{S}_{AR}(\Omega) = \hat{S}_{AR}(z)\big|_{z=e^{j\Omega}} = \frac{\hat{R}_{ee}\Delta T}{|\hat{A}(e^{j\Omega})|^2} \qquad (6.2\text{--}4)$$

The basic parameter estimation problem for the autoregressive model in the general (infinite covariance) case is given as the minimum (error) variance solution to

$$\min_{\underline{a}} J(t) = E\{e^2(t)\} \qquad (6.2\text{--}5)$$

[†]A popular method of "simulating" the spectrum at equally spaced frequencies is to find the DFT of $A(e^{j\Omega})$, multiply by its conjugate, and divide into $\hat{R}_{ee}$, as in Eq. (6.2–4).

where the estimation error is defined by

$$e(t) := y(t) - \hat{y}(t)$$

and $\hat{y}$ is the minimum-variance estimate obtained from the AR model

$$y(t) = -\sum_{i=1}^{N_a} a_i y(t-i) + e(t) \tag{6.2-6}$$

as

$$\hat{y}(t) = -\sum_{i=1}^{N_a} \hat{a}_i y(t-i) \tag{6.2-7}$$

Note that $\hat{y}$ is actually the one-step predicted estimate $\hat{y}(t \mid t-1)$ based on $[t-1]$ past data samples; hence the popular name "linear (combination of data) predictor." The estimator can easily be derived by minimizing the *prediction error*; $e(t)$ obtained from Eq. (6.2–6) as

$$e(t) = \sum_{i=0}^{N_a} a_i y(t-i) \qquad \text{with } a_0 = 1 \tag{6.2-8}$$

Minimizing Eq. (6.2–5) with respect to the AR coefficients, $\{a_j\}$ for $j = 1, \ldots, N_a$, we obtain

$$\frac{\partial J}{\partial a_j} = \frac{\partial}{\partial a_j} E\{e^2(t)\} = 2E\left\{e(t)\frac{\partial e(t)}{\partial a_j}\right\} = 0 \tag{6.2-9}$$

The prediction-error gradient is given by

$$\frac{\partial e(t)}{\partial a_j} = \frac{\partial}{\partial a_j}\left(\sum_{i=0}^{N_a} a_i y(t-i)\right)\Bigg|_{} = y(t-j) \qquad j = 1, \ldots, N_a \tag{6.2-10}$$

and substituting into Eq. (6.2–9), we obtain the orthogonality condition

$$E\{e(t)y(t-j)\} = 0 \qquad \text{for } j = 1, \ldots, N_a \tag{6.2-11}$$

Substituting for $e(t)$, we obtain

$$E\left\{\sum_{i=0}^{N_a} a_i y(t-i)y(t-j)\right\} = \sum_{i=0}^{N_a} a_i E\{y(t-i)y(t-j)\} \qquad \text{for } j = 1, \ldots, N_a$$

This leads to the *normal* or *Yule-Walker* equations, as

$$\sum_{i=0}^{N_a} a_i R_{yy}(i-j) = 0 \qquad \text{for } j = 1, \ldots, N_a \tag{6.2-12}$$

Setting $a_0 = 1$ and $R(-j) = R(j)$, we obtain the recursion

$$\sum_{i=1}^{N_a} a_i R_{yy}(i-j) = -R_{yy}(j) \qquad \text{for } j = 1, \ldots, N_a \tag{6.2-13}$$

Expanding these equations over j, we have

$$
\begin{bmatrix}
R_{yy}(0) & R_{yy}(1) & \cdots & R_{yy}(N_a - 1) \\
\vdots & \ddots & \ddots & \vdots \\
& & R_{yy}(1) & \\
R_{yy}(N_a - 1) & \cdots & & R_{yy}(0)
\end{bmatrix}
\begin{bmatrix}
a_1 \\
\vdots \\
a_{N_a-1} \\
a_{N_a}
\end{bmatrix}
= -
\begin{bmatrix}
R_{yy}(1) \\
\vdots \\
R_{yy}(N_a)
\end{bmatrix}
\tag{6.2-14}
$$

Clearly, the solution to the estimation problem is given by

$$
\hat{\underline{a}}(N_a) = -\mathbf{R}_{yy}^{-1}\underline{R}_{yy}(N_a)
\tag{6.2-15}
$$

where $\mathbf{R}_{yy}$ is an $N_a \times N_a$ Toeplitz matrix. The quality is obtained by substituting Eq. (6.2–8) into Eq. (6.2–5):

$$
J(t) = E\{e^2(t)\} = E\left\{ \left(\sum_{i=0}^{N_a} a_i y(t-i) \right) \left(\sum_{j=0}^{N_a} a_j y(t-j) \right) \right\}
$$

$$
= \sum_{i=0}^{N_a} a_i \sum_{j=0}^{N_a} a_j E\{y(t-i)y(t-j)\}
$$

$$
= \sum_{i=0}^{N_a} a_i \sum_{j=0}^{N_a} a_j R_{yy}(i-j)
$$

or, removing the term ($a_0 = 1$) in the first sum and multiplying, we obtain

$$
J(t) = \sum_{j=0}^{N_a} a_j R_{yy}(j) + \sum_{i=0}^{N_a} a_i \sum_{j=0}^{N_a} a_j R(i-j)
\tag{6.2-16}
$$

From the normal equations of Eq. (6.2–12), the second term is zero; therefore, we have

$$
R_{ee} = J(t) = R_{yy}(0) + \sum_{j=1}^{N_a} a_j R_{yy}(j)
\tag{6.2-17}
$$

which gives the prediction-error variance.

Various methods of solution can be developed, depending on the cost function selected and its associated limits. More specifically, the AR spectral estimators are calculated by replacing the ensemble average with a time average,

$$
J(T) = E_T\{e^2(t)\} \approx \sum_t e^2(t)
\tag{6.2-18}
$$

If the data record is N points, $\{y(t)\}$, $0 \le t \le N - 1$, then the following methods evolve, depending on the limits selected for the sum shown in Eq. (6.2–18):

1. The windowed or *autocorrelation method* is obtained with $0 \le t \le N-1 + N_a$ (data assumed zero-padded).

2. The *pre-windowed* autocorrelation method is obtained with $0 \le t \le N - 1$ (beginning of data assumed zero-padded).

3. The *post-windowed* autocorrelation method is obtained with $N_a \le t \le N - 1 + N_a$ (end of data assumed zero-padded).

4. The unwindowed or *covariance method* is obtained with $N_a \le t \le N - 1$ (no data assumed zero-padded).

The windowing refers to the original data record, $\{y(t)\}$, $0 \le t \le N - 1$, as depicted in Fig. 6.2–1. To see this, we must write the prediction error in vector form, first noting that

$$e(t) = \sum_{i=0}^{N_a} a_i y(t - i) = \begin{bmatrix} y(t) & y(t-1) & \cdots & y(t - N_a) \end{bmatrix} \begin{bmatrix} 1 \\ a_1 \\ \vdots \\ a_{N_a} \end{bmatrix} = \underline{y}'\underline{a} \quad (6.2\text{-}19)$$

and then expanding this expression for $0 \le t \le N - 1 + N_a$, with $y(t) = 0$, $t < 0$ and $t > N - 1$, to give

$$\underline{e} = \mathbf{Y}\underline{a}$$

where $\underline{e}' = \begin{bmatrix} e(0) & \cdots & e(N_a) & \cdots & e(N-1) & \cdots & e(N-1+N_a) \end{bmatrix}$

$\underline{a}' = \begin{bmatrix} 1 & a_1 & \cdots & a_{N_a} \end{bmatrix}$

and the data matrix $\mathbf{Y}'$ is given by

$$\mathbf{Y}' = \begin{bmatrix} y(0) & \cdots & | & y(N_a) & \cdots & y(N-1) & | & 0 & \cdots & & 0 \\ \vdots & \ddots & | & \vdots & & \vdots & | & & \ddots & & \vdots \\ 0 & & | & y(0) & \cdots & y(N-1+N_a) & | & & & y(N-1) \end{bmatrix}$$

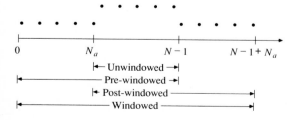

FIGURE 6.2–1
AR spectral estimation: unwindowed, pre- and post-windowed, and windowed methods

We see from Fig. 6.2–1 that the autocorrelation or windowed method corresponds to appending zeros to both the beginning and end of the data record, while the covariance or unwindowed method corresponds to using just the available data. The pre- and post-windowed methods correspond to variations of the autocorrelation method, but the covariance matrix is *not* Toeplitz. Note that we can write the cost function using the vector matrix notation as

$$J(t) = \underline{e}'\underline{e} = \underline{a}'\mathbf{Y}'\mathbf{Y}\underline{a} = \underline{a}'\mathbf{R}_{yy}\underline{a} \qquad (6.2\text{–}20)$$

where $\mathbf{Y}'\mathbf{Y}$ can be defined as $\mathbf{R}_{yy}$, an autocorrelation matrix. If the autocorrelation method is selected, then $\mathbf{R}_{yy}$ is Toeplitz and symmetric, and many efficient recursions can be used to solve for the AR parameters (see [5] for details). Next we will investigate the windowed and unwindowed solutions in more detail.

The autocorrelation method evolves by solving

$$\min_{\underline{a}} J(N - 1 + N_a) = \sum_{t=0}^{N-1+N_a} e^2(t) \qquad (6.2\text{–}21)$$

Differentiating as before, we obtain the orthogonality condition

$$2 \sum_{t=0}^{N-1+N_a} e(t)\frac{\partial e(t)}{\partial a_j} = 0 \qquad \text{for } j = 1, \ldots, N_a$$

or

$$\sum_{t=0}^{N-1+N_a} \left(\sum_{i=0}^{N_a} a_i y(t - i) \right) y(t - j) = 0$$

which gives

$$\sum_{i=0}^{N_a} a_i \left(\sum_{t=0}^{N-1+N_a} y(t - i)y(t - j) \right) = \sum_{i=0}^{N_a} a_i \hat{R}_{yy}(i - j) = 0 \qquad (6.2\text{–}22)$$

where the *sample autocorrelation* function is defined from [†]

$$\hat{R}_{yy}(i - j) := \sum_{t=0}^{N-1+N_a} y(t)y(t + i - j)$$

Expanding this set of normal equations over j gives

$$\begin{bmatrix} \hat{R}_{yy}(0) & \cdots & \hat{R}_{yy}(N_a - 1) \\ \vdots & \ddots & \vdots \\ \hat{R}_{yy}(N_a - 1) & \cdots & \hat{R}_{yy}(0) \end{bmatrix} \begin{bmatrix} a_1 \\ \vdots \\ a_{N_a} \end{bmatrix} = - \begin{bmatrix} \hat{R}_{yy}(1) \\ \vdots \\ \hat{R}_{yy}(N_a) \end{bmatrix} \qquad (6.2\text{–}23)$$

[†]We ignore the $1/N$ premultiplier, since it cancels out of the final expression.

The quality can be estimated as before from Eq. (6.2–17) as

$$\hat{J}(t) = \hat{R}_{yy}(0) + \sum_{j=1}^{N_a} a_j \hat{R}_{yy}(j) \tag{6.2–24}$$

If, however, we choose the limits of the criterion sum (in the unwindowed or covariance method) to be

$$J(t) = \sum_{t=N_a}^{N-1} e^2(t) \tag{6.2–25}$$

then no assumption is made about the data outside of the interval. Therefore, the normal equations become

$$\sum_{i=0}^{N_a} a_i \left(\sum_{t=N_a}^{N-1} y(t-i)y(t-j) \right) = \sum_{i=0}^{N_a} a_i \hat{C}_{yy}(j,i) = 0 \qquad j = 1, \ldots, N_a$$

or

$$\sum_{i=1}^{N_a} a_i \hat{C}_{yy}(j,i) = -\hat{C}_{yy}(j,0) \qquad j = 1, \ldots, N_a \tag{6.2–26}$$

Note that $\hat{C}_{yy}$ can be written recursively as

$$\hat{C}_{yy}(j+1, i+1) = \hat{C}_{yy}(j,i) + y(N_a-j-1)y(N_a-i-1) - y(N-1-j)y(N-1-i)$$

which implies that $y(t)$ is known for $N_a \le t \le N-1$, leading to

$$\begin{bmatrix} \hat{C}_{yy}(1,1) & \hat{C}_{yy}(1,2) & \cdots & \hat{C}_{yy}(1,N_a) \\ \vdots & \vdots & & \vdots \\ \hat{C}_{yy}(N_a,1) & \hat{C}_{yy}(N_a,2) & \cdots & \hat{C}_{yy}(N_a,N_a) \end{bmatrix} \begin{bmatrix} a_1 \\ a_2 \\ \vdots \\ a_{N_a} \end{bmatrix} = - \begin{bmatrix} \hat{C}_{yy}(1,0) \\ \hat{C}_{yy}(2,0) \\ \vdots \\ \hat{C}_{yy}(N_a,0) \end{bmatrix} \tag{6.2–27}$$

$$\mathbf{C}_{yy}\underline{a}(N_a) = -\underline{C}_{yy}(N_a)$$

or

$$\hat{\underline{a}}(N_a) = -\mathbf{C}_{yy}^{-1}\underline{C}_{yy}(N_a)$$

where $C \in R^{N_a \times N_a}$ is a symmetric covariance matrix.

Note that the autocorrelation and covariance methods give identical results when the data sequence is truncated so that $y(t) = 0$ for $t < N_a$ and $t > N-N_a-1$, as well as when $N \gg N_a$. It can also be shown that the AR spectral estimator is asymptotically unbiased and has variance [14]

$$\text{var}(S_{AR}(\Omega)) = \frac{2N_a}{N} S_{yy}^2(\Omega) \tag{6.2–28}$$

We summarize the AR spectral estimators in Table 6.2–1. Let us reconsider the previous example of sinusoids in bandpass noise and apply the AR spectral estimator to the problem.

TABLE 6.2–1
AR spectral estimation methods

Covariance estimation

$$\hat{R}_{yy}(k) = \sum_t y(t)y(t+k)$$

  (i) Windowed (autocorrelation) method: $t = 0, \ldots, N-1+N_a$
  (ii) Pre-windowed method: $t = 0, \ldots, N-1$
  (iii) Post-windowed method: $t = N_a, \ldots, N-1+N_a$
  (iv) Un-windowed (covariance): $t = N_a, \ldots, N-1$

Parameter estimation

$$\hat{\underline{a}} = -\mathbf{R}_{yy}^{-1}\underline{R}_{yy} \qquad \text{autocorrelation method}$$

or $\qquad\quad \hat{\underline{a}} = -\mathbf{C}_{yy}^{-1}\underline{C}_{yy} \qquad \text{covariance method}$

Noise estimation

$$\hat{R}_{ee} = \hat{R}_{yy}(0) + \sum_{j=1}^{N_a} a_j \hat{R}_{yy}(j)$$

Spectral estimation

$$S_{\mathrm{AR}}(\Omega) = \frac{\hat{R}_{ee}\Delta T}{|\hat{A}(e^{j\Omega})|^2}$$

Example 6.2–1. Using SIG [6], consider the performance of the AR spectral estimator on the bandpass sinusoidal simulation of the previous chapter (see Example 5.4–2) by choosing various orders (N_a) to produce the spectral estimates. The results of this estimator are shown in Fig. 6.2–2 for selected orders ($N_a = 100, 50, 25, 12$ and 16 [automatic selection]). We note that by decreasing the order of the model, the spectral estimate gets smoother, again at the cost of line resolution. In the high-order estimates, we see the spectral lines at 30, 50, and 90 Hz enhanced, while the lower-order estimates capture the bandpass response characteristics as well. The "automatic fit" (see following discussion) captures the lines at 35 and 50 Hz, as well as capturing the bandpass spectrum, but does not resolve the lines at 56 and 90 Hz.

We note also that the main "knob" in AR spectral estimation is the selection of the order of the model. There are a number of criteria for model-order selection [7], but there are still various opinions on which criterion is the most reasonable. If the order selected is too low, there will not be enough poles to adequately represent the underlying spectrum. Too high a choice of order will usually result in spurious peaks in the estimated spectrum. Two of the most popular criteria for order selection were introduced by Akaike [8]. The first criterion is the final prediction error (FPE), developed so that the selected order of the AR process minimizes the average error in the one-step prediction error (see Chapter 7 for more details); the FPE of order N_a is given by

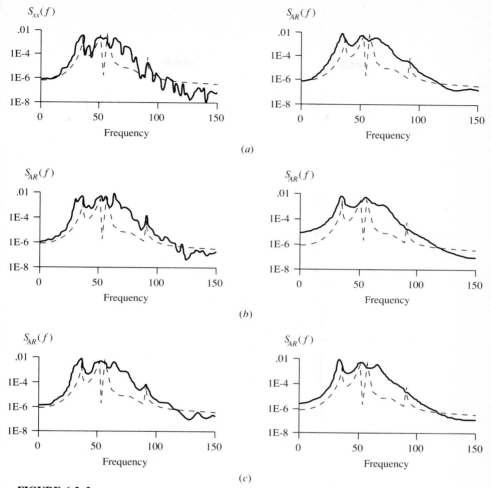

FIGURE 6.2–2
AR spectral estimation of bandpass sinusoidal simulation (*a*) Raw/true (dashed) spectra and 25th order AR spectral estimate (*b*) 100th and 12th order AR spectral estimates (*c*) 50th and 16th (automatic) order AR spectral estimates

$$\text{FPE } (N_a) = \left(\frac{N + 1 + N_a}{N - 1 - N_a} \right) \hat{R}_{ee} \qquad (6.2\text{–}29)$$

where N is the number of data samples, N_a is the predictor order, and R_{ee} is the prediction-error variance for the N_a-order predictor. Note that as $N_a \to N$, FPE increases, reflecting the increase in the uncertainty in the estimate of R_{ee} or prediction-error power. The order selected is the one for which FPE is minimum.

Akaike suggested another criterion, using a maximum-likelihood approach to derive the Akaike information criterion (AIC) [8,9]. The AIC determines the model order by minimizing a function developed from an information-theory viewpoint (see Sec. 7.7 for details). Assuming the process has gaussian statistics, the AIC is

$$\text{AIC}(N_a) = -\ln \hat{R}_{ee} + 2\left(\frac{N_a}{N}\right) \tag{6.2-30}$$

The first term is the prediction-error variance, while the second consists of two parts; $2/N$ is an additive constant accounting for the removal of the sample mean, while N_a is the penalty for the use of extra AR coefficients that do not result in a substantial reduction in prediction-error power. Again, the order selected is the one that minimizes the AIC estimate, or MAICE. As $N \to \infty$, AIC $\to$ FPE.

Let us again examine the sinusoidal bandpass noise example for various orders of the AR spectral estimates in terms of the AIC criterion.

> **Example 6.2-2.** For the sinusoids in the bandpass noise problem of the previous section, determine the optimum order of the AR spectral estimator, based on the AIC criterion. After making various runs in SIG [6] of the spectral estimator at different orders, the resulting AIC is displayed in Fig. 6.2-3. Here we see that the function reaches a minimum at 16th order and then begins to increase, indicating that this is the optimum order of the spectral estimator.

6.3   MOVING-AVERAGE (ALL-ZERO) SPECTRAL ESTIMATION

The moving-average or all-zero model is characterized by the difference equation

$$y(t) = C(q^{-1})e(t) \tag{6.3-1}$$

where y is the measured data

e is a zero-mean, white-noise sequence with variance R_{ee}

C is an N_cth order polynomial in backward-shift operator q^{-1}

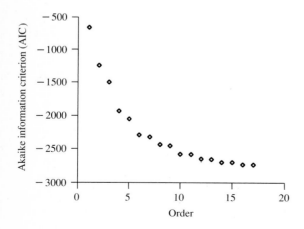

FIGURE 6.2-3
AIC order estimation of sinusoids in bandpass noise

Taking Z-transforms, we have

$$H_{MA}(z) = \frac{Y(z)}{E(z)} = C(z) \tag{6.3-2}$$

where
$$C(z) = c_0 + c_1 z^{-1} + \cdots + c_{N_c} z^{-N_c}$$

Substituting Eq. (6.3–2) into Eq. (6.3–1), we obtain

$$S_{MA}(z) := S_{yy}(z) = H_{MA}(z)H_{MA}^*(z)S_{ee}(z) = |H_{MA}(z)|^2 S_{ee}(z)$$

or
$$S_{MA}(z) = C(z)C(z^{-1})S_{ee}(z) = |C(z)|^2 R_{ee}\Delta T \tag{6.3-3}$$

which is the desired representation of the all-zero spectral density. The parametric spectral estimation procedure requires that we first estimate the MA model parameters ($\{\hat{c}_i\}, \hat{R}_{ee}$) from the data and then form the estimate

$$\hat{S}_{MA}(\Omega) = \hat{S}_{MA}(z)\big|_{z=e^{j\Omega}} = |\hat{C}(e^{j\Omega})|^2 \hat{R}_{ee}\Delta T \tag{6.3-4}$$

The parametric estimation problem can be solved, as before, by defining the minimum-error variance criterion and minimizing with respect to the moving-average model parameters; that is, we define

$$\min_{\underline{c}} J(t) = E\{e^2(t)\}$$

where the prediction error is $e(t) = y(t) - \hat{y}(t)$ and $\hat{y}(t) = \sum_{i=1}^{N_c} c_i e(t-i)$. Differentiating and setting the gradient to zero, we obtain the orthogonality condition

$$\frac{\partial J(t)}{\partial c_k} = 2E\left\{ e(t)\frac{\partial e(t)}{\partial c_k} \right\} = 0 \qquad k = 0, \ldots, N_c$$

and the parametric gradient is given by

$$\frac{\partial e(t)}{\partial c_k} = \frac{\partial}{\partial c_k}\left(y(t) - \sum_{i=0}^{N_c} c_i e(t-i) \right) = -e(t-k) \qquad k = 0, \ldots, N_c \tag{6.3-5}$$

Substituting this result into the orthogonality condition and using the MA model gives

$$E\{e(t)e(t-k)\} = E\left\{ \left(y(t) - \sum_{i=0}^{N_c} c_i e(t-i) \right)\left(y(t-k) - \sum_{j=0}^{N_c} c_j e(t-j-k) \right) \right\} = 0$$

or

$$E\{y(t)y(t-k)\} - \sum_{i=0}^{N_c} c_i E\{e(t-i)y(t-k)\} - \sum_{j=0}^{N_c} c_j E\{e(t-j-k)y(t)\}$$

$$+ \sum_{i=0}^{N_c} \sum_{j=0}^{N_c} c_i c_j E\{e(t-i)e(t-j-k)\} = 0$$

$$R_{yy}(k) - \sum_{i=0}^{N_c} c_i E\left\{ e(t-i) \left(\sum_{j=0}^{N_c} c_j e(t-k-j) \right) \right\}$$

$$- \sum_{j=0}^{N_c} c_j E\left\{ e(t-j-k) \left(\sum_{i=0}^{N_c} c_i e(t-i) \right) \right\}$$

$$+ \sum_{i=0}^{N_c} \sum_{j=0}^{N_c} c_i c_j E\{ e(t-i) e(t-j-k) \} = 0$$

$$R_{yy}(k) - \sum_{j=0}^{N_c} \sum_{i=0}^{N_c} c_j c_i R_{ee}(i-k-j) - \sum_{j=0}^{N_c} \sum_{i=0}^{N_c} c_i c_j R_{ee}(i-j-k)$$

$$+ \sum_{i=0}^{N_c} \sum_{j=0}^{N_c} c_i c_j R_{ee}(i-j-k) = 0$$

or since $e(t)$ is white, $R_{ee}(i-j-k) = R_{ee}\delta(i-j-k)$ and therefore

$$R_{yy}(k) - R_{ee} \sum_{j=0}^{N_c} c_j c_{j+k} - R_{ee} \sum_{j=0}^{N_c} c_j c_{j+k} + R_{ee} \sum_{j=0}^{N_c} c_j c_{j+k} = 0$$

or
$$R_{yy}(k) = \begin{cases} R_{ee} \sum_{j=0}^{N_c} c_j c_{j+k} & k = 0, \ldots, N_c \\ 0 & \text{otherwise} \end{cases} \tag{6.3-6}$$

We note immediately that the Wiener-Khintchine relation becomes

$$S_{yy}(z) = \sum_{k=-\infty}^{\infty} R_{yy}(k) z^{-k} = R_{ee} \sum_{k=-\infty}^{\infty} \left(\sum_{j=0}^{N_c} c_j c_{j+k} \right) z^{-k} \tag{6.3-7}$$

The classical *correlation method* of the previous section is related to the MA model through Eq. (6.3–4). In fact, from the spectral estimation viewpoint it is far simpler to estimate the covariance function and use the correlation method.

The estimation of MA parameters is related to the spectral factorization problem, since Eq. (6.3–5) can be altered by multiplying by $z^{-j} z^j$ to obtain

$$S_{yy}(z) = R_{ee} \sum_{k=-\infty}^{\infty} \sum_{j=-\infty}^{\infty} c_j c_{j+k} z^{-(j+k)} z^{+j} = R_{ee} \left(\sum_{l=-\infty}^{\infty} c_l z^{-l} \right) \left(\sum_{j=-\infty}^{\infty} c_j z^j \right)$$

or
$$S_{MA}(z) = S_{yy}(z) = C(z) C(z^{-1}) R_{ee}$$

as before. Thus, many spectral factorization algorithms exist [7] to solve this problem.

6.4 AUTOREGRESSIVE MOVING-AVERAGE OR RATIONAL (POLE-ZERO) SPECTRAL ESTIMATION

When spectra have sharp lines as well as deep notches, then both poles (AR) and zeros (MA) are required to describe the spectra adequately. The autoregressive moving-average (ARMA) or pole-zero model is characterized by the difference equation

$$A(q^{-1})y(t) = C(q^{-1})e(t) \tag{6.4-1}$$

where

y is the measured data
e is a zero-mean, white-noise sequence with variance R_{ee}
A, C are N_ath and N_cth order polynomials in the backward shift operator q^{-1}

Taking Z-transforms of Eq. (6.4–1), we have

$$H_{ARMA}(z) = \frac{Y(z)}{E(z)} = \frac{C(z)}{A(z)} \tag{6.4-2}$$

where
$$C(z) = 1 + c_1 z^{-1} + \cdots + c_{N_c} z^{-N_c},$$
and
$$A(z) = 1 + a_1 z^{-1} + \cdots + a_{N_a} z^{-N_a}$$

Substituting for H in Eq. (6.1–1), we obtain

$$S_{ARMA}(z) := S_{yy}(z) = H_{ARMA}(z)H^*_{ARMA}(z)S_{ee}(z) = |H_{ARMA}(z)|^2 S_{ee}(z)$$

or
$$S_{ARMA}(z) = \left(\frac{C(z)}{A(z)}\right)\left(\frac{C(z^{-1})}{A(z^{-1})}\right) R_{ee} = \left|\frac{C(z)}{A(z)}\right|^2 R_{ee}\Delta T \tag{6.4-3}$$

which is the desired representation of the rational (pole-zero) spectral density. The parametric spectral estimation procedure requires that we obtain the ARMA parameter estimates ($\{\hat{a}_i\}, \{\hat{c}_i\}, \hat{R}_{ee}$) from the data and then from the power spectrum

$$\hat{S}_{ARMA}(\Omega) = \hat{S}_{ARMA}(z)\big|_{z=e^{j\Omega}} = \left|\frac{\hat{C}(e^{j\Omega})}{\hat{A}(e^{j\Omega})}\right|^2 \hat{R}_{ee}\Delta T \tag{6.4-4}$$

The ARMA parameter estimation problem is nonlinear again, due to the presence of the MA polynomial. Many methods have been developed to estimate the parameters; most are based on various optimization schemes [4,7,10].

First, let us investigate the optimal solution to this problem for the ARMA model. If we define the mean-squared error criterion as before,

$$\min_{\Theta} J = E\{e^2(t)\}$$

From the ARMA model, we have

$$e(t) = \sum_{i=0}^{N_a} a_i y(t - i) - \sum_{i=1}^{N_c} c_i e(t - i) \qquad a_0, c_0 = 1 \qquad (6.4\text{--}5)$$

As before, we obtain the orthogonality conditions for both sets of parameters; that is,

$$\frac{\partial J}{\partial a_j} = 2E\left\{ e(t) \frac{\partial e(t)}{\partial a_j} \right\} = 0 \qquad j = 1, \ldots, N_a$$

and

$$\frac{\partial J}{\partial c_j} = 2E\left\{ e(t) \frac{\partial e(t)}{\partial c_j} \right\} = 0 \qquad j = 1, \ldots, N_c$$

The parametric gradients are given by

$$\frac{\partial e(t)}{\partial a_j} = \frac{\partial}{\partial a_j} \left(\sum_{i=0}^{N_a} a_i y(t - i) - \sum_{i=1}^{N_c} c_i e(t - i) \right) = y(t - j)$$

and similarly

$$\frac{\partial e(t)}{\partial c_j} = -e(t - j)$$

Substituting these expressions into the orthogonality conditions gives the relations

$$\frac{\partial J}{\partial a_j} = E\{e(t)y(t - j)\} = E\left\{ \left(\sum_{i=0}^{N_a} a_i y(t - i) - \sum_{i=1}^{N_c} c_i e(t - i) \right) y(t - j) \right\} = 0$$

$$j = 1, \ldots, N_a$$

$$0 = \sum_{i=0}^{N_a} a_i R_{yy}(i - j) - \sum_{i=1}^{N_c} c_i R_{ey}(i - j) \qquad \text{for } j = 1, \ldots, N_a$$

or removing the a_0 term from the first summation, we obtain

$$-R_{yy}(j) = \sum_{i=1}^{N_a} a_i R_{yy}(i - j) - \sum_{i=1}^{N_c} c_i R_{ey}(i - j) \qquad \text{for } j = 1, \ldots, N_a \qquad (6.4\text{--}6)$$

Similarly,

$$\frac{\partial J}{\partial c_j} = -E\{e(t)e(t - j)\} = -E\left\{ \left(\sum_{i=0}^{N_a} a_i y(t - i) - \sum_{i=1}^{N_c} c_i e(t - i) \right) e(t - j) \right\} = 0$$

$$j = 1, \ldots, N_c$$

$$= -\sum_{i=0}^{N_a} a_i E\{y(t - i)e(t - j)\} + \sum_{i=1}^{N_c} c_i E\{e(t - i)e(t - j)\} = 0$$

$$j = 1, \ldots, N_c$$

$$R_{ye}(j) = -\sum_{i=1}^{N_a} a_i R_{ye}(i-j) + \sum_{i=1}^{N_c} c_i R_{ee}(i-j) \qquad j = 1, \ldots, N_c \quad (6.4\text{--}7)$$

If we assume that the covariances in Eqs. (6.4–6,7) are *known*, then the following matrix equation results:

$$\begin{bmatrix} \mathbf{R}_{yy} & | & -\mathbf{R}_{ey} \\ \hline -\mathbf{R}_{ye} & | & \mathbf{R}_{ee} \end{bmatrix} \begin{bmatrix} a \\ \hline c \end{bmatrix} = \begin{bmatrix} -\underline{R}_{yy} \\ \hline \underline{R}_{ye} \end{bmatrix}$$

$$\mathbf{R}\underline{\Theta} = \underline{R}$$

or
$$\hat{\underline{\Theta}}_{\mathrm{LS}} = \mathbf{R}^{-1}\underline{R} \qquad (6.4\text{--}8)$$

Of course, this relation only holds if we have the required covariances; however, since we do not have $\{e(t)\}$, we cannot calculate the covariance. One technique suggested is called the *extended least-squares* (ELS) method, where the values for $\{e(t)\}$ are estimated using the current parameters of the ARMA model,

$$\hat{e}(t) = \sum_{i=0}^{N_a} \hat{a}_i y(t-i) - \sum_{i=1}^{N_c} \hat{c}_i \hat{e}(t-i) \qquad (6.4\text{--}9)$$

If we use this recursion to generate $\{\hat{e}(t)\}$, then clearly the required covariances can be computed. The solution to this problem can be calculated recursively, using the ELS algorithm, by simply writing the ARMA relations of Eq. (6.4–5) in vector form as

$$y(t) = \begin{bmatrix} -y(t-1) & \cdots & -y(t-N_a) & | & \hat{e}(t-1) & \cdots & \hat{e}(t-N_c) \end{bmatrix} \begin{bmatrix} a_1 \\ \vdots \\ a_{N_a} \\ \hline c_1 \\ \vdots \\ c_{N_c} \end{bmatrix} + e(t)$$

or
$$y(t) = \underline{c}'(t)\underline{\Theta} + e(t) \qquad (6.4\text{--}10)$$

where $\underline{c}', \Theta \in R^{N_a+N_c}$. As before, the values for $\{e(t)\}$ are estimated using the relation of Eq. (6.4–9) thereby "extending" the data vector, $\underline{c}(t)$. Once this is accomplished, the recursive least-squares or Kalman-filter identifier (see Chapters 7 and 9) can be used to solve this problem.

Let us consider a suboptimal approach to solving the ARMA spectral estimation problem. If we calculate the output covariance of the ARMA model, we obtain

$$R_{yy}(k) = E\{y(t)y(t+k)\} = -\sum_{i=1}^{N_a} a_i E\{y(t)y(t+k-i)\}$$

$$+ \sum_{i=0}^{N_c} c_i E\{y(t)e(t+k-i)\}$$

or
$$R_{yy}(k) = -\sum_{i=1}^{N_a} a_i R_{yy}(k-i) + \sum_{i=0}^{N_c} c_i R_{ye}(k-i) \qquad (6.4\text{--}11)$$

If we use the impulse-response relation,

$$y(t) = \sum_{t=0}^{\infty} h(k)e(t-k)$$

then we have (see Table 4.4–1)

$$R_{ye}(k-i) = E\{y(t)e(t+k-i)\} = E\left\{\sum_{j=0}^{\infty} h(j)e(t-j)e(t+k-i)\right\}$$

$$= \sum_{j=0}^{\infty} h(j)R_{ee}(j+k-i)$$

But
$$R_{ye}(k-i) = \begin{cases} R_{ee}h(i-k) & k \le i \\ 0 & k > i \end{cases} \qquad (6.4\text{--}12)$$

since $e(t)$ is a white-noise sequence,

$$R_{ee}(j+k-i) = R_{ee}\delta(j+k-i)$$

Substituting Eq. (6.4–12) into Eq. (6.4–11), we have the Yule-Walker equations for the ARMA process:

$$R_{yy}(k) = \begin{cases} -\sum_{i=1}^{N_a} a_i R_{yy}(k-i) + R_{ee}\sum_{i=0}^{N_c} c_i h(i-k) & k = 0,\ldots,N_c \\ -\sum_{i=1}^{N_a} a_i R_{yy}(k-i) & k = N_c+1,\ldots,N_c+N_a \end{cases}$$
$$(6.4\text{--}13)$$

Clearly, these equations are nonlinear ($k \le N_c$) in the parameters, since it follows that

$$h(i) * a_i = c_i \qquad \text{for } j = 0,\ldots,N_c$$

or
$$\sum_{i=0}^{N_a} a_i h(j-i) = \begin{cases} c_j & j = 0,\ldots,N_c \\ 0 & j > N_c \end{cases} \qquad (6.4\text{--}14)$$

Many ARMA parameter estimation techniques have been formulated which estimate the ARMA parameters jointly, as required; however, suboptimum techniques have also been developed, usually based on a least-squares criterion and the solutions of linear equations. These suboptimum methods usually estimate the AR and MA parameters separately.

If we choose the suboptimal approach, then the second set of relations of Eq. (6.4–13) result, usually called the *extended* or *modified Yule-Walker* equations:

$$R_{yy}(k) = -\sum_{i=1}^{N_a} a_i R_{yy}(k-i) \qquad k = N_c + 1, \ldots, N_c + N_a$$

This leads to the linear relations

$$\begin{bmatrix} R_{yy}(N_c) & \cdots & R_{yy}(N_c + 1 - N_a) \\ \vdots & \ddots & \vdots \\ R_{yy}(N_c + N_a - 1) & \cdots & R_{yy}(N_c) \end{bmatrix} \begin{bmatrix} a_1 \\ \vdots \\ a_{N_a} \end{bmatrix} = - \begin{bmatrix} R_{yy}(N_c + 1) \\ \vdots \\ R_{yy}(N_c + N_a) \end{bmatrix}$$

$$(6.4\text{--}15)$$

which can be solved to yield the AR part of the spectrum. The MA part can be obtained in a number of different ways. One approach is to filter the data with the estimated AR model; that is,

$$y_{MA}(t) := \hat{A}(q^{-1})y(t) = \hat{A}(q^{-1})\left[\frac{C(q^{-1})}{A(q^{-1})}\right]e(t) \approx C(q^{-1})e(t)$$

$$y(t) \rightarrow \hat{A}(q^{-1}) \rightarrow y_{MA}(t) = C(q^{-1})e(t)$$

yielding a purely MA process. The MA process can then be estimated using the correlation method or the spectral factorization method. Clearly, if we just want the spectrum then the MA parameters are not required. In this case,

$$S_{ARMA}(\Omega) = \frac{\text{DFT}\left\{\hat{R}_{y_{MA}y_{MA}}(k)\right\}}{|\hat{A}(e^{j\Omega})|^2} \qquad (6.4\text{--}16)$$

We summarize the extended ARMA method in Table 6.4–1. Another approach (see [11] for details) is to use the impulse response of the AR filter to estimate the MA model directly. Before we leave this section, let us consider applying the ELS spectral estimation algorithm (see Chapter 7) to our simulation problem.

Example 6.4–1. We again consider the sinusoidal bandpass simulation and apply the ARMA spectral estimator to this problem. We note that because the ARMA estimator inherently has both poles and zeros, it is capable of matching the spectral lines characterized by poles and the "notches" characterized by zeros. Using the ELS approach in SIG [6], the performance of this estimator is shown in Fig. 6.4–1 for orders of 35, 25, and 16. Overall, we note the sharp spectral lines and notches characteristic of this spectrum. We see the effect of holding N_a constant and decreasing the number of zeros from ARMA(35,12) to ARMA(35,8). The latter appear very much as an AR spectrum. In each case, we recognize the ability of the estimator to resolve the lines at 35, 56, and 90 Hz as well as to resolve the bandpass response. It is interesting to note that the 16th order spectrum ap-

TABLE 6.4–1
ARMA spectral estimation (extended)

Covariance estimation

$$\hat{R}_{yy}(k) = \frac{1}{N} \sum_{t=0}^{N-k+1} y(t)y(t+k)$$

AR parameter estimation

$$\hat{\underline{a}} = -\mathbf{R}_{yy}^{-1} \underline{R}_{yy}$$

Filtering

$$\hat{y}_{MA}(t) = \hat{A}(q^{-1})y(t)$$

MA parameter estimation

$$|\hat{C}(e^{j\Omega})|^2 \hat{R}_{ee} \Delta T = \text{DFT}\{R_{y_{MA}y_{MA}}(k)\}$$

Spectral estimation

$$S_{ARMA}(\Omega) = \frac{|\hat{C}(e^{j\Omega})|^2 \hat{R}_{ee} \Delta T}{|\hat{A}(e^{j\Omega})|^2}$$

pears to be a reasonable bandpass spectrum but cannot resolve the lines at 50 and 56 Hz.

6.5   LATTICE SPECTRAL ESTIMATION

Burg [3] developed the lattice spectral estimation technique to eliminate the windowing problems of the autocorrelation method, as well as the instability of the covariance method (see Sec. 6.2). The lattice technique was developed primarily as a solution to the maximum-entropy spectral estimation problem (see Appendix B); therefore, it generates an AR model based on the estimated lattice parameters ($\{k_i\}$, R_{ee}). The PSD is therefore obtained from

$$H_{LAT}(z) = \frac{1}{A(z)}$$

to give

$$S_{LAT}(z) = H_{LAT}(z)H_{LAT}^*(z)S_{ee}(z) = \frac{R_{ee}\Delta T}{|A(z)|^2}$$

The *Burg* or *lattice method* is based on minimizing the forward and backward prediction errors:

$$\min_{\underline{k}} J(t) = E\{e_f^2(t) + e_b^2(t)\} \tag{6.5–1}$$

subject to the constraint that the forward and backward prediction errors satisfy the *lattice recursions*

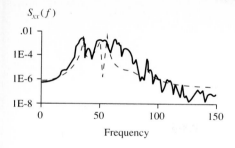

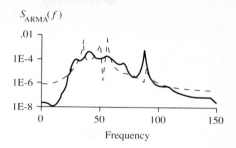

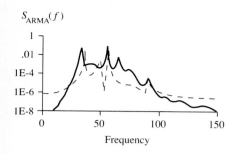

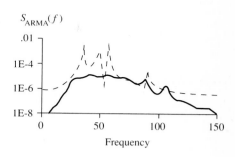

(a)

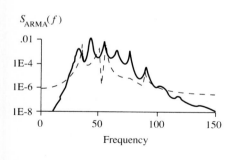

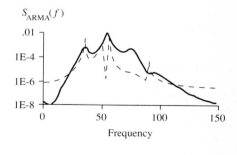

(b)

(c)

FIGURE 6.4–1

ARMA spectral estimation of sinusoidal bandpass simulation using the ELS technique (a) Raw/true (dashed) spectrum and ARMA(25,13) spectral estimate (b) ARMA(35,12) and ARMA(25,8) spectral estimates (c) ARMA(35,8) and ARMA(16,10) spectral estimates

$$e_f(t, i) = e_f(t, i - 1) - k_i e_b(t - 1, i - 1)$$

$$e_b(t, i) = e_b(t - 1, i - 1) - k_i e_f(t, i - 1) \qquad (6.5\text{–}2)$$

where $e_f(t, i)$ and $e_b(t, i)$ are the forward and backward prediction errors of the ith section at time t, and k_i is the corresponding reflection coefficient.

Performing the indicated minimization of Eq. (6.5–1), we have

$$\frac{\partial J}{\partial k_i} = \frac{\partial}{\partial k_i} E\{e_f^2(t, i) + e_b^2(t, i)\} = 2E\left\{e_f(t, i)\frac{\partial e_f(t, i)}{\partial k_i} + e_b(t, i)\frac{\partial e_b(t, i)}{\partial k_i}\right\} = 0$$

$$(6.5\text{–}3)$$

The error gradients can be calculated directly from the model of Eq. (6.5–2) as

$$\frac{\partial e_f}{\partial k_i} = -e_b(t-1, i-1)$$

and, similarly, we have

$$\frac{\partial e_b}{\partial k_i} = -e_f(t, i-1)$$

Substituting these gradients into Eq. (6.5–3) gives

$$\frac{\partial J}{\partial k_i} = -2E\{e_f(t, i)e_b(t-1, i-1) + e_b(t, i)e_f(t, i-1)\} = 0$$

Using the lattice recursions, we obtain

$$-2E\{(e_f(t, i-1) - k_i e_b(t-1, i-1))e_b(t-1, i-1)$$
$$+ (e_b(t-1, i-1) - k_i e_f(t, i-1))e_f(t, i-1)\} = 0$$

or

$$\frac{\partial J}{\partial k_i} = -2E\{2e_f(t, i-1)e_b(t-1, i-1) - k_i(e_f^2(t, i-1) + e_b^2(t-1, i-1))\} = 0$$

$$(6.5\text{–}4)$$

Solving this equation for k_i gives the relation for the reflection coefficient as

$$k_i = \frac{2E\{e_f(t, i-1)e_b(t-1, i-1)\}}{E\{e_f^2(t, i-1) + e_b^2(t-1, i-1)\}} \qquad (6.5\text{–}5)$$

If we assume the data are ergodic, then Burg's technique proposes replacing the ensemble averages of Eq. (6.5–5) with time averages using only the unwindowed data record,

$$k_i = \frac{2\sum_{t=N_a}^{N-1} e_f(t, i-1)e_b(t-1, i-1)}{\sum_{t=N_a}^{N-1} e_f^2(t, i-1) + e_b^2(t-1, i-1)} \qquad \text{for } i = 1, \ldots, N_a \qquad (6.5\text{–}6)$$

The reflection coefficients are estimated from the data, using Eqs. (6.5–2,6) for each stage, and the corresponding prediction-error variance (see Sec. 7.5 for proof) is given by:

$$\hat{R}_{ee} = \prod_{i=1}^{N_a} (1 - \hat{k}_i^2)\hat{R}_{yy}(0) \qquad (6.5\text{–}7)$$

Then the lattice spectral estimate is developed by transforming the lattice to an AR model, using the Levinson algorithm in Table 4.6–1, to give

$$\hat{S}_{\mathrm{LAT}}(z) = \frac{\hat{R}_{ee}\Delta T}{|\hat{A}(z)|^2}$$

We summarize this technique in Table 6.5–1. The Burg lattice method was designed specifically to operate over short data records and has had large impact in array signal processing [11,12,13]. Other methods based on the lattice model have evolved, distinguished by how the forward and backward prediction errors are combined to form the cost function (see [14] for details); however, these are all suboptimal.

Before we leave this section, let us reconsider our sinusoidal bandpass noise example and investigate the performance of this spectral estimator using Burg's lattice model technique.

Example 6.5–1. Again consider the performance of the lattice spectral estimator, using the Burg algorithm on the bandpass sinusoid simulation. Using SIG [6], the results are shown in Fig. 6.5–1 for various selected orders ($N_a = 100, 50, 25, 12$ and 16 (automatic)). Here we see enhanced line resolution of the high-order models with the "smoother" estimates of the low-order models. The lines at 50 and 56 Hz appear to be resolved by this method ($N_a = 100, 50$), as do the lines at 35 and 90 Hz. The bandpass spectrum is apparent in the low-order spectral estimates, again indicating the tradeoff of resolution for smoothness or variance for bias.

TABLE 6.5–1
Lattice spectral estimation

Prediction error estimation

$$e_f(t,i) = e_f(t,i-1) - k_i e_b(t-1,i-1)$$

$$e_b(t,i) = e_b(t-1,i-1) - k_i e_f(t,i-1)$$

Reflection coefficient estimation

$$k_i = \frac{2\sum_{t=N_a}^{N-1} e_f(t,i-1)e_b(t-1,i-1)}{\sum_{t=N_a}^{N-1} e_f^2(t,i-1) + e_b^2(t-1,i-1)} \qquad \text{for } i = 1,\dots,N_a$$

AR parameter calculation (Table 4.6–1)

$$a(i,i) = k_i$$

$$a(j,i) = a(j,i-1) - k_i a(i-j,i-1) \qquad j = 1,\dots,i-1$$

Noise estimation

$$\hat{R}_{ee} = \prod_{i=1}^{N}(1-k_i^2)\hat{R}_{yy}(0)$$

Spectrum estimation

$$S_{\mathrm{LAT}}(\Omega) = \frac{\hat{R}_{ee}\Delta T}{|\hat{A}(e^{j\Omega})|^2}$$

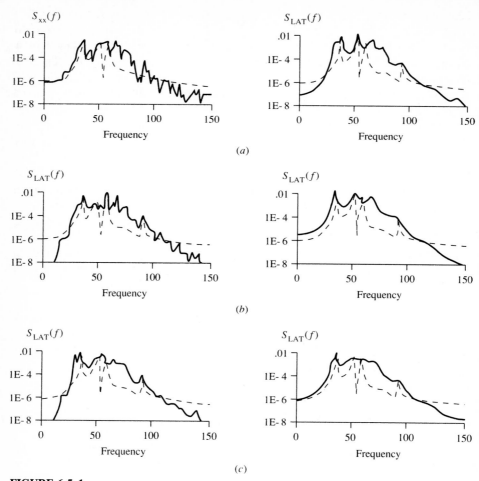

FIGURE 6.5–1
Lattice spectral estimation for bandpass sinusoidal simulation (*a*) Raw/true (dashed) spectra and 25th order lattice estimate (*b*) 100th and 12th order lattice spectral estimates (*c*) 50th and 16th (automatic) order lattice spectral estimates

6.6 MINIMUM-VARIANCE, DISTORTIONLESS RESPONSE SPECTRAL ESTIMATION

The minimum-variance, distortionless response (MVDR) spectral estimation method relies on estimating the PSD by measuring the power out of a set of narrowband filters. The MVDR method is sometimes called the maximum-likelihood method (MLM), but it is not a true maximum-likelihood estimator in the sense of maximizing the likelihood functional of Sec. 5.3. However, the maximum-likelihood name is inaccurately identified with the technique originally developed by Capon [15] for seismic arrays. The MVDR method is similar to the classical correlation/periodogram techniques, except that the narrowband filters are different for each frequency, whereas those of the classical techniques are fixed. In this way, the narrowband filters of the MVDR approach can be thought of as adapting to the process for which PSD is desired; that is, it is a "data-adaptive" technique.

The idea behind the MVDR spectral estimator is that, for a given discrete frequency Ω_m, the signal to be extracted is characterized by a complex exponential (narrow-band); that is,

$$x(t) = A e^{j\Omega_m t} \qquad (6.6\text{–}1)$$

and
$$X(\Omega) = A \delta(\Omega - \Omega_m)$$

The measured output signal is assumed to be contaminated with additive white noise as

$$y(t) = x(t) + n(t) \qquad (6.6\text{–}2)$$

An estimator is to be designed so that the output variance is minimized and the frequency under consideration, Ω_m, is passed undistorted. We assume that the estimator is a FIR operator of the form

$$\hat{y}(t) = h(t) * y(t) = \sum_{k=0}^{N} h(k) y(t - k) \qquad (6.6\text{–}3)$$

Next, we assume that the estimator is designed to pass the narrowband signal $x(t)$; that is, we want to pass $A e^{j\Omega_m t}$ with unity gain, or

$$H(\Omega)\big|_{\Omega = \Omega_m} = 1$$

We also require the estimate to be unbiased, such that

$$E\{\hat{y}(t)\} = x(t) = A e^{j\Omega_m t} = \sum_{k=0}^{N} h(k) E\{y(t - k)\}$$

but from Eq. (6.6–2) we have

$$A e^{j\Omega_m t} = \sum_{k=0}^{N} h(k) E\{x(t - k) + n(t - k)\}$$

or
$$A e^{j\Omega_m t} = \sum_{k=0}^{N} h(k)\left(A e^{j\Omega_m(t-k)}\right) \qquad (6.6\text{–}4)$$

Dividing, we obtain the desired result:

$$1 = \sum_{k=0}^{N} h(k) e^{-jk\Omega_m} = H(\Omega)\big|_{\Omega = \Omega_m} \qquad (6.6\text{–}5)$$

which implies that the estimator will pass the complex narrowband signal with unity gain when $\Omega = \Omega_m$. We can express this in vector notation by expanding the sum as

$$\begin{bmatrix} 1 & e^{-j\Omega_m} & \cdots & e^{-jN\Omega_m} \end{bmatrix} \begin{bmatrix} h(0) \\ h(1) \\ \vdots \\ h(N) \end{bmatrix} = \underline{V}'(\Omega_m)\underline{h} = 1 \qquad (6.6\text{–}6)$$

The MVDR estimator is obtained by minimizing the variance subject to the above constraint,

$$\min_{\underline{h}} J = \underline{h}'\mathbf{R}_{yy}\underline{h} \qquad (6.6\text{–}7)$$

subject to the constraint $\underline{V}'(\Omega_m)\underline{h} = 1$; we now see the reasoning behind the name. The output power or variance of Eq. (6.6–7) is minimized (minimum-variance), subject to the constraint of passing the signal at one frequency Ω_m (distortionless response) with unity gain.

This estimator can be derived using the Lagrange multiplier method of transforming the constrained problem to an unconstrained problem by augmenting the multiplier λ into the cost function and minimizing:

$$\min_{\underline{h}} J = \underline{h}'\mathbf{R}_{yy}\underline{h} + \lambda(1 - \underline{V}'(\Omega_m)\underline{h}) \qquad (6.6\text{–}8)$$

Using the chain rule of Eq. (5.1–4), we have

$$\nabla_h J = 2\mathbf{R}_{yy}\underline{h} - \lambda\underline{V}(\Omega_m) = 0$$

or

$$\underline{h} = \frac{\lambda}{2}\mathbf{R}_{yy}^{-1}\underline{V}(\Omega_m) \qquad (6.6\text{–}9)$$

Substituting this into the constrained equation above, we have

$$\underline{V}'(\Omega_m)\mathbf{R}_{yy}^{-1}\underline{V}(\Omega_m)\frac{\lambda}{2} = 1$$

or

$$\lambda = \frac{2}{\underline{V}'(\Omega_m)\mathbf{R}_{yy}^{-1}\underline{V}(\Omega_m)} \qquad (6.6\text{–}10)$$

and therefore the filter weights are

$$\underline{h}_{\text{MVDR}} = \frac{\mathbf{R}_{yy}^{-1}\underline{V}(\Omega_m)}{\underline{V}'(\Omega_m)\mathbf{R}_{yy}^{-1}\underline{V}(\Omega_m)} \qquad (6.6\text{–}11)$$

Substituting this expression into J gives the minimum output power (variance) as

$$\underline{h}'\mathbf{R}_{yy}\underline{h} = \frac{1}{\underline{V}'(\Omega_m)\mathbf{R}_{yy}^{-1}\underline{V}(\Omega_m)} \qquad (6.6\text{–}12)$$

The expression yields the required results; the MVDR *spectral estimator* at Ω_m is given by

$$S_{\text{MVDR}}(\Omega_m) = \frac{\Delta T}{\underline{V}'(\Omega_m)\mathbf{R}_{yy}^{-1}\underline{V}(\Omega_m)} = \Delta T\underline{h}'\mathbf{R}_{yy}\underline{h} \qquad (6.6\text{–}13)$$

which implies that the spectral estimate at a given frequency is obtained by scaling the output covariance appropriately at the desired frequency.

In practice, the MVDR method exhibits more resolution than the classical correlation/periodogram estimators, but less than the MEM estimator. We note from Eq. (6.6–13) that the peak of this spectral estimate for a narrowband process is proportional to the power, while for MEM (see Appendix B) it is proportional to the square of the power. In fact, it can be shown [7] that the MVDR and MEM spectra are related by

$$\frac{1}{S_{\mathrm{MVDR}}(\Omega)} = \frac{1}{N_a} \sum_{i=1}^{N_a} \frac{1}{S_{\mathrm{MEM}}(\Omega, i)} \qquad (6.6\text{–}14)$$

where $S_{\mathrm{MEM}}(\Omega, i)$ is the MEM PSD corresponding to the ith order autoregressive model. Thus, we see that the MVDR spectral estimate averages all of the lower-order (less resolution) and higher-order AR models to obtain its estimate, thus explaining its lower resolution. In fact, a suggested method of implementing this technique is to use an AR spectral estimator and perform the averaging suggested in Eq. (6.6–14). We summarize this method in Table 6.6–1. The following example demonstrates the capability of the MVDR spectral estimator.

Example 6.6–1. Consider the sinusoidal bandpass simulation and apply the MVDR spectral estimator, using Eq. (6.6–14). Using SIG [6] the performance of the MVDR spectral estimator is shown in Fig. 6.6–1. The overall spectra indicate much smoother (averaged) spectra compared to the AR estimates, with increased resolution compared to the classical methods. In (b) we see the results of averaging model orders up to 100 and 12. The 100th-order average is able to resolve the lines at 35 and 90 Hz but tends to "average" the lines at 50 and 56 Hz. The same result with less resolution occurs for the 50th-order average MVDR spectrum. Decreasing the number of model orders to 25, 12 and 16 (automatic) enhances the bandpass response, as expected, while maintaining the line resolution at 35 Hz.

TABLE 6.6–1
MVDR spectral estimation

Covariance estimation

$$\hat{R}_{yy}(k) = \frac{1}{N} \sum_{t=0}^{N-1} y(t)y(t + k)$$

Parameter estimation

$$\underline{h}_{\mathrm{MVDR}} = \frac{\mathbf{R}_{yy}^{-1} \underline{V}(\Omega_m)}{\underline{V}'(\Omega_m)\mathbf{R}_{yy}^{-1} \underline{V}(\Omega_m)}$$

Spectrum estimation

$$S_{\mathrm{MVDR}}(\Omega_m) = \Delta T(\underline{h}'\mathbf{R}_{yy}\underline{h}) = \frac{\Delta T}{\underline{V}'(\Omega_m)\mathbf{R}_{yy}^{-1}\underline{V}(\Omega_m)} \qquad \text{for } m = 0, 1, \ldots$$

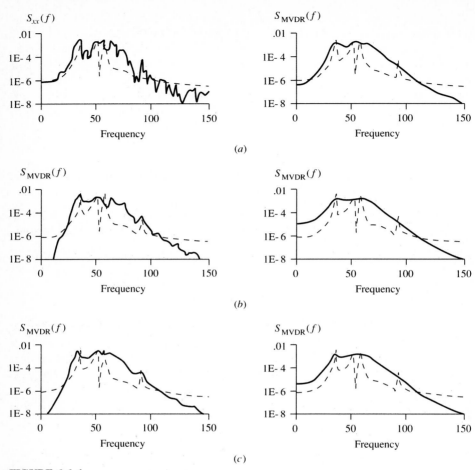

FIGURE 6.6–1
MVDR spectral estimation of bandpass sinusoidal simulation (*a*) Raw/true (dashed) spectra and 25th order average MVDR spectral estimates (*b*) 100th and 12th order MVDR spectral estimates (*c*) 50th and 16th order MVDR spectral estimates

6.7 HARMONIC METHODS OF SPECTRAL ESTIMATION

In this section we discuss spectral estimation techniques, which can be used to extract harmonics or sinusoidal (narrowband) signals from noisy data. The signal can be modeled as

$$y(t) = \sum_{i=1}^{N_s} A_i \sin(\Omega_i t + \phi_i) + n(t) \tag{6.7–1}$$

where n is zero-mean, white, with variance R_{nn}, and $\phi \sim U(-\pi, \pi)$. The covariance of this signal can be calculated analytically (see Example 4.3–1) as

$$R_{yy}(k) = \sum_{i=1}^{N_s} \frac{A_i^2}{2} \cos k\Delta T\Omega_i + R_{nn}\delta(k) \qquad (6.7\text{-}2)$$

or
$$R_{yy}(k) = \begin{cases} R_{nn} + \sum_{i=1}^{N_s} P_i & k = 0 \\ \sum_{i=1}^{N_s} P_i \cos k\Delta T\Omega_i & k \neq 0 \end{cases} \qquad (6.7\text{-}3)$$

where $P_i = A_i^2/2$ and ΔT is the sampling interval.

The corresponding PSD is given by

$$S_{\text{PHD}}(\Omega) := S_{yy}(\Omega) = \Delta T \sum_{i=1}^{N_s} 2\pi P_i \left[\delta(\Omega + \Omega_i) + \delta(\Omega - \Omega_i)\right] + R_{nn} \qquad (6.7\text{-}4)$$

The spectrum consists of N_s spectral lines located at $\Omega_i = 2\pi f_i \Delta T$, with power P_i on a base level of R_{nn}, and, therefore, we must determine $(\{P_i\}, \{\Omega_i\})$ to characterize the sinusoids.

We know from linear systems theory [16] that an individual sinusoid can be characterized by a second-order system or difference equation†

$$x(t) = -\alpha_1 x(t-1) - \alpha_2 x(t-2)$$

where $\alpha_1 = 2\cos\Omega_1$ and $\alpha_2 = 1$. In general, then, N_s sinusoids can be represented by the $2N_s$-order difference equation

$$x(t) = -\sum_{i=1}^{2N_s} \alpha_i x(t-i) \qquad (6.7\text{-}5)$$

Equivalently, we can use the model of Eq. (6.7-1):

$$y(t) = x(t) + n(t) = -\sum_{i=1}^{2N_s} \alpha_i x(t-i) + n(t) \qquad (6.7\text{-}6)$$

If we substitute for $x(t-i)$ above, we have

$$y(t) = -\sum_{i=1}^{2N_s} \alpha_i (y(t-i) - n(t-i)) + n(t)$$

or
$$y(t) = -\sum_{i=1}^{2N_s} \alpha_i y(t-i) + \sum_{i=0}^{2N_s} \alpha_i n(t-i) \qquad (6.7\text{-}7)$$

for $\alpha_0 = 1$, showing that this is an N_sth order ARMA model. If we write this expression in vector notation, we obtain

---

†This evolves from the fact that the Z-transform of a sinusoid is $z^{-1}\sin\Omega_0/(1 + 2\cos\Omega_0 z^{-1} + z^{-2})$ which gives the homogeneous solution.

$$[y(t)...y(t-2N_s)]\begin{bmatrix} \alpha_0 \\ \vdots \\ \alpha_{2N_s} \end{bmatrix} = [n(t)...n(t-2N_s)]\begin{bmatrix} \alpha_0 \\ \vdots \\ \alpha_{2N_s} \end{bmatrix}$$

or
$$\underline{Y}'\underline{\alpha} = \underline{N}'\underline{\alpha} \tag{6.7-8}$$

which also implies that Eq. (6.7-6) can be written as

$$\underline{Y} = \underline{X} + \underline{N} \tag{6.7-9}$$

If we multiply Eq. (6.7-8) by $\underline{Y}$, substitute Eq. (6.7-9), and take expected values, we have

$$E\{\underline{Y}\,\underline{Y}'\}\underline{\alpha} = E\{\underline{Y}\,\underline{N}'\}\underline{\alpha} = E\{\underline{X}\,\underline{N}'\}\underline{\alpha} + E\{\underline{N}\,\underline{N}'\}\underline{\alpha}$$

$$\mathbf{R}_{yy}\underline{\alpha} = R_{nn}\mathbf{I}\underline{\alpha} \tag{6.7-10}$$

or
$$(\mathbf{R}_{yy} - \mathbf{R}_{nn})\underline{\alpha} = (\mathbf{R}_{yy} - R_{nn}\mathbf{I})\underline{\alpha} = \underline{0}$$

since n is white, $\mathbf{R}_{nn} = R_{nn}\mathbf{I}$, and $\mathbf{R}_{yy}$ is Toeplitz. This relation is an eigenequation, where R_{nn} is the eigenvalue of the covariance matrix $\mathbf{R}_{yy}$.[†] Therefore, ARMA parameter vector $\underline{\alpha}$ is the eigenvector associated with eigenvalue R_{nn}, scaled with $\alpha_0 = 1$. So we see that for known lags, the parameters $\{\alpha_i\}$, $i = 1, \ldots, 2N_s$ can be determined from Eq. (6.7-10). It has been shown that for this process in additive white noise, R_{nn} corresponds to the *minimum* eigenvalue of $\mathbf{R}_{yy}$. This result forms the basis of the Pisarenko harmonic decomposition (PHD) procedure [17].

Once the ARMA parameters are found, it is well known [18] that they can be used to form the system characteristic equation whose roots are the required sinusoidal frequencies; that is

$$z^{2N_s} + \alpha_1 z^{2N_s-1} + \ldots + \alpha_{2N_s} = \prod_{i=1}^{N_s}(z - z_i)(z - z_i^*) = 0 \tag{6.7-11}$$

where $z_i = e^{j\Omega_i \Delta T}$.

After determining the roots of the characteristic equation, the sinusoidal frequencies are found from

$$\Omega_i = \frac{1}{\Delta T} \arctan\left(\frac{\text{Im } z_i}{\text{Re } z_i}\right) \tag{6.7-12}$$

The associated power can be determined directly by expanding the covariance relation of Eq. (6.7-3),

---

[†]Recall that an *eigenvalue* λ of a matrix $\mathbf{A}$ is defined as a root of det $(\mathbf{A} - \lambda\mathbf{I}) = 0$ with corresponding *eigenequation* $(\mathbf{A} - \lambda\mathbf{I})\underline{V} = 0$ where $\underline{V}$ is the corresponding *eigenvector*.

$$
\begin{bmatrix} \cos \Omega_1 \Delta T & \cdots & \cos \Omega_{N_s} \Delta T \\ \vdots & & \vdots \\ \cos N_s \Omega_1 \Delta T & \cdots & \cos N_s \Omega_{N_s} \Delta T \end{bmatrix} \begin{bmatrix} P_1 \\ \vdots \\ P_{N_s} \end{bmatrix} = \begin{bmatrix} R_{yy}(1) \\ \vdots \\ R_{yy}(N_s) \end{bmatrix} \qquad (6.7\text{--}13)
$$

or
$$
\mathbf{C}\underline{P} = \underline{R}_{yy}(N_s)
$$

which is solved by

$$
\hat{\underline{P}} = \mathbf{C}^{-1}\underline{R}_{yy}(N_s) \qquad (6.7\text{--}14)
$$

So we see that the PHD approach can be used to extract the parameters $(\{P_i\}, \{\Omega_i\})$ from data contaminated with noise; however, we must have precise knowledge of the number of sinusoids N_s and the covariance $R_{yy}(k)$. In practice this is not possible; therefore, other techniques have evolved. The number of sinusoids can be estimated by iterating the eigenequation and examining the minimum eigenvalue. When it does not change significantly, then the correct order has been established. Since the covariance matrix is Toeplitz, the classical power method of linear algebra (see [19] for an example) can be used to calculate the eigenvector corresponding to the minimum eigenvalue. Here the sequence of vectors

$$
\mathbf{R}_{yy}\underline{\alpha}(i) = \underline{\alpha}(i-1) \qquad \text{for } i = 0, 1 \ldots \qquad (6.7\text{--}15)
$$

converges in the limit to the eigenvector of the minimum eigenvalue for some initial guess $\underline{\alpha}'(0)$. In practice, this technique usually converges in a few iterations and the minimum eigenvalue, say $\lambda_{\min}$, is

$$
\lambda_{\min} = R_{nn} = \frac{\underline{\alpha}'\mathbf{R}_{yy}\underline{\alpha}}{\underline{\alpha}'\underline{\alpha}} \qquad (6.7\text{--}16)
$$

which is called the *Rayleigh quotient* of the eigenvector α [19]. We summarize the Pisarenko spectral estimation technique in Table 6.7–1.

Next we consider an alternate method of retrieving harmonics. This method is based on the Prony technique of extracting damped exponentials from noisy data. It offers an alternative to the Pisarenko technique and does not require the estimation of the covariance function—one of the shortcomings of the Pisarenko method. The *Prony spectral estimation* technique is based on estimating the parameters of a damped exponential representation given by

$$
\hat{y}(t) = \sum_{i=1}^{N_e} A_i e^{p_i t} \qquad (6.7\text{--}17)
$$

where A is complex and p is a pole given by

$$
p_i = d_i + j\Omega_i
$$

for d_i the damping factor and Ω_i the angular frequency. It is well-known from linear systems theory that Eq. (6.7–17) is the solution of a homogeneous linear difference equation:

TABLE 6.7–1
Pisarenko (PHD) spectral estimation

Minimum eigenvalue estimation

For $i = 1, \ldots$

$$\mathbf{R}_{yy}\underline{\alpha}(i) = \underline{\alpha}(i - 1)$$

$$\lambda_{\min}(i) = \frac{\underline{\alpha}'(i)\mathbf{R}_{yy}\underline{\alpha}(i)}{\underline{\alpha}'(i)\underline{\alpha}(i)}$$

$$\hat{R}_{nn} = \lambda_{\min}(i) \approx \lambda_{\min}(i - 1)$$

Characteristic root estimation

Set $\alpha(z) = 0$ for $\{z_i\}$, $i = 1, \ldots, 2N_s$

Frequency estimation

For $j = 1, \ldots, N_s$

$$\hat{\Omega}_j = \frac{1}{\Delta T} \arctan\left(\frac{\operatorname{Im} z_j}{\operatorname{Re} z_j}\right)$$

Power estimation

$$\underline{\hat{P}} = \mathbf{C}^{-1}\underline{R}_{yy}(N_s)$$

Spectral estimation

$$S_{\mathrm{PHD}}(\Omega) = \sum_{i=1}^{N_s} 2\pi\hat{P}_i\left[\delta\left(\Omega + \hat{\Omega}_i\right) + \delta(\Omega - \hat{\Omega}_i)\right] + \hat{R}_{nn}$$

for $\underline{\hat{P}}' = [P_1 \ldots P_{N_s}]$

$$\sum_{i=0}^{N_e} a_i\hat{y}(t - i) = 0 \tag{6.7–18}$$

Or, taking Z-transforms, we obtain the characteristic equation of the system relating coefficients of the difference equation to the roots of the characteristic equation or the complex exponential. That is,

$$A(z) = \sum_{i=0}^{N_e} a_i z^{-i} = \prod_{i=1}^{N_e}(z - p_i) \qquad a_0 = 1$$

or $A(z)$ has $\{p_i\}$ as its roots and coefficients $\{a_i\}$ when multiplied out. Based on these representations, the extended Prony technique can be developed.

We first define the prediction error as

$$e(t) = y(t) - \hat{y}(t) \qquad \text{for } 0 \leq t \leq N - 1$$

or, solving for $\hat{y}(t)$ in Eq. (6.7–18) and substituting, we obtain

$$e(t) = y(t) + \sum_{i=1}^{N_e} a_i\hat{y}(t - i) = y(t) + \sum_{i=1}^{N_e} a_i\Big(y(t - i) - e(t - i)\Big)$$

Solving for $y(t)$, we obtain the equivalent ARMA model:

$$y(t) = -\sum_{i=1}^{N_e} a_i y(t-i) + \sum_{i=0}^{N_e} a_i e(t-i) \qquad \text{for } N_e \le t \le N-1 \qquad (6.7\text{--}19)$$

or

$$A(q^{-1})y(t) = A(q^{-1})e(t)$$

If we attempt to construct an estimator for this model, it will be nonlinear, as shown in Sec. 6.4 for ARMA models; therefore, we develop a suboptimum approach called the *extended Prony technique*, where we define

$$\epsilon(t) := \sum_{i=0}^{N_e} a_i e(t-i) \qquad \text{for } t = N_e, \ldots, N-1 \qquad (6.7\text{--}20)$$

leading to an AR model,

$$A(q^{-1})y(t) = \epsilon(t) \qquad (6.7\text{--}21)$$

Minimizing the sum-squared error criterion

$$\min_{\underline{a}} J(t) = \sum_{t=N_e}^{N-1} \epsilon^2(t) \qquad (6.7\text{--}22)$$

leads to the orthogonality condition,

$$\frac{\partial J(t)}{\partial a_k} = 2 \sum_{t=N_e}^{N-1} \epsilon(t) y(t-k) = 2 \sum_{t=N_e}^{N-1} \left(\sum_{i=0}^{N_e} a_i y(t-i) \right) y(t-k) = 0 \qquad \text{for } k = 1, \ldots, N_e$$

$$(6.7\text{--}23)$$

where we have used the fact that

$$\frac{\partial \epsilon(t)}{\partial a_k} = \frac{\partial}{\partial a_k} \left(\sum_{i=0}^{N_e} a_i y(t-i) \right) = y(t-k)$$

Expanding these equations over the indices, we obtain

$$(\mathbf{Y}'\mathbf{Y})\underline{a} = \begin{bmatrix} J_{\min} \\ \vdots \\ 0 \end{bmatrix}$$

where

$$\mathbf{Y} = \begin{bmatrix} y(N_e) & \cdots & y(0) \\ \vdots & & \vdots \\ y(N-1) & \cdots & y(N-N_e-1) \end{bmatrix}, \qquad J_{\min} = \sum_{i=0}^{N_e} \hat{a}_i \left(\sum_{t=N_e}^{N-1} y(t-i)y(t) \right)$$

These equations can be solved using the unwindowed or covariance method of Sec. 6.2 to obtain the estimate of the AR parameters:

$$\underline{\hat{a}} = (\mathbf{Y}'\mathbf{Y})^{-1} \begin{bmatrix} J_{\min} \\ 0 \\ \vdots \\ 0 \end{bmatrix} \qquad (6.7\text{-}24)$$

Once the values for $\{\hat{a}_i\}$ are estimated using the AR techniques, the complex exponentials $\{p_i\}$ are determined from the roots of the characteristic equation $A(z)$. The set of complex gains $\{A_i\}$ can be determined by expanding Eq. (6.7–17), to obtain

$$\mathbf{P}\underline{A} = \hat{\underline{Y}} \qquad (6.7\text{-}25)$$

where $\mathbf{P} \in R^{N \times N_e}$ is a *Vandermonde* matrix of exponentials, given by

$$\mathbf{P} = \begin{bmatrix} 1 & \cdots & 1 \\ e^{p_1 \Delta T} & & e^{p_{N_e} \Delta T} \\ \vdots & & \vdots \\ e^{p_1 (N-1)\Delta T} & \cdots & e^{p_{N_e}(N-1)\Delta T} \end{bmatrix}$$

and

$$\underline{A}' = [A_1 \cdots A_{N_e}]$$

$$\underline{\hat{Y}}' = [\hat{y}(0) \cdots \hat{y}(N_e - 1)]$$

which can be solved using least-squares techniques to give

$$\underline{\hat{A}} = (\mathbf{P}'\mathbf{P})^{-1}\mathbf{P}'\underline{\hat{Y}} \qquad (6.7\text{-}26)$$

Once these parameters are determined, the PSD is estimated as

$$S_{\text{PRONY}}(\Omega) = |\hat{Y}(\Omega)|^2 \qquad (6.7\text{-}27)$$

where

$$\hat{Y}(\Omega) = \sum_{i=1}^{N_e} \frac{2A_i d_i}{d_i^2 + (\Omega - \Omega_i)^2}$$

This estimate yields the Prony spectrum. Note that the peaks are linearly proportional to the energy. This estimate is capable of producing both narrow or wideband spectral shapes, a function which is controlled by the damping factors, $\{d_i\}$. We summarize the Prony spectral estimation technique in Table 6.7–2.

The Prony technique can be restricted to sinusoids, yielding the Prony harmonic decomposition; this follows directly from Eq. (6.7–17) with $\{d_i\}$ set to zero:

$$\hat{y}(t) = \sum_{i=1}^{N_s} A_i e^{j\Omega_i t} + A_i^* e^{-j\Omega_i t} = \sum_{i=1}^{N_s} A_i \cos(\Omega_i t + \phi_i)$$

The roots of the characteristic equation in this case are given by

$$A(z) = \prod_{i=1}^{N_s} (z - z_i)(z - z_i^*) = \sum_{i=0}^{2N_s} a_i z^{2N_s - i} = 0$$

TABLE 6.7–2
Prony spectral estimation

Parameter estimation

$$\hat{\underline{a}} = (\mathbf{Y'Y})^{-1} \begin{bmatrix} J_{\min} \\ 0 \\ \vdots \\ 0 \end{bmatrix}$$

Root estimation

$$A(z) = \prod_{i=1}^{N_e} (z - \hat{p}_i) = \sum_{i=1}^{N_e} \hat{a}_i z^{-i}$$

Residue estimation

$$\hat{\underline{A}} = (\mathbf{P'P})^{-1} \mathbf{P'} \hat{\underline{Y}}$$

Spectral estimation

$$S_{\mathrm{PRONY}}(\Omega) = \left| \sum_{i=1}^{N_e} \left[\frac{2\hat{A}_i \hat{d}_i}{\hat{d}_i^2 + (\Omega - \hat{\Omega}_i)^2} \right] \right|^2$$

The solution then follows as before by solving a set of linear vector-matrix equations, except in this case the data matrix $\mathbf{Y}$ is both Toeplitz and Hankel (see [7,20] for details).

6.8   SUMMARY

In this chapter we have discussed *modern* methods of random signal analysis. We showed that most of the modern techniques of spectral estimation fall into the class of parametric methods. We developed the AR (all-pole), MA (all-zero), ARMA (pole-zero), and lattice-model approaches and applied them to sinusoids in our bandpass noise example for comparative purposes. We then developed the minimum-variance, distortionless response (MVDR) or so-called maximum-likelihood method (MLM). Finally, the popular harmonic methods of spectral estimation, namely the Pisarenko harmonic decomposition (PHD) and Prony techniques, were discussed.

REFERENCES

1. D. Childers, Ed., *Modern Spectral Analysis* (New York: IEEE Press, 1978).
2. S. Kesler, Ed., *Modern Spectral Analysis II* (New York: IEEE Press, 1986).
3. J. Burg, "Maximum Entropy Spectral Analysis," Ph.D. Dissertation, Stanford Univ., 1975.
4. L. Ljung and T. Soderstrom, *Theory and Practice of Recursive Identification* (Boston: MIT Press, 1983).
5. J. Makhoul, "Spectral Analysis of Speech by Linear Prediction," *IEEE Trans. Audio. Electroacoust.*, AU-21, 1973.

6. D. Lager and S. Azevedo, "SIG—A General Purpose Signal Processing Code," *Proc. IEEE*, 1987.

7. S. Kay and S. Marple, "Spectrum Analysis—A Modern Perspective," *Proc. IEEE*, Vol. 69, 1981.

8. H. Akaike, "Fitting Autoregressive Models for Prediction," *Ann. Inst. Statist. Math.*, Vol. 21, 1969.

9. C. Shannon, *The Mathematical Theory of Communications* (Urbana, Ill.: Univ. of Illinois Press, 1948).

10. J. Cadzow, "High Performance Spectral Estimator—A New ARMA Method," *IEEE Trans. Acoust., Speech, Signal Processing*, ASSP-28, 1980.

11. B. Friedlander, "Lattice Methods for Spectral Estimation," *Proc. IEEE*, Vol. 70, 1982.

12. E. Robinson and S. Treitel, *Geophysical Signal Analysis* (Englewood Cliffs, N. J.: Prentice-Hall, 1980).

13. S. Haykin, Ed., *Array Signal Processing* (Englewood Cliffs, N.J.: Prentice-Hall, 1984).

14. J. Makhoul, "Stable and Efficient Lattice Methods for Linear Prediction," *IEEE Trans. Acoustics, Speech, and Signal Processing*, Vol. ASSP-25, 1977.

15. J. Capon, "High-Resolution Frequency-Wavenumber Spectrum Analysis," *Proc. IEEE*, Vol. 57, 1969.

16. T. Kailath, *Linear Systems* (Englewood Cliffs, N. J.: Prentice-Hall, 1980).

17. V. Pisarenko, "The Retrieval of Harmonics from a Covariance Function," *Geoph. J. R. Astrom. Soc.*, Vol. 30, 1973.

18. A. Oppenheim and R. Shafer, *Digital Signal Processing* (Englewood Cliffs, N. J.: Prentice-Hall, 1975).

19. G. W. Stewart, *Introduction to Matrix Computations* (New York: Academic Press, 1973).

20. S. Marple, *Digital Spectral Analysis with Applications* (Englewood Cliffs, N. J.: Prentice-Hall, 1987).

SIG NOTES

The power spectral density estimators are also available (SD***) for random signal analysis in SIG. The classical nonparametric methods are obtained using the SDCORRELATION and SDWELCH commands for the respective correlation and periodogram (Welch's) methods. Many of the modern parametric spectral estimation methods are also available. SDAONLY implements the autocorrelation method of Sec. 6.1 using the Levinson-Durbin recursion, while SDLATTICE implements *three* of the covariance lattice [14] spectral estimators. Also available are the SDBURG, SDARMA, and SDMVDR, which implement the original Burg [3] lattice spectral estimator; the ARMA estimator uses ELS and the Capon [15] maximum-likelihood or minimum-variance distortionless response (MVDR) estimator.

EXERCISES

6.1. Simulate a $N = 2048$ point test signal sampled at $\Delta T = 0.001$ sec by passing white noise of variance 0.01 through a fourth-order Butterworth bandpass filter, with cutoff frequencies of 30 and 70 Hz respectively. Now add sinusoids at frequencies of 35, 55, 70, and 90 Hz, with respective amplitudes of 0.025, 0.015, and 0.01, and design a spectral estimator using

(a) Correlation method

(b) Periodogram method

(c) AR

(d) Lattice

(e) ARMA

Show plots of the resulting estimates.

6.2. Repeat the previous problem on a data record of $N=128$ points, sampled at $\Delta T=0.01$ sec, with unit amplitude sinusoids at 5, 7, 10, 20 through a fourth-order bandpass Butterworth filter, with cutoff frequencies at 2 and 35 Hz, with gaussian noise, $N(0,8)$.

6.3. Suppose you are asked to analyze the spectral content of a proposed system (shown below) and develop a spectral estimator to extract sinusoidal signals.

The low-pass filter which is part of the data acquisition system has a cutoff frequency of 40 Hz, while the DC/trend filter is high-pass with a cutoff of 80 Hz. It is expected that the sinusoids of interest are at 20, 60, and 100 Hz. You have limited data storage arrays ($N = 1024$) and the noise is assumed gaussian, $N(0,0.01)$. Simulate this system and design some spectral estimators to extract the sinusoids. Which estimator did you choose?

6.4. Design a spectral estimator to extract a 30 Hz sinusoidal signal contaminated with bandpass filtered noise $\sim N(0,0.1)$ for $N = 1024$ points (cutoffs: 15 and 25 Hz).

CHAPTER
7

PARAMETRIC
SIGNAL
PROCESSING

In this chapter, we develop the parametric approach to signal processing. After introducing basic concepts, we develop the essential results from estimation theory. Once this foundation is established, we then develop each of the popular parametric methods, starting with the Levinson all-pole and all-zero filters, proceeding to the lattice filters, and then investigating the popular prediction-error filters. Next, we discuss the concepts involved in estimating the order of the underlying parametric model. Finally, we discuss a case study applying the parametric approach to electromagnetic signal processing.

7.1 INTRODUCTION

Parametric methods of signal processing have evolved as the modern approach to processing both deterministic as well as random signals. Probably the catalyst that has initiated this now popular approach evolved from the application of signal-processing techniques to the analysis and synthesis of speech [1,2]. In fact, these techniques are used heavily in the modern parametric approach to spectral estimation [3]. Parametric methods have existed in other disciplines for quite some time, initially evolving from time-series analysis [4] and the control/estimation area, most notably after the development of the Kalman filter [5] (see Chapter 9 and [6]). More recently, the work of Ljung [7], Goodwin [8], and others has provided an analytic and practical framework for designing and evaluating the performance of parametric processors.

Parametric methods are more appropriately called model-based methods (see [9]), because each technique first assumes a pre-specified model set (e.g., all-

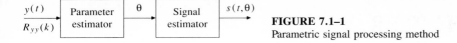

FIGURE 7.1–1
Parametric signal processing method

pole, all-zero, etc.) and then estimates the appropriate model parameters. In fact, the model-based approach consists of three essential ingredients: (1) data, (2) model set, and (3) criterion. Once we select the model set, we "fit" the model (parameter estimation) to the data according to some criterion. Parametric methods correspond to a model-based signal-processing scheme when the model is unknown.

The parametric approach to signal processing is best summarized in Fig. 7.1–1. Here we see that once the model set is selected, the estimator is used to obtain the unknown parameters that specify the model. The signal estimate is then constructed using these parameters.

The *parametric signal processing* method can be summarized in the following steps:

1. Select a representative model set (e.g., ARMAX, lattice, state-space).
2. Estimate the model parameters from the data; that is, given $\{y(t)\}$ and/or $\{u(t)\}$, find

$$\hat{\Theta} = \{\hat{A}(q^{-1}), \hat{B}(q^{-1}), \hat{C}(q^{-1}), \hat{R}_{ee}\}$$

3. Construct the signal estimate using these parameters, that is,

$$\hat{s}(t, \Theta) = f(\hat{\Theta}, y(t))$$

7.2 OPTIMAL ESTIMATION

In this section, we discuss the concept of optimal estimation—optimal, that is, in the sense that the estimator results from the optimization of some meaningful criterion function. Typically, the criterion function of most interest in signal-processing literature is that of the mean-squared error defined by

$$J(t) = E\{e^2(t)\} \tag{7.2-1}$$

where the error is derived directly from the measurement and estimate as

$$e(t) = y(t) - \hat{y}(t) \tag{7.2-2}$$

The optimal estimate is that which produces the smallest error. Geometrically, we can think of this error sequence, $\{e(t)\}$, as the result of projecting the current data onto a space of past data. Intuitively, the minimum error occurs when the projection is perpendicular or orthogonal to the space,

$$e(t) \perp Y(t - 1) \tag{7.2-3}$$

where $Y(t - 1) = \{y(t - 1), \ldots, y(0)\}$ is the past-data space. We depict this relationship in Fig. 7.2–1. We can think of the error or, more commonly, the

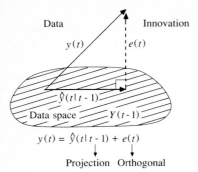

$$y(t) = \hat{y}(t \mid t - 1) + e(t)$$

Projection   Orthogonal

FIGURE 7.2–1
Orthogonal projection of the optimal estimate

innovation (new information [10]) as the orthogonal projection of the data vector $y(t)$ onto the past-data space $Y(t - 1)$, spanned by

$$\hat{y}(t \mid t - 1) = E\{y(t) \mid Y(t - 1)\} \tag{7.2–4}$$

Another approach to interpreting this result is to think of $\hat{y}(t \mid t - 1)$ as an estimate of that part of $y(t)$ correlated with past data $Y(t - 1)$ and to think of $e(t)$ as the uncorrelated or orthogonal part:

$$y(t) \quad = \quad e(t) \quad + \quad \hat{y}(t \mid t - 1) \tag{7.2–5}$$

Data   Uncorrelated   Correlated

We know from Sec. 5.2 that the optimal estimator satisfies the orthogonality condition, which is equivalent to Eq. (7.2–3):

$$E\{e(t)y(T)\} = 0 \qquad \text{for } T < t \tag{7.2–6}$$

In fact, we take the correlation-cancelling approach of Orfanidis [11] and consider the following example.

> **Example 7.2–1.** Suppose the data are composed of two parts: one correlated with the past and one not; that is,
>
> $$y(t) = y_1(t) + y_2(t)$$
>
> for $\qquad y_1(t) \in Y(t - 1) \qquad$ and $\qquad y_2(t) \perp Y(t - 1)$
>
> We are asked to analyze the performance of an estimator. The optimal estimator is an estimate only of the correlated part,
>
> $$\hat{y}(t \mid t - 1) = y_1(t)$$
>
> and therefore the innovation is
>
> $$e(t) = y(t) - \hat{y}(t \mid t - 1) = \big(y_1(t) + y_2(t)\big) - y_1(t) = y_2(t)$$
>
> the uncorrelated part of the data. Thus, we see that the results of cancelling or removing the correlated part of the measurement with the optimal estimate is the uncorrelated part or innovation.

This example emphasizes a very important point that is crucial to analyzing the performance of *any* optimal estimator. If the estimator is performing properly, and we have estimated all of the information in the currently available data, then the *innovation sequence must be zero-mean and white*. We will see when we design estimators, whether parametric, signal, spectral, or model-based, that checking the statistical properties of the corresponding innovation sequence is the first step in analyzing proper performance.

The innovation sequence can be constructed from a set or batch of data $Y(t)$, using the well-known Gram-Schmidt procedure (for examples see [12,13]), in which we describe this data space as a random vector space with proper inner product, given by

$$< a, b > := E\{ab'\} \qquad (7.2\text{--}7)$$

The construction problem becomes that of finding a nonsingular transformation L, such that

$$\underline{e} = L^{-1}\underline{Y} \qquad \text{for } \underline{e}, \underline{Y} \in R^N \qquad (7.2\text{--}8)$$

The *Gram-Schmidt* procedure recursively generates the innovation sequence as

$$e(1) = y(1)$$
$$e(2) = y(2) - \quad < y(2), e(1) >< e(1), e(1) >^{-1} e(1) = y(2) - \quad R_{ye}(2,1)R_{ee}^{-1}(1)e(1)$$
$$\vdots \qquad \vdots \qquad\qquad \vdots \qquad\qquad \vdots \qquad\qquad \vdots$$
$$e(N) = y(N) - \sum_{i=1}^{N-1} < y(N), e(i) >< e(i), e(i) >^{-1} e(i) = y(N) - \sum_{i=1}^{N-1} R_{ye}(N,i)R_{ee}^{-1}(i)e(i)$$

which implies that

$$y(t) = \sum_{i=1}^{t} L(t,i)e(i) \qquad \text{for } L(t,i) = R_{ye}(t,i)R_{ee}^{-1}(i) \qquad \text{with } L(t,t) = 1 \quad (7.2\text{--}9)$$

and, therefore, L is lower triangular:

$$L = \begin{bmatrix} 1 & \cdots & & & 0 \\ R_{ye}(2,1)R_{ee}^{-1}(1) & 1 & & & \\ \vdots & & \ddots & & \vdots \\ R_{ye}(N,1)R_{ee}^{-1}(1) & \cdots & R_{ye}(N,N-1)R_{ee}^{-1}(N-1) & 1 \end{bmatrix}$$

From Eq. (7.2–9) we see that the orthogonal decomposition of the data in Eq. (7.2–5) becomes

$$y(t) \quad = \quad e(t) \quad + \quad \sum_{i=1}^{t-1} L(t,i)e(i) \qquad (7.2\text{--}10)$$

$$\text{Data} \qquad\quad \text{Orthogonal} \qquad \text{Projection}$$

and therefore the optimal estimator is given by

$$\hat{y}(t \mid t-1) = \sum_{i=1}^{t-1} L(t,i)e(i) = \sum_{i=1}^{t-1} R_{ye}(t,i)R_{ee}^{-1}(i)e(i) \qquad (7.2\text{--}11)$$

We note that the Gram-Schmidt procedure corresponds to transforming a correlated measurement vector to an uncorrelated innovation vector through a linear transformation [14], and therefore diagonalizing the corresponding covariance matrix. That is,

$$R_{yy} = E\{\underline{Y}\,\underline{Y}'\} = E\{L\underline{e}\,\underline{e}'L'\} = LR_{ee}L'$$

or, equivalently,[†]

$$R_{ee} = L^{-1}R_{yy}L^{-T} \qquad (7.2\text{--}12)$$

where L satisfies Eq. (7.2–9) and, by construction,

$$R_{ee} = \text{diag } [R_{ee}(1) \cdots R_{ee}(N)] \qquad (7.2\text{--}13)$$

The following example emphasizes the geometric aspects of this transformation.

Example 7.2–2. For the case of $N = 3$, construct the transformation matrix L using the Gram-Schmidt procedure. Using Eq. (7.2–9), we know immediately that

$$\begin{bmatrix} y(1) \\ y(2) \\ y(3) \end{bmatrix} = \begin{bmatrix} 1 & 0 & 0 \\ L(2,1) & 1 & 0 \\ L(3,1) & L(3,2) & 1 \end{bmatrix} \begin{bmatrix} e(1) \\ e(2) \\ e(3) \end{bmatrix}$$

We depict the corresponding vectors in Fig. 7.2–2; note the projections given by Eq. (7.2–11).

[†]We use the notation $A^{-T} = [A^{-1}]'$ and diag [·] to mean diagonal matrix with entries [·].

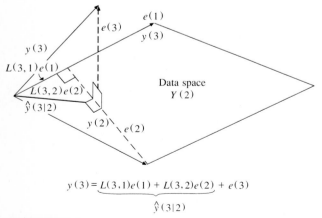

$$y(3) = \underbrace{L(3,1)e(1) + L(3,2)e(2)}_{\hat{y}(3\mid2)} + e(3)$$

FIGURE 7.2–2
Innovations construction using Gram-Schmidt orthogonalization

We also note that this result corresponds directly to the optimal Wiener solution of the "batch" linear estimation problem of Sec. 5.2, with the data vector replaced by

$$\underline{Y} = L\underline{e}$$

Using Eq. (7.2–12), we obtain the equivalent optimal estimate

$$\hat{\Theta}_{MV} = K_{MV}\underline{Y} = R_{\Theta y}R_{yy}^{-1}\underline{Y} = (R_{\Theta e}L')(L^{-T}R_{ee}L^{-1})(L\underline{e}) = R_{\Theta e}R_{ee}^{-1}\underline{e} \quad (7.2–14)$$

and we define

$$K_e := R_{\Theta e}R_{ee}^{-1} \quad (7.2–15)$$

with the corresponding quality

$$R_{\tilde{\Theta}\tilde{\Theta}} = R_{\Theta\Theta} - R_{\Theta e}R_{ee}^{-1}R_{e\Theta} \quad (7.2–16)$$

We can reconcile these results with the classical approach [16] by considering only the *scalar case*. Taking Z-transforms of Eq. (7.2–14), we obtain

$$K_{MV}(z) = \frac{S_{\Theta y}(z)}{S_{yy}(z)} \quad (7.2–17)$$

which is the frequency-domain Wiener solution. If we further restrict the optimal estimator to be causal, then we must extract the causal parts of this solution. Since $S_{yy}(z)$ is a power spectrum, it can be factored (see [16] for details) as

$$S_{yy}(z) = S_{yy}(z^+)S_{yy}(z^-)$$

where $S_{yy}(z^+)$ contains only poles and zeros that lie inside the unit circle, and $S_{yy}(z^-)$ contains only those lying outside. Thus, the causal Wiener filter is given by

$$K(z) = \left[S_{\Theta y}(z)S_{yy}^{-1}(z^-)\right]_{cp}S_{yy}^{-1}(z^+) \quad (7.2–18)$$

where the "cp" notation signifies the "causal part" of $[\cdot]$. The same results can be obtained using the innovations or whitening filter approach, yielding

$$K_e(z) = \left[S_{\Theta e}(z)S_{ee}^{-1}(z^-)\right]_{cp}S_{ee}^{-1}(z^+) = \left[S_{\Theta e}(z)R_{ee}^{-1/2}\right]_{cp}R_{ee}^{-1/2} \quad (7.2–19)$$

since $S_{ee}(z) = R_{ee}$. Consider the following example of frequency-domain Wiener filter design.

Example 7.2–3. Suppose we are asked to design a Wiener filter to estimate a zero-mean, random parameter Θ, with covariance $R_{\Theta\Theta}(k) = \alpha^{|k|}$ in a random noise unit variance; that is,

$$y(t) = \Theta(t) + \nu(t)$$

Then, we have

$$S_{yy}(z) = S_{\Theta\Theta}(z) + S_{\nu\nu}(z) \quad \text{and} \quad S_{\Theta y}(z) = S_{\Theta\Theta}(z)$$

From the sum decomposition, we have

$$S_{yy}(z) = \frac{2 - \alpha(z + z^{-1})}{(1 - \alpha z^{-1})(1 - \alpha z)}$$

Thus, the *noncausal* Wiener solution is simply

$$K(z) = \frac{S_{\Theta y}(z)}{S_{yy}(z)} = \frac{1 - \alpha^2}{2 - \alpha(z + z^{-1})}$$

while the causal solution is found by factoring $S_{yy}(z)$ as

$$S_{yy}(z) = S_{yy}(z^+)S_{yy}(z^-) = \frac{2 - \alpha(z + z^{-1})}{(1 - \alpha z^{-1})(1 - \alpha z)} = \frac{(\beta_0 - \beta_1 z^{-1})(\beta_0 - \beta_1 z)}{(1 - \alpha z^{-1})(1 - \alpha z)}$$

which gives
$$\beta_0 = \frac{\alpha}{\beta_1} \quad \text{and} \quad \beta_1^2 = 1 - \sqrt{1 - \alpha^2}$$

and therefore
$$S_{yy}(z) = \left(\frac{\alpha}{\gamma}\right)\frac{(1 - \gamma z^{-1})(1 - \gamma z)}{(1 - \alpha z^{-1})(1 - \alpha z)}$$

where
$$\gamma = \frac{1 - \sqrt{1 - \alpha^2}}{\alpha}$$

The causal Wiener filter is then found by using the residue theorem to find the causal part:

$$K(z) = \left[\frac{\sqrt{\gamma/\alpha}(1 - \alpha^2)}{(1 - \alpha z^{-1})(1 - \gamma z)}\right]_{cp}\left[\frac{(1 - \alpha z^{-1})\sqrt{\gamma/\alpha}}{(1 - \gamma z^{-1})}\right]$$

Similar results can be obtained for the innovations approach by performing the spectral factorization first.

7.3   LEVINSON (ALL-POLE) FILTERS

In this section, we develop the Levinson all-pole filters leading to the autoregressive model of Sec. 6.2. In fact, the Levinson filter is simply a recursive procedure to invert the Toeplitz matrix of Eq. (6.2–14) and obtain the set of parameters ($\{a_i\}$, R_{ee}). Levinson filters have classically been called "linear predictors" or "prediction filters" [15]. In any case, we motivate these filters as a means to produce a signal estimator and use the Levinson-Durbin recursion of inverting the Toeplitz matrix to obtain the filter parameters.

Suppose we would like to obtain an estimate of a signal $s(t)$ based on past data; that is,

$$\hat{s}(t \mid t - 1) = \hat{s}(t \mid Y_{\text{past}})$$

If we take the parametric approach, we model the signal by the linear representation

$$s(t) = -\sum_{i=1}^{N} a_i y(t - i) \qquad (7.3–1)$$

and therefore the measurement

$$y(t) = s(t) + e(t) = -\sum_{i=1}^{N} a_i y(t-1) + e(t) \qquad (7.3\text{--}2)$$

is characterized by an AR model

$$A(q^{-1})y(t) = e(t)$$

for $e(t)$ zero-mean, white, with variance R_{ee} and $a_0 = 1$.

The optimal signal-estimation problem in terms of this model becomes

$$\min_{\underline{a}} J(t) = E\{e^2(t)\} \qquad \text{for } e(t) = y(t) - \hat{s}(t) \qquad (7.3\text{--}3)$$

Following Sec. 6.2, we see that differentiating the cost function with respect to the parameters leads to the orthogonality condition

$$\frac{\partial J}{\partial a_j} = 2E\left\{ e(t) \frac{\partial e(t)}{\partial a_j} \right\} = 2E\{e(t)y(t-j)\} = 0 \qquad \text{for } j = 1, \ldots, N \quad (7.3\text{--}4)$$

Solving, we obtain the *normal equations*

$$\sum_{i=0}^{N} a_i R_{yy}(i-j) = 0 \qquad \text{for } j = 1, \ldots, N, \text{ and } a_0 = 1 \qquad (7.3\text{--}5)$$

The variance for the innovation or prediction error is obtained as before (see Eq. (6.2–17)):

$$\sum_{i=0}^{N} a_i R_{yy}(i) = J(N), \quad a_0 = 1 \qquad (7.3\text{--}6)$$

Combining Eqs. (7.3–5,6), we have

$$\sum_{i=0}^{N} a_i R_{yy}(i-j) = \begin{cases} J(N) & j = 0 \\ 0 & j = 1, \ldots, N \end{cases} \qquad \text{for } a_0 = 1 \qquad (7.3\text{--}7)$$

If we expand these equations, we obtain a Toeplitz equation:

$$\begin{bmatrix} R_{yy}(0) & \cdots & R_{yy}(N) \\ \vdots & \ddots & \vdots \\ R_{yy}(N) & \cdots & R_{yy}(0) \end{bmatrix} \begin{bmatrix} 1 \\ a_1 \\ \vdots \\ a_N \end{bmatrix} = \begin{bmatrix} J(N) \\ 0 \\ \vdots \\ 0 \end{bmatrix} \qquad (7.3\text{--}8)$$

or

$$\mathbf{R}_{yy}(N)\underline{a}(N) = \begin{bmatrix} J(N) \\ 0 \\ \vdots \\ 0 \end{bmatrix}$$

Suppose we want the solution for the $(N + 1)$th order estimator; we have

$$\mathbf{R}_{yy}(N + 1)\underline{a}(N + 1) = \begin{bmatrix} J(N + 1) \\ 0 \\ \vdots \\ 0 \end{bmatrix} \qquad (7.3\text{--}9)$$

If we assume that the optimal estimator is in fact Nth order, then

$$\underline{a}(N + 1) = \begin{bmatrix} \underline{a}(N) \\ --- \\ 0 \end{bmatrix} \quad \text{and} \quad \begin{bmatrix} J(N + 1) \\ 0 \\ \vdots \\ 0 \end{bmatrix} = \begin{bmatrix} J(N) \\ 0 \\ \vdots \\ 0 \\ \Delta_N \end{bmatrix} \qquad (7.3\text{--}10)$$

We have from Eqs. (7.3–8,9) that

$$\begin{bmatrix} \mathbf{R}_{yy}(N) & \vdots & R_{yy}(N + 1) \\ ----- & \vert & \vdots \\ R_{yy}(N + 1) & \cdots & R_{yy}(0) \end{bmatrix} \begin{bmatrix} \underline{a}(N) \\ --- \\ 0 \end{bmatrix} = \begin{bmatrix} J(N) \\ \vdots \\ \Delta_N \end{bmatrix} \qquad (7.3\text{--}11)$$

and, therefore, we must perform elementary operations on this equation to eliminate Δ_N. If we interchange the rows, multiply by $\Delta_N/J(N)$, and subtract, we obtain

$$\mathbf{R}_{yy}(N + 1)\left\{ \begin{bmatrix} \underline{a}(N) \\ --- \\ 0 \end{bmatrix} - \frac{\Delta_N}{J(N)} \begin{bmatrix} 0 \\ --- \\ \underline{a}(N) \end{bmatrix} \right\} = \left\{ \begin{bmatrix} J(N) \\ \vdots \\ \Delta_N \end{bmatrix} - \frac{\Delta_N}{J(N)} \begin{bmatrix} \Delta_N \\ \vdots \\ J(N) \end{bmatrix} \right\} \qquad (7.3\text{--}12)$$

If we define the reflection or partial correlation coefficient as

$$k_{N+1} := \frac{\Delta_N}{J(N)} \qquad (7.3\text{--}13)$$

then expanding into the components of Eq. (7.3–12) and introducing the notation $a(i,j)$ as the ith coefficient of the jth order predictor, we have

$$\mathbf{R}_{yy}(N + 1) \begin{bmatrix} 1 \\ a(1, N) - k_{N+1}a(N, N) \\ \vdots \\ a(i, N) - k_{N+1}a(N + 1 - i, N) \\ \vdots \\ -k_{N+1} \end{bmatrix} = \begin{bmatrix} (1 - k_{N+1}^2)J(N) \\ 0 \\ \vdots \\ 0 \\ \vdots \\ 0 \end{bmatrix} \qquad (7.3\text{--}14)$$

From Eq. (7.3–11) we see that Δ_N is given by

$$\Delta_N = \sum_{j=0}^{N} a(j, N)R_{yy}(N + 1 - j) \qquad (7.3\text{--}15)$$

If we equate Eqs. (7.3–9,14) we obtain the *Levinson-Durbin* recursion, which we summarize in Table 7.3–1. The following example demonstrates the recursion.

TABLE 7.3–1
Levinson-Durbin recursion

For $i = 1, \ldots, N$

$$\Delta_i = \sum_{j=0}^{i} a(j,i)R_{yy}(i+1-j)$$

$$k_{i+1} = \frac{\Delta_i}{J(i)}$$

$$a(i+1, i+1) = -k_{i+1}$$

For $j = 1, \ldots, i$

$$a(j, i+1) = a(j,i) - k_{i+1}a(i+1-j, i)$$

$$J(i+1) = (1 - k_{i+1}^2)J(i)$$

Initialization: $J(0) = R_{yy}(0)$, $\Delta_0 = R_{yy}(1)$, $k_1 = \dfrac{R_{yy}(1)}{R_{yy}(0)}$

Example 7.3–1. Suppose we have an exponentially correlated process, $R_{yy}(k) = \alpha^{|k|}$ and we would like to develop a filtered estimate, $\hat{y}(t \mid t - 1)$, using (i) the batch and (ii) the Levinson recursion.

Case (i): Batch solution

$$\underline{a} = \mathbf{R}_{yy}^{-1}\underline{R}_{yy}$$

$$\mathbf{R}_{yy} = \begin{bmatrix} R_{yy}(0) & R_{yy}(1) \\ R_{yy}(1) & R_{yy}(0) \end{bmatrix} = \begin{bmatrix} 1 & \alpha \\ \alpha & 1 \end{bmatrix}, \qquad \underline{R}_{yy} = \begin{bmatrix} R_{yy}(1) \\ R_{yy}(2) \end{bmatrix} = \begin{bmatrix} \alpha \\ \alpha^2 \end{bmatrix}$$

and therefore
$$\mathbf{R}_{yy}^{-1} = \frac{1}{1-\alpha^2}\begin{bmatrix} 1 & -\alpha \\ -\alpha & 1 \end{bmatrix}$$

and
$$\underline{a} = -\frac{1}{1-\alpha^2}\begin{bmatrix} 1 & -\alpha \\ -\alpha & 1 \end{bmatrix}\begin{bmatrix} \alpha \\ \alpha^2 \end{bmatrix} = -\frac{1}{1-\alpha^2}[\alpha(1-\alpha^2) \quad 0] = \begin{bmatrix} -\alpha \\ 0 \end{bmatrix}$$

$$J(2) = \sum_{i=0}^{2} a(i,2)R_{yy}(i) = R_{yy}(0) + a(1,2)R_{yy}(1) + a(2,2)R_{yy}(2) = 1 - \alpha^2$$

which gives $\hat{s}(t) = \hat{y}(t \mid t-1) = \alpha y(t-1)$ with $R_{ee} = 1 - \alpha^2$.

Case (ii): Levinson-Durbin recursion

$i = 0$ $J(0) = R_{yy}(0) = 1$, $\Delta_0 = R_{yy}(1) = \alpha$, $k_1 = R_{yy}(1)/R_{yy}(0) = \alpha$
$i = 1$ $a(1,1) = -k_1 = -\alpha$
           $J(1) = (1 - k_1^2)J(0) = 1 - \alpha^2$
           $\Delta_1 = \sum_{j=0}^{1} a(j,1)R_{yy}(2-j) = R_{yy}(2) + a(1,1)R_{yy}(1)$
           $= \alpha^2 - \alpha^2 = 0$
           $k_2 = \Delta_1/J(1) = 0$

$$a(2, 2) = 0$$
$$a(1, 2) = a(1, 1) = -\alpha$$
$$J(2) = (1 - k_2^2)J(1) = J(1) = 1 - \alpha^2$$

which is identical to the batch solution.

We close this section with a practical design example.

Example 7.3–2. Suppose we have a process that can be modeled by an ARMAX(2,1,1,1) model given by the difference equation

$$(1 - 1.5q^{-1} + 0.7q^{-2})y(t) = (1 + 0.5q^{-1})u(t) + \frac{1 + 0.3q^{-1}}{1 + 0.5q^{-1}}e(t)$$

where $u(t)$ is a delayed unit impulse at $t = 1.2$ sec, and $e \sim N(0, 0.01)$. We choose this type of ARMAX model because, of the algorithms we will investigate, none can exactly reproduce the structure. We simulate this process using SIG (Appendix D, [17]) for a data set of $N = 128$ at $\Delta T = 0.1$ sec, and we apply the Levinson-Durbin algorithm of Table 7.3–1 to design an all-pole filter. We utilize the AIC to automatically select the appropriate order. A third-order AR model was selected (AIC $= -278$), with the following estimated parameters:

Parameter	True value	Estimate
a_1	−1.5	−1.67
a_2	0.7	1.11
a_3	—	−0.33
b_0	1.0	0.15
b_1	0.5	—
c_1	0.3	—
d_1	0.5	—

The estimated impulse response and signal are shown in Fig. 7.3–1. The estimated impulse response appears to have captured the major resonance and the signal estimate obtained using the estimated parameters in Eq. (7.3–1) produces a reasonable "filtered" measurement, which is obtained from the estimation parameters; that is, the filtered measurement is given by

$$\hat{y}(t \mid t - 1) = \hat{s}(t \mid t - 1) = (1 - \hat{A}(q^{-1}))y(t) = -\sum_{i=1}^{N} \hat{a}_i y(t - i)$$

Other techniques are also available, based on the Cholesky decomposition [11], if we select the covariance or pre- and post-windowed methods discussed in Sec. 6.2.

7.4   LEVINSON (ALL-ZERO) FILTERS

In this section, we develop the Levinson all-zero filters, leading to a finite impulse-response model of Sec. 4.5. The technique discussed is based on the

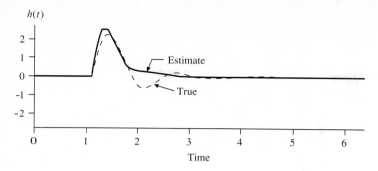

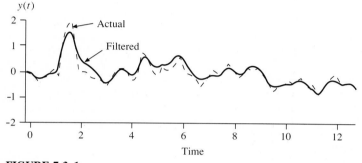

FIGURE 7.3–1
Levinson all-pole filter design for ARMAX(2,1,1,1) simulation (*a*) True and estimated impulse response (*b*) Noisy and filtered measurement

Levinson-Durbin recursion of the previous section and is sometimes called the "generalized Levinson" or Levinson-Wiggins-Robinson (LWR) algorithm, because of the improvements developed in [18]. This technique is again based on inverting a Toeplitz matrix, but now for the case where the right-hand vector is a cross rather than autocovariance. In fact, this method provides a recursive solution to the Wiener filtering problem (see Eq. (7.2–15)) for the scalar as well as multichannel cases [19]. The model set used provides the basic building block of many techniques used in adaptive signal processing [20].

We would like to obtain a linear estimate of a signal $s(t)$ based on past excitation data; that is,

$$\hat{s}(t) = \hat{s}(t \mid X_{\text{past}})$$

Taking the parametric approach, we model the signal by the FIR (all-zero) representation

$$\hat{s}(t) = \sum_{i=0}^{N} h(i)x(t - i) \tag{7.4–1}$$

If we represent the measurement $y(t)$ as

$$y(t) = s(t) + n(t) \tag{7.4–2}$$

where n is zero-mean white noise, of variance R_{nn},[†] then the optimal estimator is obtained by

$$\min_{\underline{h}} J = E\{e^2(t)\} \qquad (7.4\text{--}3)$$

where $e(t) = y(t) - \hat{s}(t)$.

Following the standard minimization technique, we differentiate with respect to $h(j)$, set to zero, and obtain the orthogonality condition

$$\frac{\partial J}{\partial h(j)} = 2E\left\{ e(t)\frac{\partial e(t)}{\partial h(j)} \right\} = -2E\{e(t)x(t-j)\} = 0 \qquad (7.4\text{--}4)$$

or

$$E\{e(t)x(t-j)\} = E\left\{ \left(y(t) - \sum_{i=0}^{N} h(i)x(t-i) \right) x(t-j) \right\} = 0 \qquad j = 0,\ldots,N \qquad (7.4\text{--}5)$$

which leads to the *normal equations*

$$\sum_{i=0}^{N} h(i)R_{xx}(i-j) = R_{yx}(j) \qquad j = 0,\ldots,N \qquad (7.4\text{--}6)$$

or

$$\begin{bmatrix} R_{xx}(0) & \cdots & R_{xx}(N) \\ \vdots & \ddots & \vdots \\ R_{xx}(N) & \cdots & R_{xx}(0) \end{bmatrix} \begin{bmatrix} h(0) \\ \vdots \\ h(N) \end{bmatrix} = \begin{bmatrix} R_{yx}(0) \\ \vdots \\ R_{yx}(N) \end{bmatrix} \qquad (7.4\text{--}7)$$

or

$$\mathbf{R}_{xx}(N)\underline{h}(N) = \underline{R}_{yx}(N) \qquad (7.4\text{--}8)$$

where $\mathbf{R}_{xx}$ is a Toeplitz matrix which can be inverted, using the "generalized Levinson" or the LWR algorithm. Note that we will have obtained the Wiener solution, since

$$\hat{\underline{h}}(N) = \mathbf{R}_{xx}^{-1}(N)\underline{R}_{yx}(N) \qquad (7.4\text{--}9)$$

The LWR recursion can be developed in two steps. The first establishes the basic recursion, and the second is the standard Levinson recursion for inverting Toeplitz matrices. We will use the notation $\{h(i,k)\}$ to denote the ith coefficient of the kth-order filter. Let us assume that we have the Nth order filter that satisfies the set of normal equations,

$$\begin{bmatrix} R_{xx}(0) & \cdots & R_{xx}(N) \\ \vdots & \ddots & \vdots \\ R_{xx}(N) & \cdots & R_{xx}(0) \end{bmatrix} \begin{bmatrix} h(0,N) \\ \vdots \\ h(N,N) \end{bmatrix} = \begin{bmatrix} R_{yx}(0) \\ \vdots \\ R_{yx}(N) \end{bmatrix} \qquad (7.4\text{--}10)$$

and we want the $(N+1)$th order solution given by

[†]Note that this problem is different from the previous in that here we have *both* input and output sequences, $\{x(t)\}$ and $\{y(t)\}$ respectively, while for the all-pole filter we just had $\{y(t)\}$. This problem is usually called the "joint-process" estimation problem.

$$
\begin{bmatrix} R_{xx}(0) & \cdots & R_{xx}(N+1) \\ \vdots & \ddots & \vdots \\ R_{xx}(N+1) & \cdots & R_{xx}(0) \end{bmatrix} \begin{bmatrix} h(0,N+1) \\ \vdots \\ h(N+1,N+1) \end{bmatrix} = \begin{bmatrix} R_{yx}(0) \\ \vdots \\ R_{yx}(N+1) \end{bmatrix} \tag{7.4-11}
$$

Suppose the optimum solution for the $(N+1)$th order filter is given by the Nth order; then $\underline{h}'(N+1) = [\underline{h}'(N) \mid 0]$, and $\underline{R}'_{yx}(N+1) = [\underline{R}'_{yx}(N) \mid \nabla_N]$ with $\nabla_N = R_{yx}(N+1)$. We can rewrite Eq. (7.4-12) as

$$
\left[\begin{array}{c|c} \mathbf{R}_{xx}(N) & R_{xx}(N+1) \\ \hline \\ R_{xx}(N+1) & \cdots & R_{xx}(0) \end{array} \right] \begin{bmatrix} \underline{h}(N) \\ \hline 0 \end{bmatrix} = \begin{bmatrix} \underline{R}_{yx}(N) \\ \hline \nabla_N \end{bmatrix} \tag{7.4-12}
$$

where

$$
\nabla_i = \sum_{j=0}^{N} h(j,N) R_{xx}(N-j+1) \tag{7.4-13}
$$

We must perform operations on Eq. (7.4-12) to assure that $\nabla_N = R_{yx}(N+1)$ for the correct solution. Let us assume there exists a solution $\{\alpha(i,N+1)\}$ such that[†]

$$
\begin{bmatrix} R_{xx}(0) & \cdots & R_{xx}(N+1) \\ \vdots & \ddots & \vdots \\ R_{xx}(N+1) & \cdots & R_{xx}(0) \end{bmatrix} \begin{bmatrix} \alpha(0,N+1) \\ \vdots \\ \alpha(N+1,N+1) \end{bmatrix} = \begin{bmatrix} J(N+1) \\ \vdots \\ 0 \end{bmatrix} \tag{7.4-14}
$$

Now, by elementary manipulations, we can reverse the order of the components, multiply by a constant, k_{N+1}, and subtract the result from Eq. (7.4-12). That is,

$$
\mathbf{R}_{xx}(N+1) \left\{ \begin{bmatrix} h(0,N) \\ \vdots \\ h(N,N) \\ 0 \end{bmatrix} - k_{N+1} \begin{bmatrix} \alpha(N+1,N+1) \\ \vdots \\ \alpha(1,N+1) \\ \alpha(0,N+1) \end{bmatrix} \right\}
$$

$$
= \left\{ \begin{bmatrix} R_{yx}(0) \\ \vdots \\ R_{yx}(N) \\ \nabla_N \end{bmatrix} - k_{N+1} \begin{bmatrix} 0 \\ \vdots \\ 0 \\ J(N+1) \end{bmatrix} \right\}
$$

or

$$
\mathbf{R}_{xx}(N+1) \left\{ \begin{bmatrix} h(0,N) - k_{N+1}\alpha(N+1,N+1) \\ \vdots \\ h(N,N) - k_{N+1}\alpha(1,N+1) \\ -k_{N+1}\alpha(0,N+1) \end{bmatrix} \right\} = \left\{ \begin{bmatrix} R_{yx}(0) \\ \vdots \\ R_{yx}(N) \\ \nabla_N - k_{N+1}J(N+1) \end{bmatrix} \right\} \tag{7.4-15}
$$

[†]Note that the solution to this set of equations results in the standard Levinson-Durbin recursion of linear-prediction theory, discussed in the previous section.

Here the multiplier k_{N+1} is selected so that

$$\nabla_N - k_{N+1}J(N + 1) = R_{yx}(N + 1)$$

By identifying the coefficients $\underline{h}(N + 1), \underline{R}_{yx}(N + 1)$ from Eq. (7.4–15) with $\alpha(0, i) = 1$, we obtain the first part of the recursion shown in Table 7.4–1.

In order to satisfy this recursion, we must also obtain the predictor, $\{\alpha(j, i + 1)\}$, and $J(i + 1)$ from the solution of Eq. (7.4–14). This can be accomplished using the Levinson-Durbin recursion of Table 7.3–1 for $\{\alpha(j, i)\}$ and $J(i)$ as shown in Table 7.4–1. This completes the solution to the Toeplitz inversion using the generalized Levinson or LWR algorithm.

Example 7.4–1. Suppose we have a signal that is exponentially correlated with $R_{xx}(k) = \alpha^{|k|}$, and it is transmitted through a medium with a velocity of 1 m/sec and attenuation A. The sensor is placed $d = 4$ m from the source. Find the optimal

TABLE 7.4–1
Generalized Levinson or LWR filters

Initialize: $J(0) = R_{xx}(0)$, $h(0, 0) = \dfrac{R_{yx}(0)}{J(0)}$, $\alpha(0, i) = 1 \forall i$

For $i = 0, \ldots, N$

$$\nabla_i = \sum_{j=0}^{i} h(j, i)R_{xx}(i - j + 1)$$

$$k_{i+1} = \frac{\nabla_i - R_{yx}(i + 1)}{J(i + 1)}$$

$$h(i + 1, i + 1) = -k_{i+1}$$

$$h(j, i + 1) = h(j, i) - k_{i+1}\alpha(i - j + 1, i + 1) \quad \text{for } j = 0, \ldots, i$$

where $\alpha(i, j)$ and $J(i)$ are obtained from the Levinson-Durbin recursion, given by the following equations:

For $i = 0, \ldots, N$

$$\Delta_i = \sum_{j=0}^{i} \alpha(j, i)R_{xx}(i - j + 1)$$

$$k^*_{i+1} = \frac{\Delta_i}{J(i)}$$

$$\alpha(i + 1, i + 1) = -k^*_{i+1}$$

$$\alpha(j, i + 1) = \alpha(j, i) - k^*_{i+1}\alpha(i + 1 - j, i) \quad \text{for } j = 1, \ldots, i$$

$$J(i + 1) = (1 - (k^*_{i+1})^2)J(i)$$

Signal estimation:

$$\hat{s}(t \mid t - 1) = \sum_{i=0}^{N} \hat{h}(i)x(t - i)$$

(FIR) estimate of the signal, that is, $\hat{s}(t \mid t - 1)$. From the characteristics of the medium we have the delay

$$\tau = \frac{d}{c} = \frac{4 \text{ m}}{1 \text{ m/sec}} = 4 \text{ sec}$$

Therefore, the received signal is

$$r(t) = Ax(t - 4) + n(t)$$

and

$$R_{rx}(k) = E\{r(t)x(t + k)\} = AE\{x(t - 4)x(t + k)\} = AR_{xx}(k - 4) = A\alpha^{|k-4|}$$

Batch approach: Using the optimal Wiener solution, we have

$$R_{xx}\underline{h} = \underline{R}_{rx}$$

$$\begin{bmatrix} 1 & \alpha \\ \alpha & 1 \end{bmatrix} \begin{bmatrix} h(0) \\ h(1) \end{bmatrix} = A \begin{bmatrix} \alpha^4 \\ \alpha^3 \end{bmatrix}$$

$$\begin{bmatrix} h(0) \\ h(1) \end{bmatrix} = \left(\frac{A}{1 - \alpha^2}\right) \begin{bmatrix} 1 & -\alpha \\ -\alpha & 1 \end{bmatrix} \begin{bmatrix} \alpha^4 \\ \alpha^3 \end{bmatrix} = \frac{A}{(1 - \alpha^2)} \begin{bmatrix} 0 \\ \alpha^3(1 - \alpha^2) \end{bmatrix} = \begin{bmatrix} 0 \\ A\alpha^3 \end{bmatrix}$$

which gives the signal estimate

$$\hat{s}(t) = A\alpha^3 x(t - 1)$$

LWR approach:

initialize $\quad J(0) = R_{xx}(0) = 1, h(0, 0) = R_{yx}(0)/J(0) = A\alpha^4, \alpha(0, i) = 1 \forall i$

$i = 0 \qquad \Delta_0 = \alpha(0, 0)R_{xx}(1) = \alpha$

$$k_1^* = \Delta_0/J(0) = \alpha$$

$$\alpha(1, 1) = -k_1^* = -\alpha$$

$$\nabla_0 = h(0, 0)R_{xx}(1) = A\alpha^4(\alpha) = A\alpha^5$$

$$J(1) = (1 - (k_1^*)^2)J(0) = 1 - \alpha^2$$

$$k_1 = \frac{\nabla_0 - R_{yx}(1)}{J(1)} = \frac{A\alpha^5 - A\alpha^3}{1 - \alpha^2} = -A\alpha^3$$

$$h(0, 1) = h(0, 0) - k_1\alpha(1, 1) = A\alpha^4 + A\alpha^3(-\alpha) = 0$$

$$h(1, 1) = -k_1 = A\alpha^3$$

which is identical to the batch approach.

Next we consider the simulation problem of the previous section.

Example 7.4–2. We would like to design a Levinson all-zero filter for the ARMAX(2,1,1,1) simulated data of the previous example. As before, we utilize SIG [17] to perform both simulation and estimation. A 32-weight FIR Levinson filter was designed using the algorithm of Table 7.4–1; the estimated parameters are given in the table on the following page. The results are shown in Fig. 7.4–1, where we see the impulse response and reconstructed measurement. Here we see that the FIR approach more closely estimates the impulse response and measurement at the cost of a much higher order.

Parameter	True	Estimated
$h(0)$	1	1
$h(1)$	2	1.944
$h(2)$	2.20	2.24
$h(3)$	2.05	1.95
$h(4)$	1.47	1.39
$h(5)$	0.76	0.68
$h(6)$	0.12	0.11

Before closing the section on FIR Wiener solutions, it is interesting to note that if we replace $\{h(i)\} \rightarrow \{b_i\}$ of the ARMAX$(1, N_b, 0)$ model, then we have these Wiener solutions as special cases of the ARMAX solution. Also, it is important to note that the LWR recursion leads to an order-recursive solution of the Toeplitz matrix. The FIR Wiener solution is the basis of most of the popular adaptive solutions (see [20]). In fact, for the stationary case the FIR adaptive solution converges precisely to $\underline{h}(N)$ of Eq. (7.4–9). We will discuss this more fully in the next chapter.

The Wiener solution also provides an optimal method of solving the decon-

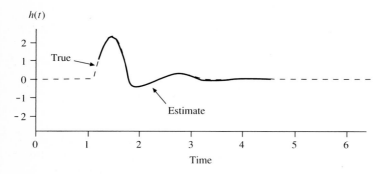

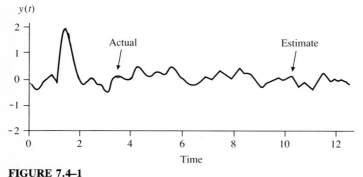

FIGURE 7.4–1
Levinson all-zero filter design for ARMAX$(2,1,1,1)$ simulation (*a*) True and estimated impulse response (*b*) Actual and estimated measurement (signal)

volution problem. It is easy to see this from the commutative property of the convolution relation; that is,

$$y(t) = h(t) * x(t) = x(t) * h(t) \qquad (7.4-16)$$

The deconvolution problem (see Chapter 9 also) is concerned with obtaining the "optimal" estimate of $x(t)$, given $(\{y(t)\}, \{h(t)\})$. Proceeding in similar fashion, the optimal estimator is given by

$$\min_{\underline{x}} J = E\{e^2(t)\} \qquad (7.4-17)$$

where $e(t) = y(t) - \hat{y}(t)$. Following the standard approach, the orthogonality conditions (with $h \to x$) are given by

$$E\{e(t)h(t - j)\} = 0 \qquad (7.4-18)$$

where $\partial e(t)/\partial x(j) = h(t - j)$. Substituting as before, we obtain the *normal equations*,

$$E\{y(t)h(t - j)\} - \sum_{i=0}^{N} x(i)E\{h(t - i)h(t - j)\} = 0 \qquad \text{for } j = 0, \ldots, N$$

or, solving, we have

$$\sum_{i=0}^{N} x(i)R_{hh}(i - j) = \underline{R}_{yh}(j) \qquad \text{for } j = 0, \ldots, N \qquad (7.4-19)$$

or

$$\mathbf{R}_{hh}(N)\underline{x}(N) = \underline{R}_{yh}(N) \qquad (7.4-20)$$

where $\mathbf{R}_{hh}$ is again Toeplitz and can be solved with the LWR recursion of Table 7.4–1.

> **Example 7.4–3.** In this example, we utilize the simulated ARMAX(2,1,1,1) measurement data, and the Levinson all-zero algorithm of Table 7.4–1 to deconvolve or estimate the input according to Eq. (7.4–20). The impulse response is that obtained in the previous example (see Fig. 7.4–1). Using SIG [17], we show the results of the solution in Fig. 7.4–2. Ideally, we should see a unit impulse at 1.2 sec. We see from the figure that the algorithm performs quite well and is able to produce a good estimate of the desired excitation.

We also show an application of this deconvolution technique in the case study of Sec. 7.8.

7.5 LATTICE FILTERS

Since its roots were established by Markel [1], the lattice filter and lattice techniques have evolved to a high degree of current popularity for many applications (see Sec. 6.5). The lattice method combines the orthogonal projections in *both*

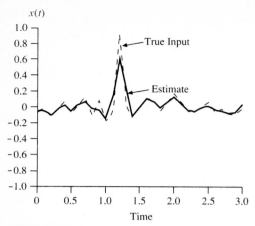

FIGURE 7.4–2
Levinson all-zero deconvolution filter design for ARMAX(2,1,1,1) simulation

time and parameter spaces [22]. Lattice filters can be developed directly from the model, or indirectly from the Levinson recursion [15]; we choose the former, since it follows the development of our previous processors.

As before, we are interested in developing an estimate of the signal $s(t)$ based on past measurements and/or past excitations; that is,

$$\hat{s}(t \mid t - 1) = \hat{s}(t \mid Y_{\text{past}} \text{ or } X_{\text{past}})$$

Again taking the parametric approach, we know from the results of Chapter 4 that, given the lattice parameters, we can construct all-pole or all-zero signal estimators. With this in mind, let us now construct an estimator for the lattice model.

The lattice recursion for the ith section at time t (see Sec. 4.6) is given by

$$e_f(t, i) = e_f(t, i - 1) - k_i e_b(t - 1, i - 1)$$
$$e_b(t, i) = e_b(t - 1, i - 1) - k_i e_f(t, i - 1)$$

(7.5–1)

where

$e_f(t, i)$ is the forward prediction error of the ith section at time t
$e_b(t, i)$ is the backward prediction error of the ith section at time t
k_i is the reflection coefficient of the ith section

The optimal parametric estimation problem is characterized by minimizing the joint-error function in terms of the lattice parameters as

$$\min_{\underline{k}} J(t) = E\{e_f^2(t) + e_b^2(t)\}$$

(7.5–2)

Performing the indicated minimization of Eq. (7.5–2) as before, we have the orthogonality condition

$$\frac{\partial J}{\partial k_i} = \frac{\partial}{\partial k_i} E\{e_f^2(t,i) + e_b^2(t,i)\} = 2E\left\{e_f(t,i)\frac{\partial e_f(t,i)}{\partial k_i} + e_b(t,i)\frac{\partial e_b(t,i)}{\partial k_i}\right\} = 0$$

$$(7.5\text{--}3)$$

The error gradients can be calculated directly from the model of Eq. (7.5–1) as

$$\frac{\partial e_f}{\partial k_i} = -e_b(t-1, i-1)$$

and, similarly, we have

$$\frac{\partial e_b}{\partial k_i} = -e_f(t, i-1)$$

Substituting these gradients into Eq. (7.5–3) gives

$$\frac{\partial J}{\partial k_i} = -2E\{e_f(t,i)e_b(t-1, i-1) + e_b(t,i)e_f(t, i-1)\} = 0$$

Using the model recursion of Eq. (7.5–1) we obtain

$$\frac{\partial J}{\partial k_i} = -2E\{2e_f(t, i-1)e_b(t-1, i-1) - k_i(e_f^2(t, i-1) + e_b^2(t-1, i-1))\} = 0$$

$$(7.5\text{--}4)$$

Solving this equation for k_i gives[†]

$$k_i = \frac{2E\{e_f(t, i-1)e_b(t-1, i-1)\}}{E\{e_f^2(t, i-1) + e_b^2(t-1, i-1)\}} \qquad (7.5\text{--}5)$$

The variance of the lattice recursion can be calculated directly from the recursion, since

$$R_{ee}(i) = E\{e_f^2(t,i)\} = E\{(e_f(t, i-1) - k_i e_b(t-1, i-1))^2\}$$

$$= E\{e_f^2(t, i-1)\} - 2k_i E\{e_f(t, i-1)e_b(t-1, i-1)\}$$

$$+ k_i^2 E\{e_b^2(t-1, i-1)\} \quad (7.5\text{--}6)$$

Solving Eq. (7.5–5) for the numerator term and substituting, we obtain

$$R_{ee}(i) = E\{e_f^2(t, i-1)\} - k_i^2(E\{e_f^2(t, i-1)\} + E\{e_b^2(t-1, i-1)\})$$

$$+ k_i^2 E\{e_b^2(t-1, i-1)\}$$

or

$$R_{ee}(i) = (1 - k_i^2)R_{ee}(i-1) \qquad (7.5\text{--}7)$$

where $R_{ee}(0) = R_{yy}(0) = \text{var}(y(t))$. Expanding this recursion over an M-stage lattice, we see that

---

[†]The reflection coefficient evolves from the Schur-Cohn stability technique and is used to test stability. If the reflection coefficient lies between ± 1, the system is *stable*; that is, $|k_i| \leq 1$ for stability (see [1] for details).

$$R_{ee} = R_{ee}(M) = \prod_{i=1}^{M}(1 - k_i^2)R_{yy}(0) \qquad (7.5\text{--}8)$$

If we assume the data are ergodic, then Burg [24] proposes replacing the ensemble averages of Eq. (7.5–5) with time averages, using only the unwindowed data record, that is,

$$k_i = \frac{2\sum_{t=N_a}^{N-1} e_f(t, i-1)e_b(t-1, i-1)}{\sum_{t=N_a}^{N-1} e_f^2(t, i-1) + e_b^2(t-1, i-1)} \qquad \text{for } i = 1, \dots, N_a \quad (7.5\text{--}9)$$

Once the reflection coefficients are estimated from the data using Eqs. (7.5–1,9) for each stage, the prediction error variance is obtained from Eq. (7.5–8).

Note that the Burg method is a block (time) or batch parameter estimation technique that is *recursive-in-order*, just as the Levinson filter of the previous sections. We summarize the technique in Table 7.5–1. It should also be noted that when implementing this block estimator, the reflection coefficient is first estimated and then *all* of the data are processed through the corresponding lattice section. The processed data is then used to estimate the next reflection coefficient and so on (see [15] for more details). In Sec. 8.5, we develop an alternative method of implementing the Burg technique.

It should also be noted that the lattice model evolves physically as a wave propagating through a layered medium. That is, if we interpret e_f and e_b as forward and backward waves (see Fig. 7.5–1), then it is possible to derive the lattice recursions directly from the wave equation ([1,19]). This model has been used quite successfully to estimate acoustic, seismic, and electromagnetic signals (waves) in layered media.

Consider the following example to see how the lattice technique can be used to estimate a signal.

Example 7.5–1. Consider the deterministic version of the previous problem. We would like to estimate the signal from a system with data set given by $Y(t) = \{1, \alpha\}$. The recursion begins with

$$i = 0 \qquad k_1 = \frac{\sum_{t=1}^{2} e_f(t, 0)e_b(t-1, 0)}{\sum_{t=1}^{2} e_f^2(t, 0) + e_b^2(t-1, 0)}$$

$$= \frac{e_f(1, 0)e_b(0, 0) + e_f(2, 0)e_b(1, 0)}{e_f^2(1, 0) + e_b^2(0, 0) + e_f^2(2, 0) + e_b^2(1, 0)}$$

$$= \frac{\alpha(1)}{1 + \alpha^2 + 1} = \frac{\alpha}{\alpha^2 + 2}$$

$$e_f(t, 1) = e_f(t, 0) - \left(\frac{\alpha}{\alpha^2 + 2}\right)e_b(t-1, 0)$$

$$e_b(t, 1) = e_b(t-1, 0) - \left(\frac{\alpha}{\alpha^2 + 2}\right)e_f(t, 0)$$

TABLE 7.5–1
Lattice filters

Prediction error estimation

$$e_f(t, i) = e_f(t, i - 1) - k_i e_b(t - 1, i - 1)$$
$$e_b(t, i) = e_b(t - 1, i - 1) - k_i e_f(t, i - 1)$$

Reflection coefficient estimation

$$k_i = \frac{2\sum_{t=N_a}^{N-1} e_f(t, i - 1)e_b(t - 1, i - 1)}{\sum_{t=N_a}^{N-1} e_f^2(t, i - 1) + e_b^2(t - 1, i - 1)} \quad \text{for } i = 1, \ldots, N_a$$

Parameter calculation (Table 4.6–1)

$$\beta(i, i) = -k_i$$
$$\beta(j, i) = \beta(j, i - 1) - k_i\beta(i - j, i - 1) \quad j = 1, \ldots, i - 1$$

Noise estimation

$$R_{ee} = \prod_{i=1}^{N_a} (1 - k_i^2)R_{yy}(0)$$

Signal estimation

$$\hat{s}(t \mid t - 1) = -\sum_{i=1}^{N_a} \beta_i y(t - i) \quad \text{(all-pole)}$$

or

$$\hat{s}(t \mid t - 1) = \sum_{i=1}^{N_a} \beta_i x(t - i) \quad \text{(all-zero)}$$

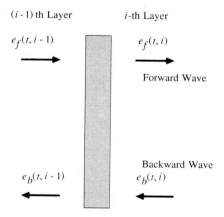

$(i - 1)$ th Layer i-th Layer

$e_f(t, i - 1)$ $e_f(t, i)$

Forward Wave

Backward Wave

$e_b(t, i - 1)$ $e_b(t, i)$

FIGURE 7.5–1
Wave propagation in layered medium—the lattice model

or

$$\{e_f(1,1), e_f(2,1)\} = \left\{1, \frac{\alpha^2 - \alpha + 2}{\alpha^2 + 2}\right\}$$

$$\{e_b(1,1), e_b(2,1)\} = \left\{-\frac{\alpha}{\alpha^2 + 2}, \frac{\alpha^2 - \alpha + 2}{\alpha^2 + 2}\right\}$$

$$k_2 = \frac{e_f(1,1)e_b(0,1) + e_f(2,1)e_b(1,1)}{e_f^2(1,1) + e_b^2(0,1) + e_f^2(2,1) + e_b^2(1,1)}$$

$$= \frac{-\alpha(\alpha^2 - \alpha + 2)}{(\alpha^2 + 2)^2 + (\alpha^2 - \alpha + 2)^2 + \alpha^2}$$

For the all-pole model, we have

$$\hat{s}(t) = \hat{y}(t \mid t - 1) = -\left(\frac{\alpha}{\alpha^2 + 2}\right) y(t - 1)$$

and a second-order predictor can be developed using $\{k_1, k_2\}$ and the Levinson recursion.

Next we consider applying the algorithm to a simulated set of data to see how well it can produce a signal estimate.

Example 7.5–2. Consider the simulated data set of Example 7.3–2, where the data are simulated from an ARMAX(2,1,1,1) model with $e \sim N(0, 0.01)$. Using SIG [17] to simulate this process, and the algorithm of Table 7.5–1 we find the results shown in Fig. 7.5–2. The AIC criterion indicates a third-order model with estimated parameters as shown in the following table:

Parameter	True	Estimated	Reflection
a_1	-1.5	-1.47	-0.89
a_2	0.7	0.73	0.59
a_3	—	0.10	-0.10
b_0	1.0	0.16	
b_1	0.5	—	
c_1	0.3	—	
d_1	0.5	—	

The filtered output is shown in Fig. 7.5–2(b). Here we see that the lattice method appears to perform better than the Levinson-Durbin approach, as indicated by the estimated impulse response and signal.

7.6   PREDICTION-ERROR FILTERS

In this section, we develop the prediction-error filters, which evolved from the control/estimation area [6,7]. In contrast to the previous estimators, the predic-

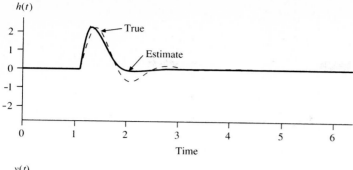

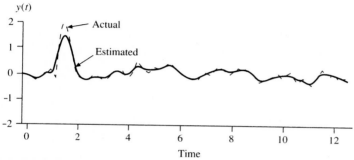

FIGURE 7.5-2

Lattice all-pole filter design for ARMAX(2,1,1,1) simulation (*a*) True and estimated impulse response (*b*) Actual and filtered measurement (signal)

tion-error filter is *recursive-in-time* rather than *recursive-in-order*, and therefore it is not a block (of data) processor. Prediction-error filters are used for adaptive control, but they have also been applied successfully in adaptive signal processing problems as well [7,8].

In the development of the parameter estimator for this case, many different parameter estimation schemes have evolved. We choose the Recursive Prediction-Error Method (RPEM) developed by Ljung [7], since it can be shown that most recursive parameter estimators are special cases of this approach.

The prediction-error approach assumes we would like to obtain an estimate of the signal based on past measurement data,

$$\hat{s}(t \mid t - 1) = \hat{s}(t \mid Y_{\text{past}})$$

Taking the parametric approach, we model the signal by the relation

$$s(t) = -\sum_{i=1}^{N_a} a_i y(t - i) + \sum_{i=0}^{N_b} b_i u(t - i) \qquad (7.6-1)$$

and, therefore, the measurement is given by

$$y(t) = s(t) + n(t) \qquad (7.6-2)$$

If we choose to represent the noise as correlated, that is,

$$n(t) = \sum_{i=1}^{N_c} c_i e(t - i) + e(t) \qquad (7.6\text{--}3)$$

then substituting Eqs. (7.6–1,3), we obtain

$$y(t) = -\sum_{i=1}^{N_a} a_i y(t-i) + \sum_{i=0}^{N_b} b_i u(t-i) + \sum_{i=0}^{N_c} c_i e(t-i), \qquad c_0 = 1 \quad (7.6\text{--}4)$$

or more succinctly, in backward shift operator notation q^{-1},

$$A(q^{-1})y(t) = B(q^{-1})u(t) + C(q^{-1})e(t) \qquad (7.6\text{--}5)$$

Thus, the ARMAX(N_a, N_b, N_c) evolves as the underlying model.

In developing the RPEM based on the ARMAX model, we first note that Eq. (7.6–4) can be rewritten in vector form as

$$y(t) = \phi'(t)\Theta + e(t) \qquad (7.6\text{--}6)$$

where

$$\phi' := [-y(t-1) \cdots - y(t-N_a) \mid u(t) \cdots u(t-N_b) \mid e(t-1) \cdots e(t-N_c)]$$

and

$$\Theta' := [a_1 \cdots a_{N_a} \mid b_0 \cdots b_{N_b} \mid c_1 \cdots c_{N_c}]$$

We define the *prediction-error criterion*[†] as

$$\min_{\Theta} J_t(\Theta) = \frac{1}{2}\sum_{k=1}^{t} e^2(k, \Theta) \qquad (7.6\text{--}7)$$

where the prediction error is given (as before) by

$$e(t, \Theta) := y(t) - \hat{y}(t \mid t - 1) \qquad (7.6\text{--}8)$$

and

$$\hat{y}(t \mid t - 1) = \hat{s}(t \mid t - 1) = \left(1 - \hat{A}(q^{-1})\right)y(t) + \hat{B}(q^{-1})u(t) \qquad (7.6\text{--}9)$$

We derive the parameter estimator in the standard manner (see Appendix C for details) by performing the indicated minimization in Eq. (7.6–7).

Starting with the *Gauss-Newton* parameter estimator,

$$\hat{\Theta}(t) = \hat{\Theta}(t-1) - \left[\frac{\partial^2}{\partial \Theta^2} J_t(\hat{\Theta})\right]^{-1} \frac{\partial}{\partial \Theta} J_t(\hat{\Theta}) \qquad (7.6\text{--}10)$$

where Θ is the $N_\Theta \times 1$ parameter vector

$\dfrac{\partial^2 J}{\partial \Theta^2}$ is an $N_\Theta \times N_\Theta$ Hessian matrix

$\dfrac{\partial J}{\partial \Theta}$ is an $N_\Theta \times 1$ gradient vector

---

[†]We explicitly show the dependence of the prediction error on the unknown parameter vector Θ.

we must construct the gradient and Hessian. The gradient is given by

$$\frac{\partial}{\partial\Theta}J_t(\Theta) = -\sum_{k=1}^{t}\psi(k,\Theta)e(k,\Theta) \tag{7.6-11}$$

where we define the negative innovation gradient vector as

$$\psi(t,\Theta) := -\frac{\partial}{\partial\Theta}e(t,\Theta) \tag{7.6-12}$$

Since we are interested in a *recursive-in-time* technique, we find the *gradient recursion*

$$\frac{\partial}{\partial\Theta}J_t(\hat{\Theta}(t-1)) = -\psi(t,\hat{\Theta}(t-1))e(t,\hat{\Theta}(t-1)) \tag{7.6-13}$$

The Hessian can also be determined in a similar manner, using this recursion and some approximations (see Appendix C for details) to obtain

$$R(t) = R(t-1) + \psi(t,\hat{\Theta}(t-1))\psi'(t,\hat{\Theta}(t-1)) \tag{7.6-14}$$

where we define the Hessian matrix as

$$R(t) := \frac{\partial^2}{\partial\Theta^2}J_t(\Theta)$$

and the Gauss-Newton parameter estimator becomes

$$\hat{\Theta}(t) = \hat{\Theta}(t-1) + R^{-1}(t)\psi(t,\hat{\Theta})e(t,\hat{\Theta}) \tag{7.6-15}$$

These equations make up the RPEM algorithm; all that is required is to determine $e(t,\Theta)$ and $\psi(t,\Theta)$. Using the ARMAX model, we have[†]

$$C(q^{-1})\psi(t) = \phi(t) \tag{7.6-16}$$

which is equivalent to inverse-filtering to obtain $\psi(t)$; that is,

$$\psi'(t) = [y'_f(t) \mid u'_f(t) \mid e'_f(t)] \tag{7.6-17}$$

where y_f, u_f, and e_f are the vector outputs of the specified filtering operation. For instance,

$$y_f(t,i) = -y(t-i) - (C(q^{-1}) - 1)y_f(t,i) \qquad \text{for } i = 1,\ldots,N_a \tag{7.6-18}$$

Similar relations are given for u_f and e_f above.

Finally, the prediction error is calculated using the most current available parameter estimate:

$$e(t) = y(t) - \phi'(t)\hat{\Theta}(t-1) \tag{7.6-19}$$

The RPEM algorithm can be implemented at this point; however, it requires the inversion of the $N_\Theta \times N_\Theta$ Hessian matrix at each time-step. This calculation

---

[†]We drop the Θ-dependent notation, $e(t,\Theta) = e(t)$, $\psi(t,\Theta) = \psi(t)$.

can be reduced by applying the matrix inversion lemma[†] to $R^{-1}(t)$; defining $P(t) := R^{-1}(t)$, we obtain

$$P(t) = P(t - 1) - \frac{P(t - 1)\psi(t)\psi'(t)P(t - 1)}{1 + \psi'(t)P(t - 1)\psi(t)} \qquad (7.6\text{--}20)$$

which completes the algorithm. We summarize the RPEM in Table 7.6–1. From the table we note that the algorithm is implemented with the *residuals* (as well as $\overline{\phi}$),

$$r(t) = y(t) - \overline{\phi}'(t)\hat{\Theta}(t) \qquad (7.6\text{--}21)$$

as soon as the new parameter update is available. It is also important to note that $A(z)$ must have roots inside the unit circle for convergence and, therefore, a stability test is also incorporated in the algorithm. We refer the interested reader to Ljung [7] for more details.

Most other recursive (in-time) parameter estimation techniques are special cases of the RPEM. For instance, the Extended Least-Square (ELS) technique is implemented as a special case of RPEM with

$$\psi(t) = \phi(t)$$

and the filtering operation (y_f, u_f, e_f) eliminated (see Table 7.8–1). Consider the following example of applying the RPEM to a simulated problem.

Example 7.6–1. Again using the ARMAX(2,1,1,1) simulated data (256 points), we applied both the RPEM and the ELS algorithm in SIG [17]. The prediction-error filter was selected as ARMAX(2,1,1,0), using the FPE criterion (FPE = 0.0125) for ELS, while the RPEM indicated an ARMAX(2,1,2,0) model (FPE = 0.015). The estimated parameters are listed below:

Parameter	True	ELS	RPEM
a_1	-1.5	-1.49	-1.29
a_2	0.7	0.7	0.66
b_0	1.0	1.95	1.23
b_1	0.5	-0.22	0.37
c_1	0.3	-0.12	-0.41
c_2		$-$	-0.64
c_3	$-$	$-$	0.632
d_1	0.5	$-$	$-$

The results of this run are shown in Fig. 7.6–1. Here we see the true and estimated impulse responses. The ELS and RPEM estimators give better results than the previous recursive filters because of the zeros available in both B and C polynomials. The "filtered" measurement or estimated signal is also shown, and it again confirms a reasonable estimate. For this problem, we see that the ELS algorithm was able to

---

[†]The matrix inversion lemma is given by $[A + BCD]^{-1} = A^{-1} - A^{-1}B[DA^{-1}B + C^{-1}]^{-1}DA^{-1}$

TABLE 7.6–1
Recursive prediction-error method

Prediction error

$$e(t) = y(t) - \hat{y}(t \mid t - 1) = y(t) - \phi'(t)\hat{\Theta}(t - 1) \quad \text{(Prediction error)}$$

$$R_{ee}(t) = \psi'(t)P(t - 1)\psi(t) + \lambda(t) \quad \text{(Prediction error variance)}$$

Update

$$\Theta(t) = \Theta(t - 1) + \mu K(t)e(t) \quad \text{(Parameter update)}$$

$$r(t) = y(t) - \overline{\phi}'(t)\hat{\Theta}(t) \quad \text{(Residual update)}$$

$$P(t) = (I - K(t)\psi'(t))P(t - 1)/\lambda(t) \quad \text{(Covariance update)}$$

$$\lambda(t) = \lambda_0\lambda(t - 1) + (1 - \lambda_0) \quad \text{(Forgetting factor update)}$$

Filtering

$$y_f(t, i) = y(t - i) - [\hat{C}(q^{-1}) - 1]y_f(t, i) \quad \text{(Measurement filter)}$$

$$u_f(t, i) = u(t - i) - [\hat{C}(q^{-1}) - 1]u_f(t, i) \quad \text{(Input filter)}$$

$$r_f(t, i) = r(t - i) - [\hat{C}(q^{-1}) - 1]r_f(t, i) \quad \text{(Residual filter)}$$

Gain

$$K(t) = P(t - 1)\psi(t)R_{ee}^{-1}(t) \quad \text{(Gain or weight)}$$

$$\overline{\phi}'(t) = [-y(t - 1) \cdots - y(t - N_a) \mid u(t) \cdots u(t - N_b) \mid r(t - 1) \cdots r(t - N_c)]$$
(Phi-update)

$$\psi'(t) = [-y_f(t - 1) \cdots - y_f(t - N_a) \mid u_f(t) \cdots u_f(t - N_b) \mid r_f(t - 1) \cdots r_f(t - N_c)]$$
(Gradient update)

obtain superior parameter estimates and predicted signals than the RPEM technique, possibly indicating a slight "tuning" problem in the RPEM algorithm.

7.7 ORDER ESTIMATION

In this section, we discuss order estimation, that is, the estimation of the order (N_a) of the number of poles of the system under investigation from noisy data. Knowledge of system order is required before parameter estimation can be performed, as we have seen in the spectral estimation as well as parametric signal processing problems. Order estimation has evolved specifically from the identification problem, where the first step is to estimate or guess the system order and then attempt to validate (innovations whiteness testing) the estimated model. Much effort has been devoted to this problem [6,7], with the most significant results evolving from the point of view of information theory [25,26,27,28].

Perhaps the simplest form of order estimation resulted from the technique called repeated least-squares. Here the least-squares parameter estimator is used

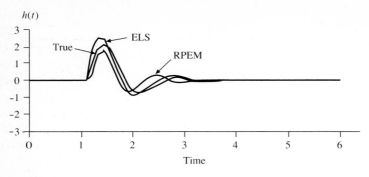

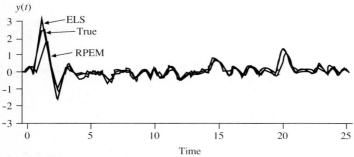

FIGURE 7.6–1

RPEM pole-zero filter design for ARMAX(2,1,1,1) simulation (*a*) True and estimated (ELS, RPEM) impulse response (*b*) Actual and filtered measurement (signal)

to fit various assumed model orders, and the estimated mean-squared error is calculated as a function of order. The model order selected was the one with minimal mean-squared error, that is,

$$J(N_i) \geq J_{\min}(N_a)$$

where $J(N_i) := E\{e^2(t, N_i)\}$, N_i the model order. The problem with this technique was that the mean-squared error monotonically decreases as model order increases; therefore, it is difficult to determine the true order, since there is *no* penalty for choosing a large order. Akaike [28] showed that this can be improved when criteria are developed from the point of view of information theory. This is the approach we will take.

Recall from Sec. 4.2 that the *average mutual information* for distribution function [28,29] is given by

$$I(p(x_i), p(x_j)) = E\left\{ \ln \frac{p(x_i)}{p(x_j)} \right\} = \sum_{i=1}^{N} p(x_i) \ln \frac{p(x_i)}{p(x_j)} \qquad (7.7\text{–}1)$$

This is called the *Kullback-Leibler* measure of dissimilarity between distributions, with the property that

$$I(p(x_i), p(x_j)) \geq 0 \qquad (7.7\text{–}2)$$

with equality, if and only if $p(x_i) = p(x_j)$. Recall also that the *joint entropy* is given simply by

$$H_{ij} := H(p(x_i), p(x_j)) = -I(p(x_i), p(x_j)) \qquad (7.7\text{--}3)$$

with properties evolving from Eq. (7.7–2) that

$$H_{ij} \leq 0 \qquad (7.7\text{--}4)$$

with equality, if and only if $p(x_i) = p(x_j)$.

For the moment, let us investigate the maximum-likelihood estimate of Θ and show how it is related to the entropy function. If the set of innovations are given by $\{e(t)\}$, then the log-likelihood function is specified by the following relations, since the values for $\{e(t)\}$ are independent:

$$L(N, \Theta) = \ln \left(\prod_{t=1}^{N} p(e(t) \mid \Theta) \right) = \sum_{t=1}^{N} \ln p(e(t) \mid \Theta) \qquad (7.7\text{--}5)$$

which in the limit approaches the *mean log-likelihood* function

$$\lim_{N \to \infty} L(N, \Theta) \to E_e\{ \ln p(e(t) \mid \Theta) \} \qquad (7.7\text{--}6)$$

If we decompose the *average information* function, we have

$$I(p(e(t)), p(e(t) \mid \Theta)) = \sum_{t=1}^{N} p(e(t)) \ln \frac{p(e(t))}{p(e(t) \mid \Theta)}$$

or, since this is a measure of dissimilarity of the true distribution $p(e(t))$ and the estimated $p(e(t) \mid \Theta)$, from Eq. (7.7–5) we have

$$I(p(e(t)), p(e(t) \mid \Theta)) = E_e\{p(e(t))\} - E_e\{p(e(t) \mid \Theta)\} \qquad (7.7\text{--}7)$$

Now in terms of the parameter estimation problem, we assume that the true distribution is obtained when $\Theta = \Theta_{\text{TRUE}}$; then we can replace $p(e(t)) \to p(e(t) \mid \Theta_{\text{TRUE}})$. In terms of this representation, we define the *Akaike Information Criterion* (AIC) as

$$\text{AIC} \approx H_{ij} = -\left[E_e\{p(e(t) \mid \Theta_{\text{TRUE}})\} - E_e\{p(e(t) \mid \Theta)\} \right] \qquad (7.7\text{--}8)$$

Under these asymptotic assumptions ($\Theta \to \Theta_{\text{TRUE}}$), there is a bias in the AIC which is equal to the number of parameters in the estimated model (N_Θ) and a variance of the estimated model.

Both must be corrected. The AIC for an N_Θth-order model is therefore given by

$$\text{AIC}(N_\Theta) = -\ln R_{ee} + 2\frac{N_\Theta}{N} \qquad (7.7\text{--}9)$$

where N_Θ is the number of free parameters and N the number of data.

In the parameter estimation problem, several families of $p(e(t) \mid \Theta)$, with different forms of $p(e(t) \mid \Theta)$, are given, and we are required to select the best "fit." The AIC still provides a useful estimate of the entropy function in this

TABLE 7.7–1
Order tests

Test	Criterion	Remarks
Prediction error fit	$\sum_{k=1}^{N_\Theta} e^2(k) / \sum_{k=1}^{N_\Theta} y^2(k)$	Locate knee of curve
Whiteness test	$\left\| \dfrac{R_{ee}^2(k)}{R_{ee}^2(0)} \right\| \leq \dfrac{\pm 1.96}{\sqrt{N}}$	Check 95% of covariances lie within bounds
Signal-error test	$\hat{y}(t) = \dfrac{\hat{B}(q^{-1})}{\hat{A}(q^{-1})} u(t)$	Check for fit
Final prediction error	$\mathrm{FPE}(N_\Theta) = \left(\dfrac{N + 1 + N_\Theta}{N - 1 - N_\Theta} \right) R_{ee}$	Penalties for overparameterization reduced for large N
Akaike information	$\mathrm{AIC}(N_\Theta) = -\ln R_{ee} + 2 \dfrac{N_\Theta}{N}$	Penalties for overparameterization
Minimum description length	$\mathrm{MDL}(N_\Theta) = -\ln R_{ee} + \dfrac{N_\Theta}{2} \ln N$	Penalties for overparameterization
Criteria for AR transfer function	$\mathrm{CAT} = \dfrac{1}{N} \sum_{i=0}^{N_\Theta} \dfrac{1}{R_{ee}(i)} - \dfrac{1}{R_{ee}(N_\Theta)}$	All N_Θth-order models must be estimated

case. Here the AIC is calculated for various model orders, and the value with the minimum AIC is selected. The estimate is called the MAICE.

It can also be shown that the Final Prediction Error (FPE) criterion [28] given by

$$\mathrm{FPE}(N_\Theta) = \left(\frac{N + 1 + N_\Theta}{N - 1 - N_\Theta} \right) R_{ee} \qquad (7.7\text{–}10)$$

is asymptotically equivalent to the $\mathrm{AIC}(N_\Theta)$ for a large number of data.

Various criteria have been developed; and many of the early methods are based on checking the whiteness of the prediction errors. Theoretically, we know this is valid from the orthogonality properties of the optimal estimator (see Sec. 7.2). In any case, we list some of the more popular order tests in Table 7.7–1 and refer the interested reader to [26] for more details.

7.8   CASE STUDY: ELECTROMAGNETIC SIGNAL PROCESSING

In this section, we consider the development of parametric signal processors to improve the signal processing capability of an Electromagnetic Test Facility (EMTF). The EMTF was developed primarily to determine the transient response characteristics of scale models subjected to various EM excitations. The facility consists of three major subsystems: (1) incident field simulator, (2) test object, and (3) measurement system. The system diagram is depicted in Fig. 7.8–1 (see [30,31,32] for details).

There are two classes of signal-processing problems to consider in this diagram; subsystem interaction and the introduction of noise and distortions into the system. The major subsystems interact with each other. This interaction must be taken into account when characterizing the test facility. There are several points where noise may be present in the facility. Internal noise and distortion may be introduced by the pulse generator, transmitting antennas, cables, sensors, and digitizers. External noise may enter the system at the test object from reflections or from environmental electromagnetic noise. The purpose of noise characterization is to remove the noise and compensate for the distortions.

The actual physical diagram of Fig. 7.8–1 can be transformed into a signal-processing model. We choose to use linear models, allowing us to utilize a wide range of signal-processing and identification techniques. Fig. 7.8–2 shows the model. The incident pulse simulator subsystem is modeled as a pulse-generator transfer function, antenna-transfer function, and propagation equation. The incident field generator is modeled with an impulse as the input and a noisy incident field pulse as output. The noise sources in the field generator include measurement noise on the output of the pulse-generator and environmental electromagnetic noise and reflections on the output of the antenna. The output of the incident field generator is the input to the test object. One of the goals of this work is to provide a low-noise input to the test object. The test object is modeled as a linear system according to the Singularity Expansion Method [33]. The output of the object is sensed by a probe with a transfer function and digitized. The identification of the object transfer function has most chance of

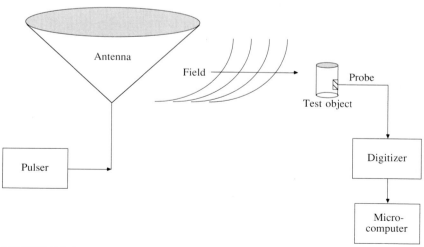

FIGURE 7.8–1
EMTF system diagram

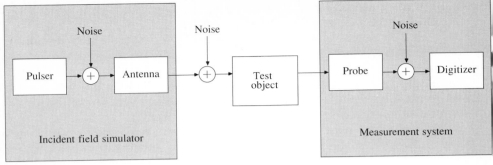

FIGURE 7.8–2
Signal processing model of EMTF

success when low-noise input and undistorted object response can be achieved— this is the goal of signal processing.

Unlike typical electronic circuits, where signals can be extracted with a simple probe, electromagnetic systems require special probes. Electromagnetic probes are small antennas excited by a field (the signal). When the field probe is placed a radial distance from the source antenna and no object is present, the probe measures the *free field response*. When a test object is present and the probe is located on the object, the probe measures the *object response*. Unfortunately, the free field (input) and object response (output) cannot be measured simultaneously, because the field is disturbed by the presence of the object. Thus, input and output data relative to a test object must be obtained by separate measurements. This practice brings up the problem of an output sequence generated by an input sequence different from the measured one.

The primary application of the EMTF is to obtain the transient response of scale models. This response is classically characterized by the spectral content of the signals using fast Fourier transforms. Good transforms imply long data records; however, for fast transients, this implies faster digitization to increase sampling rates. Limited sampling capability, coupled with noisy, uncertain measurements, suggests that parametric identification techniques may offer a reasonable alternative. Recall that the task of the *identification problem* is simply to find the best fit of a pre-selected parametric model to the data, given a set of noisy input and output data.

Identification is most successful when both input and output are available simultaneously; however, because of the nature of the EM measurements, this is not possible. Therefore, we choose to identify the free-field response or input to the object a priori, and then make object response measurements. The object identification is then performed using the identified (and filtered) free-field and measured-object responses. In order to satisfy this objective, we must characterize the various component models depicted in Fig. 7.8–2. Once these models are identified, they can be used to estimate the free-field response. Since the models are to be used only to predict the data, we use a "black-box" approach. Black-box modeling fits parametric models to measured data with no attempt to obtain the physical meaning of the estimated parameters.

The solution to the identification problem consists of three basic ingredients: (1) criterion, (2) model set, and (3) algorithm. Once the model set and criterion are selected, various algorithms can be developed. For our problem, we choose the *prediction-error criterion* given by

$$J(t) = E\{e^2(t)\}$$

where $e(t) = y(t) - \hat{y}(t \mid t - 1)$ is the *prediction error* and $\hat{y}(t \mid t - 1)$ is the best estimate of $y(t)$, given data up to $t - 1$.

For "black-box" modeling, we choose the well-known autoregressive moving average model, with exogenous inputs (ARMAX) given by:

$$A(q^{-1})y(t) = B(q^{-1})u(t) + C(q^{-1})e(t)$$

where y, u, and e are the process output, input, and noise sequences, and A, B, and C are the respective polynomials in backward-shift operator q^{-1} dimensioned N_a, N_b, and N_c.

The ARMAX model was selected for a number of reasons. First, from Fig. 7.8–2, we see that system or driving noise contaminates the object responses from various sources—the pulser, extraneous sources within the laboratory, and reflections. This noise is then filtered along with the actual plane-wave excitation by the object; therefore, the coloring filter $C(q^{-1})$, inherent in the ARMAX model, is a key element to produce unbiased parameter estimates. Since the antenna excitation is independently estimated prior to the object testing, an exogenous input $u(t)$ must also be included in the model structure, again justifying the applicability of the ARMAX model.

Since much of the facility work is performed in real time, we use recursive prediction-error algorithms (Sec. 7.6) to identify the ARMAX models, namely, the Recursive Extended Least-Squares (RELS) and Recursive Maximum-Likelihood (RML) methods (see Tables 7.6–1, 7.8–1). These algorithms are easily implemented on the microcomputer and employed for pre-processing and calibration runs.

A practical approach to perform the identification consists of (1) designing the test (input selection, etc.), (2) preprocessing the raw data (averaging, filtering, etc.), (3) estimating the ARMAX model order (N_a, N_b, N_c), (4) estimating the ARMAX model parameters ($\{a_i\}$, $\{b_i\}$, $\{c_i\}$), (5) performing prediction-error tests for "fit" validation, and (6) performing ensemble tests for model validation. Using this approach, each component of the EMTF is characterized.

The crucial part of the identification process is estimating the order of the ARMAX model. Various order tests have been suggested to solve this problem (see Table 7.7–1) and a number of them are employed to initially screen models. Perhaps the most popular methods we employ are the FPE,

$$\text{FPE}(N_\Theta) = \left(\frac{N + N_\Theta + 1}{N - N_\Theta - 1} \right) R_{ee} \qquad (7.8\text{–}1)$$

TABLE 7.8–1
Recursive extended least-squares (RELS)

Prediction error

$$e(t) = y(t) - \hat{y}(t \mid t - 1) = y(t) - \phi'(t)\Theta(t - 1) \qquad \text{(Prediction error)}$$

$$R_{ee}(t) = \phi'(t)P(t - 1)\phi(t) + \lambda(t) \qquad \text{(Prediction-error variance)}$$

Update

$$\Theta(t) = \Theta(t - 1) + \mu K(t)e(t) \qquad \text{(Parameter update)}$$

$$r(t) = y(t) - \overline{\phi}'(t)\Theta(t) \qquad \text{(Residual update)}$$

$$P(t) = (I - K(t)\phi'(t))P(t - 1)/\lambda(t) \qquad \text{(Covariance update)}$$

$$\lambda(t) = \lambda_0\lambda(t - 1) + (1 - \lambda_0) \qquad \text{(Forgetting-factor update)}$$

Gain

$$K(t) = P(t - 1)\phi(t)R_{ee}^{-1}(t) \qquad \text{(Gain or weight)}$$

$$\overline{\phi}'(t) = [-y(t - 1)\ldots - y(t - N_a) \mid u(t)\ldots u(t - N_b) \mid r(t - 1)\ldots r(t - N_c)]$$

$$\text{(Phi-update)}$$

---

and the AIC, by

$$\text{AIC}(N_\Theta) = -\ln R_{ee} + 2\frac{N_\Theta}{N} \qquad (7.8\text{–}2)$$

where N is the number of data samples, N_Θ is the total number of model parameters, and R_{ee} is the (estimated) prediction error variance.

Model order is determined by selecting the order corresponding to the minimum value of the criteria. Other tests are also utilized to aid in the selection process. The *Signal-Error Test* consists of identifying ARMAX models for a sequence of increasing orders, exciting each model with the exogenous input to produce an estimated response, $\hat{y}(t)$, and overlaying the estimated and the actual response, $y(t)$ versus $\hat{y}(t)$, to select a "reasonable" fit to the data. That is, these tests are performed by exciting the identified ARMAX model with the exogenous input and producing the estimated (filtered or smoothed) response,

$$\hat{y}(t) = \frac{\hat{B}(q^{-1})}{\hat{A}(q^{-1})}u(t)$$

Once the order is determined and the parameters of the selected $\text{ARMAX}(N_a, N_b, N_c)$ are estimated, then the adequacy of the fit is determined by examining the statistical properties of the corresponding prediction-error sequence, $\{e(t)\}$. The prediction errors must be zero-mean and statistically white; that is, 95 percent of the sample correlation estimates must lie within the bounds established by the interval, I_e,

$$I_e = \left[0.0 \pm 1.96\frac{\hat{R}_{ee}(0)}{\sqrt{N}}\right]$$

The first component characterized is the pulse-generator unit. Here, the pulser is initiated and an ensemble of measurements generated. SIG [17] is used to identify the average pulser response. The identification process is initiated by performing a series of order tests, using the FPE and AIC criteria to select the model order, ARMAX(N_a, N_b, N_c), estimating the model parameters for the selected order, and finally performing signal-error and prediction-error tests to assure a reasonable fit to the measured data. The results for the pulser identification are shown in Fig. 7.8–3. Here we see the signal-error test overlaying the estimated (solid) and measured (dashed) responses. Note that the fit is quite reasonable and smooth, compared to the original measurement. We also note the results of the order test displaying the AIC versus order. Here we see that the minimum occurs for the ARMAX(5,4,1) model.[†] Note also that the FPE criteria yielded the same results (order) throughout all of the runs as well. These results

---

[†]The order tests are performed over all of the parameters, starting with the a-parameters to determine N_a, then varying N_b and N_c respectively, in search of a minimum.

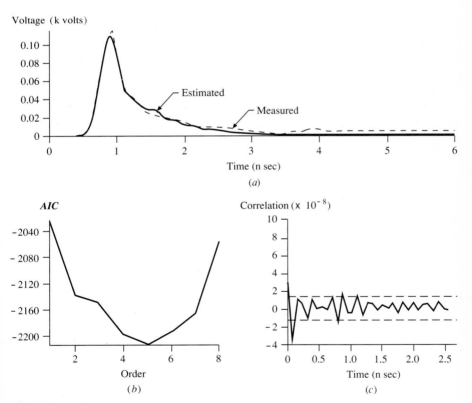

FIGURE 7.8–3
Pulser identification: (a) Signal error (b) Order (c) Whiteness tests

are further validated from the prediction-error whiteness test, also shown in the figure. Aside from an early lag correlation, the test indicates that the errors are statistically white.

The model is also validated for prediction by calculating the residual error series over the ensemble of measurements, that is,

$$\epsilon_i(t) = y_i(t) - \hat{y}(t)$$

where i corresponds to the ith record in the ensemble and $\hat{y}$ is obtained from the model using the estimated parameters. Once determined, the corresponding standard errors can be calculated to ascertain the adequacy of the model; that is, the standard error, bias error, and RMS error are given by

$$\sigma_\epsilon = \sqrt{E\{\epsilon^2(t)\} - E^2\{\epsilon(t)\}}$$

$$\beta_\epsilon = E\{\epsilon(t)\} - 0$$

$$\text{RMS}_\epsilon = \sqrt{\sigma_\epsilon^2 + \beta_\epsilon^2}$$

The results of this validation and the particular fit are depicted in Table 7.8–2. The next component to consider is the bicone antenna shown in Fig.

TABLE 7.8–2
EMTF models and validation

Type	Pulse generator	Probe
	Order	
	5	2
	Coefficient estimates	
a_1	−1.262	−0.904
a_2	1.101	0.231
a_3	−0.926	
a_4	0.445	
a_5	−0.169	
b_0	0.110	0.529
b_1	0.218	−0.530
b_2	0.324	
b_3	0.399	
b_4	0.243	
c_1	0.559	0.519
	Prediction error	
μ_e	0.244×10^{-5}	1.83×10^{-5}
σ_e	5.60×10^{-4}	5.38×10^{-4}
	Validation	
β_ϵ	1.14×10^{-2}	-4.9×10^{-5}
σ_ϵ	9.8×10^{-3}	1.008×10^{-3}
RMS_ϵ	1.5×10^{-2}	1.11×10^{-3}

7.8–1. From first principles, the antenna model can be derived [34] as

$$E_\Theta(r, t) = \frac{2V_p}{\ln \cot (\Psi/2)} \left[\frac{Z_{\text{bicone}}}{Z_{\text{bicone}} + 2Z_L} \right] \left(\frac{1}{r} \right)$$

where E_Θ and V_p are the electric field produced by the bicone and pulser voltage; Z_{bicone} and Z_L are the bicone and pulser terminal impedances; and Ψ and r are the bicone half angle and radial distance from the bicone ground-plane center line. For the EMTF, we know that $Z_{\text{bicone}} = 375.8 \ \Omega$, $\Psi = 45°$, $Z_L = 50 \ \Omega$, and r is determined by the particular facility configuration used in the given experiment.

Next, the D-dot (sensor) probe is modeled. It transduces the derivative of the electric field intensity into an output voltage. The first-principles model obtained from the manufacturer's specifications did not match the measurement data well; therefore, a model was identified from the data. The results of the probe identification are shown in Fig. 7.8–4. Here we again see the smooth estimated response, $\hat{y}(t)$, obtained in the *Signal-Error Test*, as well as the corresponding order test, which indicates that the ARMAX(2,1,1) model satisfies the minimum value of the AIC and FPE tests. The prediction-error whiteness test is clearly white; the validation statistics are shown in Table 7.8–2 as well.

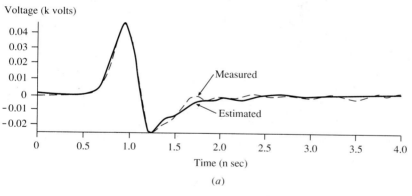

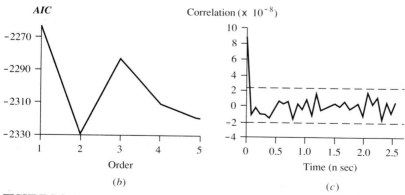

FIGURE 7.8–4
D-dot probe identification (*a*) Signal error (*b*) Order (*c*) Whiteness tests

Finally, we investigate the sampling scope, a prime instrument used in measuring EMP response. This scope has an essentially flat frequency response in the region of most interest ($\sim$ 1 GHz); thus, no modeling was necessary.

Since the measurement probes are bandlimited, their effect on the input and output data must be removed or deconvolved prior to identification. Deconvolution of sensor probes from the data can be achieved in a variety of ways. A simple (suboptimal) procedure is to use the inverse of the identified model and excite it with the sensor measurement to obtain an estimate of the deconvolved signal, say $\hat{u}(t)$. That is,

$$\hat{u}(t) = \hat{H}^{-1}(q^{-1})y(t)$$

where $\hat{H}^{-1}(q^{-1}) = \hat{A}(q^{-1})/\hat{B}(q^{-1})$, the inverse model. In this approach, we are merely filtering the data with the inverse of the identified model.

Another and more desirable approach is to actually solve the corresponding optimal (mean-squared error) deconvolution problem (see Sec. 7.4), which gives the solution of the following set of normal equations,

$$\hat{\underline{u}} = \mathbf{R}_{hh}^{-1}\underline{R}_{yh}$$

where $h(t)$ is the impulse response obtained from the identified model, and $\mathbf{R}_{hh}$ and $\underline{R}_{yh}$ are the corresponding auto- and cross-covariance matrices.

Since $\mathbf{R}_{hh}$ has a Toeplitz structure, it can be solved efficiently using the LWR. Applying this recursion to the free-field response measurement should yield the actual antenna response. The results are shown in Fig. 7.8–5. Here we see that the deconvolver (128-weight finite-impulse response filter) performs reasonably well, as indicated by the similarity between the true modeled (dashed) response and the estimated response. The overall EMTF model was validated using an ensemble of free-field measurements, and the average error was quite reasonable. This completes the characterization of the EMTF for this particular configuration. The model developed is used for experimental design, reconfig-

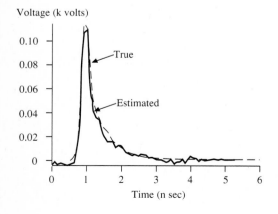

FIGURE 7.8–5
Deconvolved free-field antenna response (probe deconvolution)

uration studies, and, of course, primarily for use in identifying the EMP response of an object.

Next we discuss the application of the EMTF to a thin-wire dipole—a fundamental electromagnetic object. The purpose of this experiment is to develop a model of the dipole to predict that dipole's response when subjected to various excitations. It has been shown that a dipole can be represented in the time domain by a series of damped exponentials using the Singularity Expansion Method (SEM) [33]. In general, the EM response to input $u(t)$ can be represented as

$$y(\underline{r}, t) = \sum_i \nu_i(\underline{r}) e^{s_i t} \eta_i(\underline{e}, s_i) u(t) \qquad (7.8\text{--}3)$$

where

s_i is the natural complex frequency dependent on object parameters ($s_i = d_i \pm j2\pi f_i$)

ν_i is the natural mode, a nontrivial solution of field equations at s_i

$\underline{r}$ values are spatial coordinates of position over the body

η_i is the coupling coefficient vector, the strength of the oscillation

$\underline{e}$ values are exciting field characteristics

The most important feature of this model is the set of complex frequencies and natural modes ($\{s_i\}, \{\nu_i(\underline{r})\}$), which are excitation independent (not a function of $\underline{e}$). The effect of the exciting wave is in the coupling coefficients. To completely characterize the electromagnetic interaction, we need only s_i, ν_i, and η_i.

In most EMP problems, the quantities of interest are the poles and residues, which can be represented in the frequency domain by

$$Y(\underline{r}, s) = \sum_i \frac{R_i(\underline{r})}{s + s_i} \qquad (7.8\text{--}4)$$

where R_i is a complex residue such that $R_i(\underline{r}) = \nu_i(\underline{r}) \, \eta_i(\underline{e}, s_i)$.

Thus, the response of an EM object can be predicted by its poles and residues ($\{s_i\}, \{R_i(\underline{r})\}$). Since the residues are dependent on characteristics of the excitation due to η, the researcher of electromagnetics is primarily interested in the poles. The EM response-prediction problem becomes that of finding the object poles. From the signal-processing viewpoint, this is a system-identification problem, solved in discrete time and transformed to the s-plane.

To study dipole scattering and validate EMTF models, a 0.4-meter, thin-wire dipole was mounted on the EMTF ground plane and excited with a transient electromagnetic field. Its response was measured by means of an electric-field probe mounted at its center and connected directly to a sampling scope. These data were low-pass filtered by an anti-aliasing filter, decimated by a factor of two, and deconvolved. Furthermore, they were obtained from an ensemble average of ten records. From the measured data shown in Fig. 7.8–6,

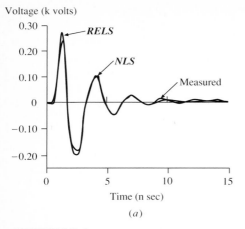

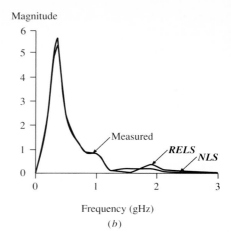

FIGURE 7.8–6
Identified dipole response: (a) Temporal (b) Spectra

the first two resonances appear to be adequately excited, but the others are not and are probably contaminated with noise. Therefore, we should expect to be able to identify the first two easily but to have trouble with the others because of the low signal levels. Two algorithms were used to estimate the dipole parameters: the RELS algorithm for prediction-error modeling and a nonlinear least-squares NLS algorithm for output-error modeling [35]. The RELS algorithm, which was discussed in the previous section, was iterated a number of times, enabling it to achieve performance equivalent to the off-line maximum likelihood [7].

The NLS algorithm is simply a nonlinear least-squares numerical optimization algorithm which determines the parameters of an ARX model, that is, ARMAX($N_a, N_b, 0$). It produces parameter estimates Θ that minimize the mean-squared output error,

$$J(\Theta) = E\{\epsilon^2(t)\} \approx \sum_i \epsilon^2(i) \qquad (7.8\text{–}5)$$

where $$\epsilon(t) = y(t) - \hat{y}_m(t)$$

and $\hat{y}_m(t)$ is obtained by exciting the identified model with the exogenous input, that is,

$$A_m(q^{-1})\hat{y}_m(t) = B_m(q^{-1})u(t) \qquad (7.8\text{–}6)$$

The parameter vector $\underline{\Theta}$ is updated using the classical gradient iteration equation from numerical optimization,

$$\underline{\Theta}(i + 1) = \underline{\Theta}(i) + \Delta(i)\underline{d}(i) \qquad (7.8\text{–}7)$$

where $\Delta(i)$ is the step size and $\underline{d}(i)$ is a descent direction.

The algorithm uses Newton's method and has proved very useful for transient data. Note that this is a computationally intensive (off-line) output-error algorithm, rather than a recursive (one-step) prediction-error algorithm. For Newton's method, the direction vector is given by

$$\underline{d}(i) = -\left[\frac{\partial^2 J(\Theta_i)}{\partial \Theta^2}\right]^{-1}\left[\frac{\partial J(\Theta_i)}{\partial \Theta}\right]'$$

and the step size is varied according to the Armijo rule [35].

Models were tried for several different orders and for several different starting points. Many local minima were observed for orders ranging from five to fifteen. To obtain better starting points, the recursive least-squares (RLS) algorithm was used. This algorithm was easily programmed to run on a small minicomputer at the EMTF, but the more intensive off-line RELS and NLS algorithms require a larger machine.

Since the dipole theoretically contains an infinite number of poles, large-models were selected to approximate as many of the theoretical poles as possible. Order tests performed with the RLS algorithm revealed that either a 12th- or 15th-order model contained approximations to most of the theoretically predicted resonances and provided reasonable fits to the data, as indicated by the response and Fourier spectra in Fig. 7.8–6. Furthermore, the prediction-error sequence satisfied a statistical test for whiteness. We decided that an ARMAX(15,12,0) model that we estimated using RLS was a reasonable starting point for the RELS and NLS algorithms.

The RELS and NLS algorithms were applied to the data records. The RELS algorithm was used to estimate an ARMAX(15,12,2) model, and the result, according to the FPE and AIC, was found to be superior to the model estimated with RLS. Similarly, an output-error model, with $N_a = 15$ and $N_b = 12$, was estimated with NLS. The identified response and spectrum are shown in Fig. 7.8–6; the fits were reasonable.

The identified parameters are transformed to the continuous domain using the impulse-invariant transformation [16], which first requires the partial fraction expansion of the identified model to produce

$$H(z) = \sum_i \frac{R_i}{1 + z^{-1}z_i} + \frac{R_i^*}{1 + z^{-1}z_i^*} \qquad (7.8\text{–}8)$$

where $z_i = e^{s_i \Delta T}$, $s_i = -d_i \pm j2\pi f_i$ is a complex pole, and R_i is a complex residue.

Once this is accomplished, the impulse-invariant mapping is given by

$$\left(\left\{\frac{1}{\Delta T}\ln z_i\right\}, \{R_i\}\right) \longmapsto \left(\{s_i\}, \{R_i(\underline{r})\}\right)$$

We note that, using the impulse-invariant transformation, we have implicitly assumed that the continuous system or the EM object response in this application is characterized by a bandlimited response or the transformation will produce aliased results. This is in fact a valid assumption, since bandlimiting occurs during both data acquisition and the decimation processing.

TABLE 7.8-3
Identified poles vs. analytical poles for 40 cm dipole antenna

Pole	Theoretical		RELS (on-line) – ARMAX (15,12,2)				NLS (off-line) – ARMAX (15,12,0)			
	$-d(\times 10^8)$	f(Ghz)	$-\hat{d}(\times 10^8)$	$\hat{d}_{err}$	$\hat{f}$(Ghz)	$\hat{f}_{err}$(Ghz)	$-\hat{d}(\times 10^8)$	$\hat{d}_{err}$	$\hat{f}$(Ghz)	$\hat{f}_{err}$(Ghz)
(1)	0.30	0.35	0.78	1.57	0.33	0.04	0.79	1.59	0.33	0.04
(3)	0.53	10.87	1.05	0.99	10.77	0.01	0.84	0.59	10.87	0
(5)	0.66	18.25	0.40	0.30	19.35	0.06	1.43	1.17	17.17	0.06
(7)	0.79	25.65	2.38	2.01	28.19	0.10 *	1.03	0.31	18.99	0.26
(9)	0.99	32.76	1.75	0.76	37.83	0.16 *	1.02	0.02	35.90	0.10
(11)	1.32	40.04	1.77	0.347	48.13	0.20	0.50	0.62	49.15	0.23

$a_{err} = \frac{a\text{TRUE} - \hat{a}}{a\text{TRUE}}$

* Residue order of magnitude low

Tesche [36] has calculated theoretical values for the poles of a dipole; these results are dependent on both its length and its diameter. Typically, the results are normalized to remove the effects of length. Table 7.8–3 shows the theoretical and the estimated poles. Since the response is measured at the center of the dipole, only the odd-numbered poles are excited. It is clear that neither algorithm gives good estimates of the damping factors (real part of the poles). Because of the limited amount of data in a transient signal, this result is not unexpected. Both algorithms perform much better on the frequency estimates (imaginary parts) and are even able to extract some of the higher-frequency resonances. We conclude that for a simple object, such as a dipole, the singularity-expansion method coupled with the appropriate identification technique can be a useful tool for evaluating transient responses.

7.9 SUMMARY

In this chapter, we have discussed the fundamental concepts underlying modern or parametric signal-processing techniques combining the equivalent methods from signal-processing, time-series analysis, and system-identification areas. After introducing the basic methodology, we developed the concepts of optimal (linear) estimation from the geometric viewpoint and reconciled these results with the classical. Next, we developed the Levinson all-pole and all-zero filters and showed how they could be used to solve the deconvolution problem. We then developed the lattice and prediction-error filters from the system-identification viewpoint. We analyzed the problem of model-order estimation and developed some of the popular criteria. Finally, we discussed a case study involving applications of the various aspects of parametric signal processing.

REFERENCES

1. J. Markel and A. Gray, *Linear Prediction of Speech* (New York: Springer-Verlag, 1976).
2. L. Rabiner and R. Shafer, *Digital Processing of Speech Signals* (Englewood Cliffs, N.J.: Prentice-Hall, 1978).
3. D. Childers, Ed., *Modern Spectral Analysis*, (New York: IEEE Press, 1978).
4. G. Box and G. Jenkins, *Time Series Analysis: Forecasting and Control* (San Francisco: Holden-Day, 1976).
5. R. Kalman, "A New Approach to Linear Filtering and Prediction Problems," *J. Basic Eng.* Vol. 82, 1960.
6. K. Astrom and P. Eykhoff, "System Identification—A Survey," *Automatica*, Vol. 7, 1971.
7. L. Ljung and T. Soderstrom, *Theory and Practice of Recursive Identification* (Boston: MIT Press, 1983).
8. G. Goodwin and K. Sin, *Adaptive Filtering, Prediction and Control* (Englewood Cliffs, N.J.: Prentice-Hall, 1984).
9. J. Candy, *Signal Processing: The Model-Based Approach* (New York: McGraw-Hill, 1986).
10. T. Kailath, *Lectures on Kalman and Wiener Filtering Theory* (New York: Springer-Verlag, 1981).
11. S. Orfanidis, *Optimum Signal Processing* (New York: Macmillan, 1985).
12. B. Noble, *Applied Linear Algebra* (Englewood Cliffs, N.J.: Prentice-Hall, 1967).

13. R. Deutsch, *Estimation Theory* (Englewood Cliffs, N.J.: Prentice-Hall, 1965).
14. G. Bierman, *Factorization Methods of Discrete Sequential Estimation* (New York: Academic Press, 1977).
15. S. Haykin, *Introduction to Adaptive Filters* (New York: Macmillan, 1984).
16. S. Tretter, *Introduction to Discrete-Time Signal Processing* (New York: Wiley, 1976).
17. D. Lager and S. Azevedo, "SIG—A General Purpose Signal Processing Code," *Proc. IEEE*, 1987.
18. R. Wiggins and E. Robinson, "Recursive Solution to the Multichannel Filtering Problem," *J. Geophysics, Res.*, Vol. 20, 1965.
19. M. Silvia and E. Robinson, *Deconvolution of Geophysical Time Series in the Exploration of Oil and Natural Gas* (New York: Elsevier, 1979).
20. B. Widrow and S. Stearns, *Adaptive Signal Processing* (Englewood Cliffs, N.J.: Prentice-Hall, 1984).
21. B. Friedlander, "Lattice Filters for Adaptive Processing," *Proc. IEEE*, Vol. 70, 1982.
22. M. Honig and D. Messerschmitt, *Adaptive Filters: Structures, Algorithms, and Applications* (Boston: Kluwer, 1984).
23. C. Cowen and P. Grant, Eds., *Adaptive Filters* (Englewood Cliffs, N.J.: Prentice-Hall, 1985).
24. A. Giordano and F. Hsu, *Least Squares Estimation with Applications to Digital Signal Processing* (New York: Wiley, 1985).
25. N. Mohanty, *Random Signals, Estimation, and Identification* (New York: Van Nostrand, 1986).
26. R. Jategaonkar, J. Raol, and S. Balakrishna, "Determination of Model Order for Dynamical Systems," *IEEE Trans. Sys., Man, Cyber.*, Vol. 12, 1982.
27. K. Astrom, Ed. "System Identification—A Tutorial," *Automatica*, 1981.
28. H. Akaike, "A New Look at the Statistical Model Identification," *IEEE Trans. Autom. Control*, Vol. 19, 1974.
29. W. Gersch "Lecture Notes on Time Series," Short course at Lawrence Livermore National Lab, 1985.
30. J. Candy and J. Zicker, "Electromagnetic Signal Processing: An Estimation/Identification Application," *Automatica*, 1987.
31. R. Bevensee, F. Deadrick, E. Miller, and J. Okada, "Validation and Calibration of the LLL Transient Electromagnetic Measurement Facility," *LLNL Report*, UCRL-52225, 1977.
32. E. Miller, Ed., *Transient Electromagnetic Measurements* (New York: Van Nostrand, 1986).
33. C. Baum, "On the Singularity Expansion Method for the Solution of EM Interaction Problems," *AFWL Interaction Note 88*, 1971.
34. B. Steinberg, *Principles of Aperture and Array System Design* (New York: Wiley, 1976).
35. D. Goodman, "NLS: A System Identification Package for Transient Signals," *LLNL Report*, UCID-19767, 1983.
36. F. Tesche, "On the Analysis of Scattering and Antenna Problems Using the Singularity Expansion Technique," *IEEE Trans. Antennas Propagation*, Vol. 21, 1973.

SIG NOTES

SIG can be used to design parametric signal processors using the LSI*** commands. The Levinson all-pole and all-zero filters can be designed using the LSIAONLY and IMPULSERESPONSE commands, respectively. Lattice-filter designs can be accomplished using the covariance lattice procedures through the LSILATTICE command or the LSIBURG (SHARE) for the original Burg block design method. Finally, prediction-error filters can be developed, using the LSIABC command, which employ the ELS method (see Secs. 7.6 and 7.8). Automatic order estimation is available using the AIC and FPE procedures. SIG also features a set of deconvolution commands characterized by the LSD***. The suboptimal all-pole approach (see Secs. 7.3 and 7.8) is accomplished using the LSDAONLY command, while the optimal all-zero approach (see Sec. 7.4) is available in the LSDBONLY command. Finally, the lattice approach to all-pole deconvolution is available using the LSDLATTICE command.

EXERCISES

7.1. Given a sequence of random variables $\{y(t)\}$, define the minimum-variance estimate:

$$e(t) = y(t) - \hat{y}(t \mid t - 1) \qquad \hat{y}(1 \mid 0) = 0$$

$$\hat{y}(t \mid t - 1) = E\{y(t) \mid Y(t - 1)\}$$

(a) Show that the values for $\{e(t)\}$ are orthogonal.

(b) Show that if $E\{e^2(t)\} > 0$, then

$$Y(N) = \begin{bmatrix} y(1) \\ \vdots \\ y(N) \end{bmatrix} \qquad \text{and} \qquad \underline{e}(N) = \begin{bmatrix} e(1) \\ \vdots \\ e(N) \end{bmatrix}$$

are related by $\qquad Y(N) = L(\underline{N})\underline{e}(N) \qquad$ and $\qquad N := 1, 2, \ldots, N$

(c) If $W(\underline{N})$ is a linear operation that yields $\hat{Y}(N)$ from $Y(N)$, show that

$$W(\underline{N}) = I - L^{-1}(\underline{N}) \qquad \text{and} \qquad \hat{Y}(N) = Y(N) - \underline{e}(N)$$

(d) Show that $L(\underline{N}) = R_{ye}(\underline{N})R_{ee}^{-1}(\underline{N})$

(e) Show that $R_{yy}(\underline{N}) = L(\underline{N})R_{ee}(\underline{N})L'(\underline{N})$

(f) If x is a related vector, show that

$$\hat{x} = R_{xy}(\underline{N})R_{yy}^{-1}(\underline{N})Y(N) - R_{xe}(\underline{N})R_{ee}^{-1}(\underline{N})\underline{e}(N)$$

(g) If $R_{yy}(i,j) = \alpha^{|i-j|}$ for $0 < \alpha < 1$, find $L(\underline{N})$ and $R_{yy}^{-1}(\underline{N})$.

7.2. Construct an innovation sequence $\{e(1), e(2)\}$ from the measurement sequence

$$\begin{bmatrix} 1 \\ 4 \end{bmatrix}, \quad \begin{bmatrix} 2 \\ 3 \end{bmatrix} \quad \text{if } R_{ee} = \text{diag}\{1, 2\} \quad \text{and} \quad R_{ye}(2, 1) = \begin{bmatrix} 2 & 1 \\ 1 & 6 \end{bmatrix}$$

7.3. Construct the Gram-Schmidt orthogonalization procedure, starting with $e(N) = y(N)$ rather than $e(1) = y(1)$, as in Eq. (7.2–9).

(a) Determine the transformation matrix L^*; how does it differ from L in Eq. (7.2–9)?

(b) Construct the covariance equivalence transformation as in Eq. (7.2–12); how does is differ?

7.4. Suppose we are given a measurement characterized by

$$y(t) = x(t) + n(t)$$

where x is exponentially correlated $R_{xx}(k) = (1/2)^{|k|}$ and n is zero-mean, white noise with variance $R_{nn} = 0.5$.

(a) Determine the optimal realizable Wiener filter.

(b) Determine the optimal realizable Wiener filter for one-step prediction, that is, $x(t + 1 \mid Y(t))$.

(c) Simulate the process and obtain the Wiener solution in both time and frequency domain.

7.5. We are asked to design a Wiener filter to estimate a random signal with covariance $R_{ss}(k) = (1/2)^{|k|}$ from a noisy measurement contaminated with unit-variance white noise.

(a) Determine the optimal Wiener "noncausal" solution.

(b) Determine the causal Wiener solution.

7.6. Suppose that a random signal s is zero-mean with covariance

$$R_{ss}(k) = a \exp(-b|k|)$$

and is to be estimated using an instrument with the following model:

$$y(t) = cs(t) + n(t)$$

with n zero-mean and white with variance R_{nn}, $a = b = 1$, $c = 2$, and $R_{nn} = 4$.
(a) Find the noncausal Wiener solution $K_{NC}(z)$ and $k_{NC}(t)$.
(b) Find the causal Wiener solution $K(z)$ and $k(t)$.
(c) Verify the results using the whitening filter approach.

7.7. Given the covariance of $R_{yy}(k) = (1/2)^{|k|}$, then
(a) Calculate the second-order, all-pole Wiener filter. Explain the result.
(b) Assuming that $R_{yy}(k) = G(1/2)^{|k|}$, what effect does G have on the parameter estimates in part (a)?
(c) Explain how to estimate the gain G in part (b) and in general.

7.8. Suppose we have the autoregressive model

$$y(t) = -a_1 y(t-1) - a_2 y(t-2) + e(t)$$

where $\{e(t)\}$ is white.
(a) Derive $\hat{y}(t \mid t-1)$.
(b) Derive $\hat{y}(t \mid t-2)$.

7.9. Suppose we have the moving-average model

$$y(t) = e(t) + c_1 e(t-1)$$

where $\{e(t)\}$ is white.
(a) Derive $\hat{y}(t \mid t-1)$.
(b) Derive $\hat{y}(t \mid t-2)$.

7.10. We are given a sequence $\{y(t)\}$ generated by passing white noise, with variance R_{ee} through a second-order, all-pole digital filter with transfer function

$$H(z) = \frac{1}{1 + a_1 z^{-1} + a_2 z^{-2}}$$

Find estimates for $\{\hat{a}_1, \hat{a}_2, \hat{R}_{ee}\}$ in terms of the statistics of $y(t)$ using (a) the batch approach and (b) the Levinson-Durbin recursion.

7.11. Repeat Exercise 7.10 using a second-order FIR model

$$H(z) = 1 + b_1 z^{-1} + b_2 z^{-2}$$

using (a) the batch approach and (b) the LWR recursion.

7.12. Starting with the Levinson recursion

$$a(i+1, i+1) = -k_{i+1} \qquad\qquad\quad \text{for } i = 1, \ldots, N$$
$$a(j, i+1) = a(j, i) - k_{i+1} a(i+1-j, 1) \qquad \text{for } j = 1, \ldots, i$$

expand over the noted indices and take Z-transforms of both sides to show
(a) $A_{i+1}(z) = A_i(z) - k_{i+1} z^{-(i+1)} A_i(z^{-1})$ where $A_i(z) = 1 + a(i,1)z^{-1} + \cdots + a(i,i)z^{-i}$
(b) Define the *reverse* polynomial $A_i^R(z) = z^{-i} A_i(z^{-1})$ and using (a) show that

$$A_{i+1}^R(z) = z^{-1} A_i^R(z) - k_{i+1} A_i(z)$$

(c) Combine (a) and (b) and show that the predictor polynomials satisfy the recursion

$$\begin{bmatrix} A_{i+1}(z) \\ A_{i+1}^R(z) \end{bmatrix} = \begin{bmatrix} 1 & -k_{i+1}z^{-1} \\ -k_{i+1} & z^{-1} \end{bmatrix}\begin{bmatrix} A_i(z) \\ A_i^R(z) \end{bmatrix}$$

which is initialized with $i = 0$, $A_0(z) = A_0^R(z) = 1$. Compare with Eq. (4.6–5). This is the *Levinson forward recursion* on both forward and backward prediction polynomials.

(d) Starting with the AR models and the forward recursion, derive Eq. (4.6–3).

(e) Suppose we are given the reflection coefficient sequence $\{k_1, \ldots, k_4\} = \{1/2, -1/2, 1/2, -1/2\}$; find $A_4(z)$.

(f) The mean-squared error can also be found recursively, using the recursion $J(i + 1) = (1 - k_{i+1}^2)J(i)$ with initial conditions $J(0) = R_{yy}(0) = 40.5$; find $J(4)$.

(g) Using the fact that the fourth-order predictor satisfies the relation

$$R_{yy}(i) = -\sum_{j=1}^{i} a(i,j)R_{yy}(i - j)$$

find the corresponding covariances $\{R_{yy}(1) \cdots R_{yy}(4)\}$, using the results of (e) above.

7.13. The *Levinson reverse recursion* can be derived directly from the forward recursion of the previous exercise. Show that the recursion is given by

$$\begin{bmatrix} A_i(z) \\ A_i^R(z) \end{bmatrix} = \frac{1}{1 - k_{i+1}^2}\begin{bmatrix} 1 & k_{i+1} \\ k_{i+1}z & z \end{bmatrix}\begin{bmatrix} A_{i+1}(z) \\ A_{i+1}^R(z) \end{bmatrix}$$

(a) Starting with the AR models and the backward recursion, derive Eq. (4.6–4).

(b) Given the fourth-order predictor and corresponding mean-squared error

$$A_4(z) = 1 - 1.25z^{-1} + 1.3125z^{-2} + 0.5z^{-4}, \qquad J(4) = 0.81$$

find the corresponding reflection coefficients $\{k_1, \ldots, k_4\}$, mean-squared errors, $\{J(0), \cdots, J(3)\}$, and lower-order predictors.

7.14. Show that the following properties of lattice filters hold true:

(a) Both forward and backward prediction errors are orthogonal to the input sequence $\{y(t)\}$; that is,

$$E\{e_f(t, i)y(t - j)\} = 0 \qquad 1 \le j \le i$$

and
$$E\{e_b(t, i)y(t - j)\} = 0 \qquad 0 \le j \le i - 1$$

(b) The cross-covariance of both forward and backward prediction errors with the input satisfies

$$E\{e_f(t, i)y(t)\} = E\{e_b(t, i)y(t - i)\} = J(i)$$

for $J(i)$ the corresponding mean-squared error (power).

(c) The backward prediction errors are uncorrelated (white); that is,

$$E\{e_b(t, i)e_b(t, j)\} = \begin{cases} J(i) & i = j \\ 0 & \text{otherwise} \end{cases}$$

(d) The forward prediction errors satisfy

$$E\{e_f(t,i)e_f(t,j)\} = J(i) \qquad i \geq j$$

therefore, they are correlated (colored).

(e) The forward and backward prediction errors are correlated,

$$E\{e_f(t,i)e_b(t,j)\} = \begin{cases} k_j J(i) & i \geq j \\ 0 & i < j \end{cases}$$

7.15. Using the recursion for the backward prediction errors in terms of the predictor coefficients, that is,

$$e_b(t,i) = \sum_{j=0}^{i} a(i-j,i)y(t-j)$$

(a) Show that $\underline{e}_b(t) = L\underline{y}(t)$ with L lower triangular.

(b) Show that

$$\mathbf{R}_{e_b e_b} = L\mathbf{R}_{yy}L'$$

with $\mathbf{R}_{e_b e_b} = \text{diag } [J(0) \cdots J(i)]$.

(c) Calculate the inverse of (b), and show that this is equivalent to a Cholesky decomposition,

$$\mathbf{R}_{yy}^{-1} = \mathbf{S}'\mathbf{S}$$

with $\mathbf{S} = \mathbf{D}^{-1/2}L$.

7.16. Simulate the following AR system:

$$(1 + 1.5q^{-1} + 0.5625q^{-2})y(t) = e(t) + \delta(t-1.2)$$

where $e \sim N(0, 0.01)$, and $\delta(t-d)$ is a unit impulse function at delay-time d for $\Delta T = 0.1$ sec and $N = 64$. Design the following sets of parametric processors:

(a) Levinson all-pole filter

(b) Levinson all-zero filter

(c) Lattice all-pole filter

In each case, determine the impulse response of the designed filter and compare it to the true (signal estimate). Also obtain the filtered output $\hat{y}$ and compare to the measured.

7.17. Simulate the "standard" system

$$y(t) - 1.5y(t-1) + 0.7y(t-2)$$
$$= u(t-1) + 0.5u(t-2) + e(t) - e(t-1) + 0.2e(t-2)$$

where e is zero-mean with $R_{ee} = 0.25$, and u is pseudorandom binary (± 1) over 500 samples.

(a) Develop the RELS estimator for this problem. What are the final parameter estimates?

(b) Show the parameter estimate plots.

CHAPTER

8

ADAPTIVE
SIGNAL
PROCESSING

In this chapter, we develop the adaptive approach to signal processing. After introducing the basic principles behind adaptive processing, we investigate some of the fundamental concepts involved in adaptive algorithms. We then investigate various processors based on the model set selected: the all-zero, all-pole, pole-zero, and lattice adaptive filters. Next, we investigate the application of the adaptive techniques to various problems, including noise cancelling, prediction, and spectral estimation. Finally, we investigate the utilization of these techniques in a case study—transient pulse estimation.

8.1 INTRODUCTION

Adaptive methods of signal processing have evolved from various disciplines, with their roots firmly embedded in numerical optimization theory. Adaptive techniques have been very popular in such areas as signal processing [1,2], array design [3], control [4], and system identification [5,6]. These techniques find application in a variety of problems in which, in contrast to optimal estimation (see Sec. 7.2), signal characteristics are not known very well. Specifically, if the signal has unknown statistics, is nonstationary or time-varying, or if the underlying phenomenology is unknown or only partially known, then adaptive processors offer a potential solution. In fact, if the signal is stationary but the statistics are unknown, the adaptive processor will converge to the optimal estimator. The main problem in adaptive processing is to find an algorithm to adjust parameters (statistics, gain, coefficients, etc.) where incomplete knowledge of the signal characteristics exist.

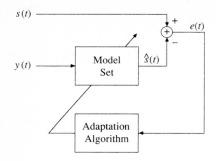

FIGURE 8.1–1
Adaptive signal-processing method

The adaptive approach to signal processing is summarized in Fig. 8.1–1. Once the appropriate model set is selected, an adaptation algorithm must be developed based on a particular criterion to produce the desired signal estimate. We summarize the *adaptive signal processing* method in the following steps:

1. Select a model set (e.g., ARMAX, lattice, state-space).
2. Select an adaptation algorithm (e.g., gradient, Newton) to estimate model parameters as

$$\underline{\hat{\Theta}}(i + 1) = \underline{\hat{\Theta}}(i) + \Delta_i \underline{d}_i$$

for the ith iteration of step size Δ with direction vector $\underline{d}_i$.
3. Construct the processor to adaptively estimate the signal, that is,

$$\hat{s}(t, \hat{\Theta}(i)) = f(\hat{\Theta}(i), y(t))$$

8.2 ADAPTATION ALGORITHMS

In this section, we briefly discuss some of the popular numerical optimization techniques that have been applied to the design of adaptive processors. We make no attempt to develop the underlying theory, which is quite extensive, but refer the reader to Luenberger [7,8] for details.

Most popular adaptive algorithms are based on iteratively updating the N_Θ-parameter vector, $\underline{\Theta}(i)$, that is,

$$\underline{\Theta}(i + 1) = \underline{\Theta}(i) + \Delta\underline{\Theta}(i) \tag{8.2--1}$$

where $\Delta\underline{\Theta}$ is the correction vector added to the previous iteration to obtain the update. The gradient algorithm approach, to which we will limit our discussion, adjusts the parameters iteratively, to minimize the criterion function by descending along this performance surface to reach the minimum. The iteration of Eq. (8.2–1) is adjusted so that the criterion is decreased; that is,

$$J(\underline{\Theta}(i) + \Delta\underline{\Theta}(i)) \leq J(\underline{\Theta}(i)) \tag{8.2--2}$$

must decrease until a minimum is achieved. Suppose we expand the criterion in a Taylor series,

$$J\big(\underline{\Theta}(i) + \Delta\underline{\Theta}(i)\big) = J\big(\underline{\Theta}(i)\big) + \Delta\underline{\Theta}'(i)\frac{\partial}{\partial\Theta}J\big(\underline{\Theta}(i)\big) + h.o.t.$$

where *h.o.t.* means higher order terms. Assuming $\Delta\underline{\Theta}(i)$ is small, then we can neglect the *h.o.t.* and

$$J\big(\underline{\Theta}(i) + \Delta\underline{\Theta}(i)\big) \approx J\big(\underline{\Theta}(i)\big) + \Delta\underline{\Theta}'(i)\frac{\partial}{\partial\Theta}J\big(\underline{\Theta}(i)\big) \le J\big(\underline{\Theta}(i)\big) \qquad (8.2\text{--}3)$$

It remains to choose $\Delta\underline{\Theta}(i)$ such that this inequality holds. Choosing the negative gradient direction with step size Δ_i,

$$\Delta\underline{\Theta}(i) = -\Delta_i\frac{\partial}{\partial\Theta}J\big(\underline{\Theta}(i)\big) \qquad (8.2\text{--}4)$$

guarantees this inequality, since

$$J\big(\underline{\Theta}(i)\big) - \left(\Delta_i\frac{\partial}{\partial\Theta}J'\big(\underline{\Theta}(i)\big)\right)\frac{\partial}{\partial\Theta}J\big(\underline{\Theta}(i)\big) \le J\big(\underline{\Theta}(i)\big) \qquad (8.2\text{--}5)$$

which gives the *steepest descent* or *gradient* algorithm

$$\underline{\Theta}(i+1) = \underline{\Theta}(i) - \Delta_i\frac{\partial}{\partial\Theta}J\big(\underline{\Theta}(i)\big) \qquad (8.2\text{--}6)$$

In addition, the step size Δ_i is a positive scalar chosen in a suitable way. The gradient method becomes inefficient, when the parameter estimates approach the optimum. We illustrate the gradient technique in Fig. 8.2–1.

If the direction vector is selected as an inverse Hessian matrix multiplied by the gradient, that is,

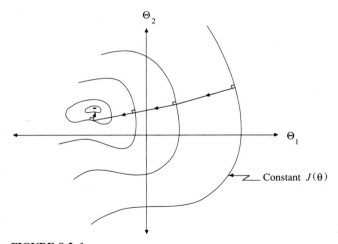

FIGURE 8.2–1
Typical gradient solution for a two-parameter surface

$$\Delta \underline{\Theta}(i) = \Delta_i \underline{d}_i = -\Delta_i \left[\frac{\partial^2}{\partial \Theta^2} J(\underline{\Theta}(i)) \right]^{-1} \left[\frac{\partial}{\partial \Theta} J(\underline{\Theta}(i)) \right] \qquad (8.2\text{--}7)$$

then this is termed the Newton direction. These Newton or quasi-Newton methods perform much better than the gradient algorithms closer to the minimum. In fact, if the criterion function is quadratic, then this algorithm will converge in one step to the optimum solution $\underline{\Theta}_{opt}$. Far from the minimum, these methods may be inefficient; the Hessian is replaced by a positive definite approximation to assure a downward point-search direction [7]. Other techniques based on different choices of step size evolve as well (e.g., conjugate gradient method), but we will confine our discussion to these two cases.

If our criterion includes an expectation to account for randomness, then we obtain stochastic variants of these adaptation schemes. *Stochastic gradient techniques* are based on replacing the deterministic criterion function with its stochastic counterpart, which incorporates the expectation operation; that is,

$$J(\underline{\Theta}) = f(\underline{\Theta}, y) \rightarrow J(\underline{\Theta}) = E\{f(\underline{\Theta}, y)\}$$

and the algorithms remain identical, except for the above changes to the cost function.

The following simple example demonstrates these algorithms.

Example 8.2–1. Suppose we are to estimate a constant c from a measurement y, and we would like to obtain an estimate which is a linear function of the measurement, say,

$$\hat{c} = \Theta y \qquad \text{for a constant } \Theta$$

We choose the criterion

$$J = \frac{1}{2} e^2 = \frac{1}{2} (c - \hat{c})^2$$

and would like to develop the (i) gradient and (ii) Newton iterates.

(i) Gradient estimator:

$$\frac{\partial J}{\partial \Theta} = -(c - \Theta y) y = -(cy - \Theta y^2)$$

or, defining $r = cy$ and $R = y^2$, we have

$$\Theta(i + 1) = \Theta(i) - \Delta \frac{\partial J}{\partial \Theta} = \Theta(i) + \Delta(r - \Theta(i) R)$$

or, combining terms,

$$\Theta(i + 1) = (1 - R\Delta) \Theta(i) + r\Delta$$

(ii) Newton estimator:

$$\frac{\partial^2 J}{\partial \Theta^2} = \frac{\partial}{\partial \Theta} \left(\frac{\partial J}{\partial \Theta} \right) = \frac{\partial}{\partial \Theta} (-(r - \Theta R)) = R$$

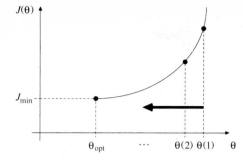

FIGURE 8.2–2
Gradient solution to constant parameter
estimation problem

Thus, the Newton estimator for this problem is

$$\Theta(i + 1) = \Theta(i) - \Delta \left[\frac{\partial^2 J}{\partial \Theta^2} \right]^{-1} \left[\frac{\partial J}{\partial \Theta} \right] = \Theta(i) + \Delta R^{-1}(r - \Theta(i)R)$$

or $\qquad \Theta(i + 1) = (1 - \Delta)\Theta(i) + \Delta R^{-1}r$

If we assume that $\hat{\Theta} = \Theta_{\text{opt}}$, then the criterion can be written

$$J = J_{\min} + (\Theta - \Theta_{\text{opt}})^2 R$$

which clearly shows that the minimum is achieved when $\hat{\Theta} = \Theta_{\text{opt}}$. We depict the gradient algorithm for this problem in Fig. 8.2–2.

Although these gradient techniques are limited, they have become very popular in the signal-processing area because of their simplicity and real-time applicability.

8.3 ALL-ZERO ADAPTIVE FILTERS

In this section, we consider the development of the all-zero or FIR adaptive (Wiener) filter. We develop the FIR Wiener solution first, since it will be necessary for steady-state analysis of stationary signals. Then, we show how these results can be used to develop and analyze an adaptive filter using the stochastic gradient algorithm. Finally, we develop the stochastic approximation or "instantaneous gradient" version of the algorithm.

Following the adaptive signal-processing method outlined in Sec. 8.1, we first select the model set to be all-zero; that is, we choose the model

$$s(t) = \sum_{i=0}^{L-1} h(i)x(t - i) = \underline{h}'\underline{X} \tag{8.3–1}$$

for $\qquad \underline{h}' = [h(0) \cdots h(L - 1)], \qquad \underline{X}' = [x(t) \cdots x(t - L + 1)]$

Next, we select the adaptation algorithm such that the mean-squared error criterion is to be minimized:

$$\min_{\underline{h}} J = E\{e^2(t)\} \tag{8.3–2}$$

where $e(t) = s_d(t) - \hat{s}(t)$ and $s_d(t)$ and $\hat{s}(t)$ are the respective desired and estimated signals, using the FIR model of Eq. (8.3–1).

Using this model, we can expand the criterion function as

$$J(\underline{h}) := E\{e^2(t)\} = E\{(s_d(t) - \underline{h}'\underline{X})(s_d(t) - \underline{X}'\underline{h})\}$$

or $\quad\quad J(\underline{h}) = E\{s_d^2(t)\} - 2\underline{h}'E\{s_d(t)\underline{X}\} + \underline{h}'E\{\underline{X}\underline{X}'\}\underline{h}$

where we have used the fact that $\underline{h}'\underline{X} = \underline{X}'\underline{h}$. If we recognize these terms as variances and see that $\underline{h}'R = R'\underline{h}$, we obtain

$$J(\underline{h}) = R_{ss} - 2R'_{sx}\underline{h} + \underline{h}'\mathbf{R}_{xx}\underline{h} \tag{8.3–3}$$

Before we design the adaptive processor, let us first develop the steady-state or Wiener solution to this problem. As usual, the Wiener solution is found by minimizing the cost function, setting it to zero, and solving; that is

$$\frac{\partial}{\partial h}J(\underline{h}) = \frac{\partial}{\partial h}\left(E\{e^2(t)\}\right) = \frac{\partial}{\partial h}\left(R_{ss} - 2R'_{sx}\underline{h} + \underline{h}'\mathbf{R}_{xx}\underline{h}\right) = 0$$

Using the chain rule of Eq. (5.1–4), we obtain

$$\frac{\partial}{\partial h}J(\underline{h}) = -2R_{sx} + 2\mathbf{R}_{xx}\underline{h} = 0 \tag{8.3–4}$$

Solving, we have the optimal Wiener solution,

$$\underline{h}_{\text{opt}} = \mathbf{R}_{xx}^{-1}\underline{R}_{sx} \tag{8.3–5}$$

with corresponding mean-squared error from Eq. (8.3–3),

$$J(\underline{h}_{\text{opt}}) = R_{ss} - 2R'_{sx}(\mathbf{R}_{xx}^{-1}\underline{R}_{sx}) + (\mathbf{R}_{xx}^{-1}\underline{R}_{sx})'\mathbf{R}_{xx}(\mathbf{R}_{xx}^{-1}\underline{R}_{sx})$$

which gives $\quad J_{\min} = J(\underline{h}_{\text{opt}}) = R_{ss} - \underline{R}'_{sx}\mathbf{R}_{xx}^{-1}\underline{R}_{sx} = R_{ss} - \underline{h}'_{\text{opt}}\underline{R}_{sx} \tag{8.3–6}$

Using this result, it is possible to show that an alternate expression for the criterion (see [1] for details) is given by

$$J(\underline{h}) = J_{\min} + (\underline{h} - \underline{h}_{\text{opt}})'\mathbf{R}_{xx}(\underline{h} - \underline{h}_{\text{opt}}) \tag{8.3–7}$$

Next we select the adaptation algorithm. Choosing the stochastic gradient technique of the previous section, we have the correction term (see Eq. (8.3–4))

$$\Delta\underline{\Theta}(i) := \Delta\underline{h}(i) = -\frac{\Delta_i}{2}\frac{\partial}{\partial h}J(\underline{h}(i)) = -\frac{\Delta_i}{2}\frac{\partial}{\partial h}\left[E\{e^2(t)\}\right] \tag{8.3–8}$$

which leads to the parameter iterator

$$\underline{h}(i+1) = \underline{h}(i) - \frac{\Delta_i}{2}\frac{\partial}{\partial h}\left[E\{e^2(t)\}\right] \tag{8.3–9}$$

Substituting the expression for the gradient of Eq. (8.3–4), we have

$$\underline{h}(i+1) = \underline{h}(i) - \frac{\Delta_i}{2}\left[-2\underline{R}_{sx} + 2\mathbf{R}_{xx}\underline{h}(i)\right]$$

or
$$\underline{h}(i + 1) = \underline{h}(i) + \Delta_i[\underline{R}_{sx} - \mathbf{R}_{xx}\underline{h}(i)]$$

which can be written as

$$\underline{h}(i + 1) = (I - \Delta_i\mathbf{R}_{xx})\underline{h}(i) + \Delta_i\underline{R}_{sx} \qquad (8.3\text{–}10)$$

with corresponding mean-squared error at the ith iteration from Eq. (8.3–6):

$$J(\underline{h}(i)) = R_{ss} - 2\underline{R}'_{sx}\underline{h}(i) + \underline{h}'(i)\mathbf{R}_{xx}\underline{h}(i) \qquad (8.3\text{–}11)$$

We summarize the stochastic gradient algorithm in Table 8.3–1 and note that implementing it will require that we have an estimate of the desired signal a priori, in order to determine the required variances. We must also have an ensemble to determine or at least estimate these variances. So, theoretically the stochastic gradient algorithm cannot be implemented as presented; however, most practical algorithms make approximations of this form. Therefore, it is important to analyze its performance to see the "best" it can achieve under ideal conditions.

The parameter estimator of the stochastic gradient algorithm can be analyzed by first subtracting the optimal solution from both sides of Eq. (8.3–10),

$$\underline{h}(i + 1) - \underline{h}_{\text{opt}} = (I - \Delta_i\mathbf{R}_{xx})\underline{h}(i) + \Delta_i(\mathbf{R}_{xx}\underline{h}_{\text{opt}}) - \underline{h}_{\text{opt}}$$

where we have solved Eq. (8.3–5) for $\underline{R}_{sx}$ and substituted. Grouping terms, we have

$$[\underline{h}(i + 1) - \underline{h}_{\text{opt}}] = (I - \Delta_i\mathbf{R}_{xx})[\underline{h}(i) - \underline{h}_{\text{opt}}]$$

or defining $\tilde{\underline{h}}(i) := \underline{h}(i) - \underline{h}_{\text{opt}}$, we have

$$\tilde{\underline{h}}(i + 1) = (I - \Delta_i\mathbf{R}_{xx})\tilde{\underline{h}}(i) \qquad (8.3\text{–}12)$$

But this is a homogeneous state-space equation (see Sec. 4.7),

$$\underline{x}(i + 1) = \mathbf{A}\underline{x}(i)$$

TABLE 8.3–1
Stochastic gradient adaptive all-zero algorithm

For $i = 0, 1, \ldots$

Gradient estimation

$$\frac{\partial}{\partial h}J(\underline{h}) = -2\underline{R}_{sx} + 2\mathbf{R}_{xx}\underline{h}(i)$$

Parameter estimation

$$\underline{h}(i + 1) = \underline{h}(i) - \frac{\Delta_i}{2}\frac{\partial}{\partial h}J(\underline{h}(i))$$

Criterion estimation $(\underline{h}(i) \rightarrow \underline{h}_{\text{opt}})$

$$J(\underline{h}) = R_{ss} - \underline{h}'_{\text{opt}}\underline{R}_{sx}$$

Initial conditions: $\underline{h}(0), \Delta_i$

with solution

$$\underline{x}(i) = \mathbf{A}^i \underline{x}(0)$$

Therefore, for our problem we have solution

$$\underline{\tilde{h}}(i) = (I - \Delta \mathbf{R}_{xx})^i \underline{\tilde{h}}(0) \tag{8.3-13}$$

using $\underline{\tilde{h}} \rightarrow \underline{x}$, $\Delta \rightarrow \Delta_i$, and $(I - \Delta \mathbf{R}_{xx}) \rightarrow \mathbf{A}$, the state transition matrix.

To facilitate the investigation of the convergence properties of this algorithm, it is best to diagonalize the transition matrix using eigenvalue-eigenvector techniques. Recall from linear algebra [9] that the *eigenvalues* of the matrix $\mathbf{A}$, $\lambda(\mathbf{A})$ are the solutions of

$$(\mathbf{A} - \lambda I)\underline{v} = \underline{0} \tag{8.3-14}$$

with corresponding *characteristic equation* found from the determinant

$$\det (\mathbf{A} - \lambda I) = |\mathbf{A}(\lambda)| = \prod_{i=1}^{N_a} (\lambda - \lambda_i) = 0 \tag{8.3-15}$$

where $A(\lambda) = \lambda^{N_a} + a_1 \lambda^{N_a - 1} + \ldots + \lambda a_{N_a - 1} + a_{N_a}$ and the roots of $A(\lambda)$ are $\lambda(\mathbf{A}) = \{\lambda_i\}$, $i = 1, \ldots, N_a$.

Corresponding to each eigenvalue there exists at least one vector solution of the above equation determined by solving

$$\mathbf{A}\underline{v}_i = \lambda_i \underline{v}_i \tag{8.3-16}$$

where $\underline{v}_i$ is the ith eigenvector of A with corresponding eigenvalue λ_i. Expanding this equation over i, we have

$$\mathbf{A}[\underline{v}_1 \cdots \underline{v}_{N_a}] = [\underline{v}_1 \cdots \underline{v}_{N_a}] \begin{bmatrix} \lambda_1 & & 0 \\ & \ddots & \\ 0 & & \lambda_{N_a} \end{bmatrix}$$

or

$$\mathbf{A}\mathbf{V} = \mathbf{V}\mathbf{\Lambda}$$

which gives the so-called "normal" form of A as

$$\mathbf{A} = \mathbf{V}\mathbf{\Lambda}\mathbf{V}^{-1} \tag{8.3-17}$$

with $\mathbf{\Lambda}$ the diagonal matrix of eigenvalues and $\mathbf{V}$ the *modal* (eigenvector) matrix.

For our problem, since $\mathbf{R}_{xx}$ is a symmetric, positive definite, covariance matrix, we know the eigenvalues are real with orthogonal eigenvectors [9]

$$\underline{v}_i' \underline{v}_j = 0 \qquad \text{for } i \neq j$$

If they are also normalized, then $\underline{v}_i' \underline{v}_j = \delta(i,j)$, and the modal matrix is orthonormal:

$$\mathbf{V}\mathbf{V}' = I$$

implying $\mathbf{V}' = \mathbf{V}^{-1}$.

Using these results, we can investigate the convergence properties of the stochastic gradient algorithm by first transforming the state-space representation to modal form, using the modal matrix of $\mathbf{R}_{xx}$; that is, we define the modal transformation

$$\tilde{\underline{h}}_M(i) := \mathbf{V}_M^{-1}\tilde{\underline{h}}(i)$$

and substitute into Eq. (8.3–12) to obtain

$$\tilde{\underline{h}}_M(i + 1) = \mathbf{V}_M^{-1}(I - \Delta\mathbf{R}_{xx})\mathbf{V}_M\tilde{\underline{h}}_M(i) = (I - \Delta\mathbf{V}_M^{-1}\mathbf{R}_{xx}\mathbf{V}_M)\tilde{\underline{h}}_M(i)$$

or $\qquad \tilde{\underline{h}}_M(i + 1) = (I - \Delta\Lambda_{xx})\tilde{\underline{h}}_M(i)$ $\qquad\qquad$ (8.3–18)

where Λ_{xx} is the eigenvalue matrix of $\mathbf{R}_{xx}$ using Eq. (8.3–17). Solving the state-space equation as before, but in modal coordinates, we have

$$\tilde{\underline{h}}_M(i) = (I - \Delta\Lambda_{xx})^i\tilde{\underline{h}}_M(0) \qquad\qquad (8.3\text{–}19)$$

For convergence to the optimal solution, we require that

$$\lim_{i\to\infty}\tilde{\underline{h}}_M(i) = \lim_{i\to\infty}(I - \Delta\Lambda_{xx})^i\tilde{\underline{h}}_M(0) \to 0 \qquad\qquad (8.3\text{–}20)$$

or $\qquad\qquad \tilde{\underline{h}}_M(i) = \underline{h}_M(i) - \underline{h}_{\text{opt}} \to 0$

which implies that $\underline{h}_M(i) \to \underline{h}_{\text{opt}}$. Therefore, we require that

$$\lim_{i\to\infty}(I - \Delta\Lambda_{xx})^i = \begin{bmatrix} \lim_{i\to\infty}(1 - \Delta\lambda_1)^i & & 0 \\ & \ddots & \\ 0 & & \lim_{i\to\infty}(1 - \Delta\lambda_L)^i \end{bmatrix} \to 0$$

since Λ_{xx} is the eigenvalue matrix, or equivalently,

$$\lim_{i\to\infty}(1 - \Delta\lambda_j)^i \to 0 \qquad j = 1,\dots,L \qquad\qquad (8.3\text{–}21)$$

which implies further that

$$|1 - \Delta\lambda_j| < 1 \qquad j = 1,\dots,L$$

or the step size must be selected such that

$$0 < \Delta < \frac{2}{\lambda_j} \qquad j = 1,\dots,L \qquad\qquad (8.3\text{–}22)$$

for convergence. *Stability* of the stochastic gradient algorithm is determined by the largest eigenvalue, since it must lie within the unit circle; therefore, the smallest allowable step size is ($\lambda_j = \lambda_{\max}$).[†]

$$0 < \Delta < \frac{2}{\lambda_{\max}}$$

[†]Another approach is to use the fact that $\lambda_{\max} < \text{Trace } \mathbf{R}_{xx} = (L)R_{xx}(0)$, which leads to the choice of step size $\Delta < 2/(L)R_{xx}(0)$.

The *speed of convergence* of the algorithm is determined by the smallest eigenvalue, since it represents the "slowest" converging mode:

$$\tilde{h}_k(i) = (1 - \Delta\lambda_k)^i \tilde{h}_k(0) \qquad \text{for } k = \min$$

where $\tilde{h}_k$ is the kth component of $\underline{\tilde{h}}_M$ and the relation follows from Eq. (8.3–21).

Suppose we choose the step size to insure stability. Then

$$\tilde{h}_k(i) = (1 - \Delta\lambda_{\min})^i \tilde{h}_k(0) = \left(1 - \left(\frac{1}{\lambda_{\max}}\right)\lambda_{\min}\right)^i \tilde{h}_k(0) = \left(1 - \frac{\lambda_{\min}}{\lambda_{\max}}\right)^i \tilde{h}_k(0)$$

$$(8.3–23)$$

then the speed of convergence is determined by the ratio of the smallest to the largest eigenvalue of $\mathbf{R}_{xx}$. Thus, the speed of convergence of the stochastic gradient algorithm is determined by the ratio

$$\frac{\lambda_{\min}}{\lambda_{\max}} = \begin{cases} 1 & \text{fast } (\lambda_{\min} \approx \lambda_{\max}) \\ 0 & \text{slow } (\lambda_{\min} \ll \lambda_{\max}) \end{cases} \qquad (8.3–24)$$

It can be shown that the optimum choice of step size is

$$\Delta = \frac{2}{\lambda_{\max} + \lambda_{\min}} \qquad (8.3–25)$$

which insures that all of the eigenvalues converge at the same rate [10], since

$$\tilde{h}_k(i) = (1 - \Delta\lambda_{\min})^i \tilde{h}_k(0) = \left(\frac{\lambda_{\max} - \lambda_{\min}}{\lambda_{\max} + \lambda_{\min}}\right)^i \tilde{h}_k(0)$$

or

$$\tilde{h}_k(i) = \left(\frac{1 - \lambda_{\min}/\lambda_{\max}}{1 + \lambda_{\min}/\lambda_{\max}}\right)^i \tilde{h}_k(0) \qquad (8.3–26)$$

We can analyze the mean-squared criterion in similar fashion. Using Eq. (8.3–7), we have

$$J(\underline{h}) - J_{\min} = (\underline{h} - \underline{h}_{\text{opt}})'\mathbf{R}_{xx}(\underline{h} - \underline{h}_{\text{opt}}) = \underline{\tilde{h}}'\mathbf{R}_{xx}\underline{\tilde{h}}$$

Transforming to modal coordinates, we obtain

$$J(\underline{\tilde{h}}_M) - J_{\min} = \underline{\tilde{h}}_M'(\mathbf{V}_M^{-1}\mathbf{R}_{xx}\mathbf{V}_M)\underline{\tilde{h}}_M = \underline{\tilde{h}}_M'\mathbf{\Lambda}_{xx}\underline{\tilde{h}}_M = \text{constant} \qquad (8.3–27)$$

which is the equation of a hyperellipse whose principal axes are defined by the eigenvectors of $\mathbf{R}_{xx}$. We show an example for the two-parameter case in Fig. 8.3–1. Here we see that the algorithm converges to the minimum directly for equal eigenvalues, while for unequal eigenvalues, the gradient can diverge. Through the use of eigenvalue-eigenvector and state-space techniques, we can evaluate the performance of the stochastic gradient algorithm.

One of the major problems with the stochastic gradient algorithm is that the gradient covariance matrices are rarely available, and therefore we cannot

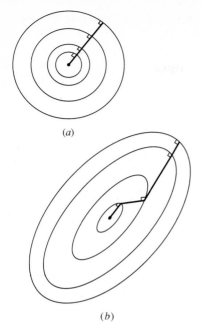

(a)

(b)

FIGURE 8.3–1
Stochastic gradient performance surface: (a) Two eigenvalues equal (b) Two eigenvalues unequal

implement the algorithm. Next, we consider an approximation to this approach, the least mean-squared (LMS) algorithm.

The LMS algorithm employs the stochastic approximation technique to estimate the gradient; that is,

$$\frac{\partial}{\partial h} J(\underline{h}) = \frac{\partial}{\partial h} \left[E\{ e^2(t) \} \right] = E \left\{ \frac{\partial}{\partial h} e^2(t) \right\} \approx 2e(t) \frac{\partial e(t)}{\partial h}$$

where we have ignored the ensemble expectation operation.

Thus, from Eq. (8.3–2) we have[†]

$$\frac{\partial e}{\partial h} = \frac{\partial}{\partial h} \left(s_d(t) - \underline{h}' \underline{X}(t) \right) = -\underline{X}(t)$$

and the gradient is estimated using

$$\frac{\hat{\partial J}}{\partial h} = -2e(t)\underline{X}(t) \qquad (8.3\text{–}28)$$

Clearly, this is an *unbiased* estimate, since

$$E \left\{ \frac{\hat{\partial J}}{\partial h} \right\} = -2E\{ e(t)\underline{X}(t) \} = \frac{\partial}{\partial h} J(\underline{h})$$

[†]We now introduce time or iteration notation for the data vector, that is, $\underline{X} \rightarrow \underline{X}(t)$.

In the algorithm we replace the stochastic gradient by its instantaneous approximation, leading to the parameter iterator; that is, we replace $i \rightarrow t$ to emphasize the instantaneous iteration and obtain

$$\underline{h}(t + 1) = \underline{h}(t) - \frac{\Delta}{2}(-2e(t)\underline{X}(t)) = \underline{h}(t) + \Delta e(t)\underline{X}(t) \qquad (8.3-29)$$

Expanding further, we have

$$\underline{h}(t + 1) = (I - \Delta\underline{X}(t)\underline{X}'(t))\underline{h}(t) + \Delta s_d(t)\underline{X}(t) \qquad (8.3-30)$$

To analyze the performance of this algorithm, let us take the expected value of the iterator,

$$E\{\underline{h}(t + 1)\} = E\{(I - \Delta\underline{X}(t)\underline{X}'(t))\,\underline{h}(t)\} + \Delta E\{s_d(t)\underline{X}(t)\}$$

Assuming that the data and parameter vectors are independent, we can separate the first term and obtain

$$\underline{m}_h(t + 1) = (I - \Delta\mathbf{R}_{xx})\underline{m}_h(t) + \Delta\underline{R}_{sx} \qquad (8.3-31)$$

where $\underline{m}_h := E\{\underline{h}\}$. Comparing this result to Eq. (8.3–10), we see that the LMS algorithm approximates the stochastic gradient algorithm "on the average." Convergence of the *average trajectory*, therefore, is identical to the stochastic gradient algorithm, since subtracting the optimal value from both sides of Eq. (8.3–31) leads to

$$\tilde{\underline{m}}_h(t + 1) = (I - \Delta\mathbf{R}_{xx})\tilde{\underline{m}}_h(t) \qquad (8.3-32)$$

and therefore $$\tilde{\underline{m}}_h(t) = (I - \Delta\mathbf{R}_{xx})^t\tilde{\underline{m}}_h(0)$$

where $\tilde{\underline{m}}_h := \underline{m}_h - \underline{m}_{h_{opt}}$ and $0 < \Delta < 2/\lambda_k$, $k = 1, \ldots, L$ as before. Note that this implies only that the *average* of *all* of the LMS trajectories converge to the optimum solution. Therefore, we expect the LMS parameter estimates to fluctuate about the optimum randomly, because the estimated or instantaneous gradient is random. These fluctuations can be controlled by the step size Δ, since it has been shown for large data lengths [11] that

$$E\{e^2(t)\} \approx \Delta J_{\min}(\underline{h}) \qquad (8.3-33)$$

and therefore, by choosing the step size,

$$\Delta = \begin{cases} \text{Large—large fluctuations, rapid convergence} \\ \text{Small—small fluctuations, slow convergence} \end{cases}$$

Before we close this discussion, let us examine properties of the "average" mean-squared error,

$$\bar{J}(\underline{h}) := E\{J(\underline{h})\}$$

Performing the required analysis, it can be shown [1] that for the LMS algorithm to *converge* (in mean square), the step size must satisfy

$$0 < \Delta < \frac{2}{\text{Trace } \mathbf{R}_{xx}} = \frac{2}{\sum_{j=1}^{N} \lambda_j} = \frac{2}{(\text{Total input power})} \qquad (8.3\text{--}34)$$

From Eq. (8.3–32), we see that for *convergence* (in the mean) the step size must satisfy

$$0 < \Delta < \frac{2}{\lambda_{\max}}$$

Clearly, since $\lambda_{\max} <$ Trace $\mathbf{R}_{xx}$, the convergence in *both* the mean and mean-squared are satisfied by the step size choice specified in Eq. (8.3–34).

Another common measure of performance is that of *misadjustment*, defined (in practice) by

$$M = \frac{\text{average excess mean-squared error}}{\text{minimum mean-squared error}} = \Delta \text{ Trace } \mathbf{R}_{xx} \qquad (8.3\text{--}35)$$

The corresponding "time constant" or decay of $\bar{J}(\underline{h})$ can be approximated by

$$\tau_{\text{mse}} \approx \frac{1}{4\Delta\lambda_j}$$

for the jth eigenvalue or, on the *average*,

$$\bar{\tau}_{\text{mse}} \approx \frac{L}{4\Delta\text{Trace } \mathbf{R}_{xx}} \qquad (8.3\text{--}36)$$

One practical aspect that should be used is the normalized least mean-squared (NLMS) technique, which employs a step size normalization computation to speed up convergence. The standard LMS algorithm is a function of the input power, since the estimated gradient is a function of $\underline{X}$; therefore, the correction term normalizes by the variance of the data:

$$\underline{h}(t + 1) = \underline{h}(t) + \Delta_t e(t)\underline{X}(t) \qquad (8.3\text{--}37)$$

where
$$\Delta_t = \frac{\alpha}{V_{xx}(t) + \beta}, \qquad 0 < \Delta_t < 2$$

for $V_{xx}(t) \approx \hat{R}_{xx}(0)$ and α and β are chosen to insure convergence (see [12] for details). We summarize the NLMS algorithm in Table 8.3–2 and use a recursive estimator for $V_{xx}(t)$.

Before we close this section on FIR adaptive filter design, let us briefly investigate the stochastic Newton-like algorithm to understand its relationship to the stochastic gradient and LMS approaches. Recall that the stochastic-Newton algorithm employs the Hessian and gradient for its direction vector; that is,

$$\Delta\underline{h}(i) = \Delta_i\underline{d}_i = -\frac{\Delta_i}{2}\left[\frac{\partial^2}{\partial h^2}J(\underline{h})\right]^{-1}\left[\frac{\partial}{\partial h}J(\underline{h})\right]$$

TABLE 8.3–2
Normalized LMS adaptive all-zero algorithm

For $t = 1, 2, \ldots$

Signal estimation

$$\hat{s}(t) = \sum_{i=0}^{L-1} h(i)x(t-i) = \underline{h}'\underline{X}(t)$$

Prediction-error estimate

$$e(t) = s_d(t) - \hat{s}(t)$$

Step-size update

$$\hat{V}_{xx}(t) = \gamma\hat{V}_{xx}(t-1) + (1-\gamma)x^2(t)$$

$$\Delta_t = \frac{\alpha}{\hat{V}_{xx}(t) + \beta}$$

Parameter estimate

$$\underline{h}(t+1) = \underline{h}(t) + \Delta_t\, e(t)\underline{X}(t)$$

Initial conditions: $h(0), \hat{V}_{xx}(0),\ \alpha,\ \beta,\ \gamma$

and, therefore, for our problem we have from Eq. (8.3–4)

$$\frac{\partial^2 J}{\partial h^2} = \frac{\partial}{\partial h}(-2\underline{R}_{sx} + 2\mathbf{R}_{xx}\underline{h}) = 2\mathbf{R}_{xx} \tag{8.3-38}$$

The Newton parameter iterator becomes

$$\underline{h}(i+1) = \underline{h}(i) - \frac{\Delta}{2}[2\mathbf{R}_{xx}]^{-1}[-2\underline{R}_{sx} + 2\mathbf{R}_{xx}\underline{h}(i)]$$

or

$$\underline{h}(i+1) = \underline{h}(i) + \frac{\Delta}{2}[\mathbf{R}_{xx}^{-1}\underline{R}_{sx} - \underline{h}(i)] = \left(1 - \frac{\Delta}{2}\right)\underline{h}(i) + \frac{\Delta}{2}\mathbf{R}_{xx}^{-1}\underline{R}_{sx} \tag{8.3-39}$$

which clearly shows a one-step convergence when $\underline{h}_{\text{opt}} = \mathbf{R}_{xx}^{-1}\underline{R}_{sx}$. This leads us to a complete class of stochastic "Newton-like" algorithms, which we specify by the iteration

$$\underline{h}(i+1) = \underline{h}(i) - \Delta_i W[-\underline{R}_{sx} + \mathbf{R}_{xx}\underline{h}(i)] \tag{8.3-40}$$

where we have

$$W = \begin{cases} [2R_{xx}]^{-1} & \text{for stochastic Newton} \\ I & \text{for stochastic gradient (and LMS)} \end{cases}$$

A final approach to this adaptive problem is to use the recursive least-squares (RLS) approach, which is essentially a special case of the RPEM of Sec.

6.5 ($\Psi = \Phi$, $A = C = 1$, $B = H$). As before, suppose we define the cost function as the sum-squared error, with

$$J_t(\underline{h}) = \sum_{t=1}^{N} \gamma_t e^2(t) \tag{8.3-41}$$

where $e(t) = s_d(t) - \hat{s}(t)$, and we have the all-zero model set

$$\hat{s}(t) = \sum_{i=0}^{L-1} h(i)x(t - i) = \underline{h}'\underline{X}(t)$$

Minimizing Eq. (8.3–41) with respect to $\underline{h}$, we obtain the usual orthogonality conditions

$$\frac{\partial J}{\partial h} = -2 \sum_{t=1}^{N} \gamma_t e(t)\underline{X}'(t) = -2 \sum_{t=1}^{N} \gamma_t \left(s_d(t) - \underline{h}'\underline{X}(t) \right) \underline{X}'(t) = 0$$

or

$$\underline{h}'\left(\sum_{t=1}^{N} \gamma_t \underline{X}(t)\underline{X}'(t) \right) = \sum_{t=1}^{N} \gamma_t s_d(t)\underline{X}'(t)$$

Or transposing and noting the scalars, we obtain

$$\left(\sum_{t=1}^{N} \gamma_t \underline{X}(t)\underline{X}'(t) \right) \underline{h} = \sum_{t=1}^{N} \gamma_t s_d(t)\underline{X}(t) \tag{8.3-42}$$

Now if we define

$$\mathbf{R}(N) := \sum_{t=1}^{N} \gamma_t \underline{X}(t)\underline{X}'(t)$$

$$\underline{R}(N) := \sum_{t=1}^{N} \gamma_t s_d(t)\underline{X}(t)$$

Then we have

$$\underline{h}(N) = \mathbf{R}^{-1}(N)\underline{R}(N) \tag{8.3-43}$$

which is the least-squares version of the Wiener solution; compare with Eq. (8.3–5). Both of these relations can be placed in recursive form by removing the tth term from the sum and identifying the remaining term; therefore,

$$\mathbf{R}(t) = \sum_{i=1}^{t-1} \gamma_i \underline{X}(i)\underline{X}'(i) + \gamma_t \underline{X}(t)\underline{X}'(t) = \mathbf{R}(t - 1) + \gamma_t \underline{X}(t)\underline{X}'(t) \tag{8.3-44}$$

and similarly,

$$\underline{R}(t) = \underline{R}(t - 1) + \gamma_t s_d(t)\underline{X}(t) \tag{8.3-45}$$

But from Eq. (8.3–43), we have

$$\underline{R}(t-1) = \mathbf{R}(t-1)\underline{h}(t-1)$$

Solving Eq. (8.3–44) for $\mathbf{R}(t-1)$ and substituting, this becomes

$$\underline{R}(t-1) = [\mathbf{R}(t) - \gamma_t \underline{X}(t)\underline{X}'(t)]\underline{h}(t-1)$$

Next substituting Eq. (8.3–45) into Eq. (8.3–43), we obtain

$$\underline{h}(t) = \mathbf{R}^{-1}(t)[\underline{R}(t-1) + \gamma_t s_d(t)\underline{X}(t)]$$

Now using the previous relation for $\underline{R}(t-1)$, we have

$$\underline{h}(t) = \mathbf{R}^{-1}(t)[\mathbf{R}(t)\underline{h}(t-1) - \gamma_t \underline{X}(t)\underline{X}'(t)\underline{h}(t-1) + \gamma_t s_d(t)\underline{X}(t)]$$

or multiplying through, we obtain the recursion

$$\underline{h}(t) = \underline{h}(t-1) + \gamma_t \mathbf{R}^{-1}(t)[s_d(t)\underline{X}(t) - \underline{X}(t)\underline{X}'(t)\underline{h}(t-1)] \quad (8.3\text{–}46)$$

We can clearly interpret this algorithm as a "Newton-like" technique (compare with Eq. (8.3–40)) when we identify

$$\gamma_t \to \Delta_t, \qquad \mathbf{W} \to \mathbf{R}^{-1}(t) \to \mathbf{R}_{xx}^{-1}$$

with appropriate gradients. In practice, it is possible to replace this algorithm and avoid inverting $\mathbf{R}(t)$ by using the so-called "matrix inversion lemma" (see [5] for details); that is, we define

$$\mathbf{P}(t) = \mathbf{R}^{-1}(t)$$

and, as in Sec. 7.4, we obtain the RLS recursion

$$\underline{h}(t) = \underline{h}(t-1) + \underline{K}(t)e(t)$$

where

$$\underline{K}(t) = \frac{\mathbf{P}(t-1)\underline{X}(t)}{1/\gamma_t + \underline{X}'(t)\mathbf{P}(t-1)\underline{X}(t)}$$

and

$$\mathbf{P}(t) = \mathbf{P}(t-1) - \frac{\mathbf{P}(t-1)\underline{X}(t)\underline{X}'(t)\mathbf{P}(t-1)}{1/\gamma_t + \underline{X}'(t)\mathbf{P}(t-1)\underline{X}(t)} \quad (8.3\text{–}47)$$

As before, it is possible to analyze the performance of the RLS algorithm in terms of its convergence properties [2]. Defining the *average* sum-squared error as

$$\bar{J}(\underline{h}) := E\{J(\underline{h})\}$$

with $\gamma_t = 1$, then it can be shown (see [2] for details) that

$$\bar{J}_t(\underline{h}) \approx J_{\min}\left(1 + \frac{L}{t}\right) \quad (8.3\text{–}48)$$

This implies that as the number of data samples gets larger ($t \to \infty$), then

$$\lim_{t \to \infty} \bar{J}_t(\underline{h}) \approx J_{\min}$$

and, therefore, it can achieve "zero" misadjustment,

$$\lim_{t \to \infty} M_{\text{RLS}} \to 0$$

Comparing the RLS to LMS approaches, we see that the RLS algorithm is an order of magnitude faster in convergence and can actually achieve the optimum (minimum) mean-squared error, or "zero misadjustment." However, the cost is in computational complexity since RLS requires $(3L/2)(L + 3)$ multiplications, compared to $2L + 1$ for the LMS algorithm.

We summarize the adaptive RLS algorithm in Table 8.3–3. Note that this is not the preferred form of implementation; the "square-root" or U-D factorized form (see [14] for details) is the popular approach. We summarize these ideas with a simple example.

> **Example 8.3–1.** Suppose we are asked to design a single-weight adaptive FIR filter using the (i) stochastic gradient, (ii) stochastic Newton, (iii) LMS algorithms, and (iv) RLS algorithms, and to analyze their performance (speed and stability). The structure of the filter is simply
>
> $$\hat{s}(t) = hx(t)$$

TABLE 8.3–3
RLS adaptive all-zero algorithm

Signal estimate

$$\hat{s}(t) = \sum_{i=0}^{L-1} h(i)x(t - i) = \underline{h}'\underline{X}(t)$$

Prediction-error estimate

$$e(t) = s_d(t) - \hat{s}(t)$$

Covariance estimate

$$\mathbf{P}(t) = \mathbf{P}(t - 1) - \frac{\mathbf{P}(t - 1)\underline{X}(t)\underline{X}'\mathbf{P}(t - 1)}{1/\gamma_t + \underline{X}'(t)\mathbf{P}(t - 1)\underline{X}(t)}$$

Gain update

$$\underline{K}(t) = \frac{\mathbf{P}(t - 1)\underline{X}(t)}{1/\gamma_t + \underline{X}'(t)\mathbf{P}(t - 1)\underline{X}(t)}$$

Parameter estimate

$$\underline{h}(t) = \underline{h}(t - 1) + \underline{K}(t)e(t)$$

Forgetting factor update

$$\gamma_t = \gamma(0)\gamma_{t-1} + (1 - \gamma(0))$$

Initial conditions: $\mathbf{P}(0) = \alpha I, \underline{h}(0) = \underline{0}, 0 < \gamma(0) < 1, \alpha$ large and the steady-state Wiener solution is

$$h_{\text{opt}} = \frac{R_{sx}}{R_{xx}}$$

The corresponding gradient and Hessian are

$$J = E\{e^2(t)\}$$

given by

$$\frac{\partial J}{\partial h} = 2E\left\{e(t)\frac{\partial e}{\partial h}\right\} = -2E\{(s_d(t) - hx(t))x(t)\} = -2R_{sx} + 2R_{xx}h$$

and $$\frac{\partial^2 J}{\partial h^2} = 2R_{xx}$$

The adaptation algorithms are as follows:

   (i) Stochastic gradient: $h(i + 1) = h(i) + \Delta(R_{sx} - R_{xx}h(i))$
   (ii) Stochastic Newton: $h(i + 1) = h(i) + \Delta(R_{sx}/R_{xx} - h(i))$
   (iii) LMS: $h(t + 1) = h(t) + \Delta e(t)x(t)$
   (iv) RLS: $h(t + 1) = h(t) + K(t)e(t)$

We can determine the speed of convergence from the following relations

   (i) Gradient: $\tilde{h}(i) = (1 - \Delta R_{xx})^i \tilde{h}(0) \Rightarrow 0 < \Delta < 2/R_{xx}$
   (ii) Newton: $\tilde{h}(i) = (1 - \Delta)^i \tilde{h}(0) \Rightarrow 0 < \Delta < 1$
   (iii) LMS: $\tilde{m}_h(t) = (1 - \Delta R_{xx})^t \tilde{m}_h(0) \Rightarrow 0 < \Delta < 2/R_{xx}$
   (iv) RLS: $J_t(\underline{h}) = J_{\min}(1 + 1/t)$

We now generate a simulation using SIG [13]; for our desired signal, we choose a discrete exponential

$$s_d(t) = 0.95s_d(t - 1) + e(t) \qquad \text{with } s_d(0) = 1, \ e \sim N(0, 0.01)$$

so that $h_d(t) = (0.95)^t$ (ignoring noise)

Next, we assume that our measured signal is scaled and with zero-mean, unit-variance white noise,

$$x(t) = 0.5s_d(t) + n(t)$$

The simulation results are shown in Fig. 8.3–2. Here we see the results of the adaptive LMS solution and the corresponding errors for two choices of step size, $\Delta = 0.01, 0.001$. We see the simulated (noisy) signal and measurement, as well as the signal estimated for each choice. In both cases, the processor tracks the signal but, as expected from Eq. (8.3–3), the larger step size yields noisier results.

Example 8.3–2. Suppose we consider the design of an adaptive processor to solve the signal-estimation problem of the previous chapter. We have an ARMAX(2,1,1,1) model with difference equation

$$(1 - 1.5q^{-1} + 0.7q^{-2})y(t) = (1 + 0.4q^{-1})u(t) + \frac{1 + 0.3q^{-1}}{1 + 0.5q^{-1}}e(t)$$

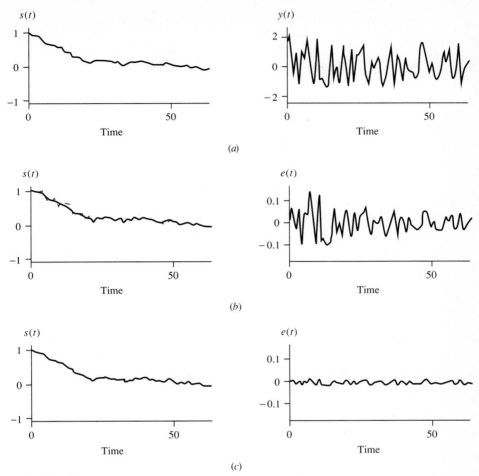

FIGURE 8.3–2
Adaptive signal weight processing example (*a*) Simulated signal and measurement (*b*) Adaptive processor for signal and error for $\Delta = 0.01$ (*c*) Adaptive processor for signal and error for $\Delta = 0.001$

where $u(t)$ is a delayed unit impulse with $e \sim N(0, 0.01)$, as in Example 6.3–2. Using SIG [13], we developed (i) the optimal Wiener (stationary) solution, (ii) the LMS adaptive solution, and (iii) the RLS adaptive solution. The results are shown in Fig. 8.3–3. Note that in this case, the best we can hope to achieve is the Wiener solution, since the data are stationary. Thus, we expect the adaptive results to converge to the Wiener solution. The estimated signals and corresponding residual errors are shown. In each case a 32-weight FIR processor was designed; we see that each processor "tracks" the measurement, as indicated by the resulting errors. From the tracking, it appears that the RLS solution approaches the optimal. This is confirmed in Fig. 8.3–3(*d*), where we have the true signal (impulse response), and the optimal Wiener and RLS solutions.

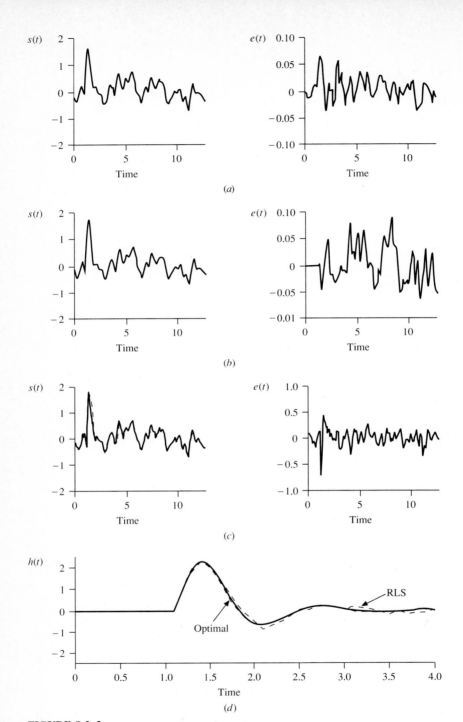

FIGURE 8.3–3
Adaptive FIR processor design for ARMAX(2,1,1,1) simulation (*a*) Optimal Wiener (stationary) solution and error (*b*) Adaptive (LMS) solution and error (*c*) Adaptive (RLS) solution and error (*d*) True impulse response and optimal Wiener and RLS solutions

8.4 POLE-ZERO ADAPTIVE FILTERS

In this section, we consider the design of the pole-zero, or IIR adaptive, Wiener filters. Adaptive IIR filters offer advantages over FIR filters, in that they require significantly fewer weights to achieve equivalent performance. For instance, it is well-known [1] that large-order FIR filters are required to estimate sinusoids in noise, while the IIR designs theoretically require only $2L$ parameters to estimate L sinusoids. IIR processors also offer the usual features of sharp cutoffs and resonances, which FIR designs can realize only with high orders. However, IIR processors have disadvantages as well. First, their performance surfaces are not quadratic and can have local minima. The processor can become unstable during adaptation as well. The work of Ljung [5], Goodwin [4] and others [14] has greatly aided in understanding the convergence of these techniques and improved performance by incorporating stability tests, enabling them to serve as alternatives to FIR adaptive techniques.

Following the prescription outlined in the introduction, the underlying model set is selected to be pole-zero with corresponding signal model

$$s(t) = -\sum_{i=1}^{N_a} a_i y(t - i) + \sum_{i=0}^{N_b} b_i u(t - i) \tag{8.4-1}$$

which can be conveniently expressed in vector form as

$$s(t) = \underline{\Theta}'(t)\underline{X}(t) \tag{8.4-2}$$

where the data vector is

$$\underline{X}'(t) = [-y(t - 1) \cdots - y(t - N_a) \mid u(t) \cdots u(t - N_b)]$$

and the parameter vector

$$\underline{\Theta}'(t) = [a_1(t) \cdots a_{N_a}(t) \mid b_0(t) \cdots b_{N_b}(t)]$$

Note that the underlying measurement model is given by

$$y(t) = s(t) + v(t) \tag{8.4-3}$$

which, substituting Eq. (8.4-2) leads to the ARMAX(N_a, N_b, 0) or, more succinctly, ARX model as a special case used in the parametric processor ($\underline{X} = \underline{\phi}$) of Sec. 7.6. Clearly, this leads to the various RPEM methods or "exact least-squares" methods of adaptive processing. In fact, Friedlander [15] shows how to construct various adaptive processors using this approach.

We will take the iterative gradient approach to develop our adaptive algorithm. As before, we select the algorithm such that the mean-squared error criterion is to be minimized:

$$\min_{\underline{\Theta}} J(t) = E\{e^2(t)\} \tag{8.4-4}$$

where $e(t) = s_d(t) - \hat{s}(t)$ and $s_d(t)$ and $\hat{s}(t)$ are the respective desired and estimated signals, using Eq. (8.4-2).

Performing the minimization in the usual manner, we have

$$\frac{\partial}{\partial \Theta} J(\Theta) = 2E\left\{ e(t)\frac{\partial}{\partial \Theta}e(t) \right\} = -2E\left\{ e(t)\frac{\partial}{\partial \Theta}y(t) \right\}$$

but now the error gradient is not as simple as before because of the IIR model structure governing $y(t)$. Thus, expanding we have

$$\frac{\partial}{\partial \Theta} J(\Theta) = -2E\left\{ e(t)\begin{bmatrix} \dfrac{\partial y(t)}{\partial a_i} \\ --- \\ \dfrac{\partial y(t)}{\partial b_j} \end{bmatrix} \right\} \quad i = 1, \ldots, N_a \quad j = 0, \ldots, N_b \quad (8.4\text{--}5)$$

Using the model, we obtain

$$\nabla_a y(t) := \frac{\partial y(t)}{\partial a_i} = -y(t-i) + \sum_{j=1}^{N_a} a_j \nabla_a y(t-j) \quad i = 1, \ldots, N_a$$

and $$\nabla_b y(t) := \frac{\partial y(t)}{\partial b_i} = u(t-i) + \sum_{j=1}^{N_a} a_j \nabla_b y(t-j) \quad i = 0, \ldots, N_b$$

or, defining $\nabla_\Theta y(t) = [\nabla_a y(t) \mid \nabla_b y(t)]'$, we have the vector recursion

$$\nabla_\Theta y(t) = \underline{X}(t) + \sum_{j=1}^{N_a} a_j \nabla_\Theta y(t-j) \quad (8.4\text{--}6)$$

Therefore, the gradient is given by

$$\frac{\partial}{\partial \Theta} J(\Theta) = -2E\{ e(t)\nabla_\Theta y(t)\} \quad (8.4\text{--}7)$$

Clearly, expanding and taking expectations leads to a complex expression for the stochastic gradient; therefore, we limit our discussion to the instantaneous gradient or LMS approach, given by the iteration

$$\underline{\Theta}(t+1) = \underline{\Theta}(t) - \frac{\Delta_t}{2}[-2e(t)\nabla_\Theta y(t)]$$

where the estimated gradient is obtained using the recursion of Eq. (8.4–6). We summarize the algorithm in Table 8.4–1. Note that the instantaneous gradient estimate is obtained by using $\nabla_\Theta y(t)$ and allowing $\{a_j(t)\}$ to vary with time. Thus, it is crucial that the step size ∇_t selected be very small, implying an overall time-averaging of the gradient. An entire class of hyperstable algorithms, called the hyperstable adaptive recursive filter (HARF), has been developed. This class "filters" the prediction errors used in the update in the same manner as the RPEM algorithm of the previous chapter [1,14].

As a special case of this approach, consider the all-pole linear predictor of

TABLE 8.4–1
LMS adaptive IIR algorithm

Signal estimate

$$\hat{s}(t) = \sum_{i=1}^{N_a} a_i y(t-i) + \sum_{i=0}^{N_b} b_i u(t-i) = \underline{\Theta}'(t)\underline{X}(t)$$

Prediction-error estimate

$$e(t) = s_d(t) - \hat{s}(t)$$

Gradient estimate

$$\nabla_{\Theta} y(t) = \underline{X}(t) + \sum_{j=1}^{N_a} a_j(t)\nabla_{\Theta} y(t-j)$$

Parameter estimate

$$\underline{\Theta}(t+1) = \underline{\Theta}(t) + \Delta e(t)\nabla_{\Theta} y(t)$$

Initial conditions: $\underline{\Theta}(0)$, $\nabla_{\Theta} y(0)$

where $\quad \underline{X}(t) = [-y(t-1) \cdots -y(t-N_a) \,|\, u(t) \cdots u(t-N_b)]'$

$$\underline{\Theta}(t) = [a_1(t) \cdots a_{N_a}(t) \,|\, b_0(t) \cdots b_{N_b}(t)]'$$

Sec. 7.3. Recall the predictor signal model

$$s(t) = -\sum_{i=1}^{N_a} a_i y(t-i)$$

and the corresponding gradient of the criterion

$$J(t) = E\{e^2(t)\}$$

given by
$$\frac{\partial J}{\partial a_j} = 2E\{e(t)\underline{Y}(t)\} \qquad j = 1, \ldots, N_a$$

where $\underline{Y}'(t) := [y(t-1) \cdots y(t-N_a)]$.

The corresponding instantaneous gradient leads to the LMS parameter estimator,

$$\underline{a}(t+1) = \underline{a}(t) - \Delta_t e(t)\underline{Y}(t) \tag{8.4–8}$$

with the prediction error given by

$$e(t) = y(t) - \hat{s}(t) = \sum_{i=1}^{N_a} a_i y(t-i) = \underline{Y}'(t)\underline{a}$$

Substituting this result into Eq. (8.4–8), we obtain the parameter update

$$\underline{a}(t+1) = \underline{a}(t) - \Delta_t\underline{Y}(t)\underline{Y}'(t)\underline{a}(t) = \left(I - \Delta_t\underline{Y}(t)\underline{Y}'(t)\right)\underline{a}(t)$$

TABLE 8.4–2
LMS adaptive all-pole algorithm

For $t = 1, 2, \ldots$

Signal estimate

$$\hat{s}(t) = \sum_{i=1}^{N_a} a_i y(t - i)$$

Prediction-error estimate

$$e(t) = \sum_{i=1}^{N_a} a_i(t) y(t - i) = \underline{Y}'(t)\underline{a}(t)$$

Gradient estimate

$$\frac{\partial}{\partial a} J(t) = -2e(t)\underline{Y}(t)$$

Parameter estimate

$$\underline{a}(t + 1) = \underline{a}(t) - \frac{\Delta_t}{2} \frac{\partial}{\partial a} J(t)$$

$$\text{Initial conditions: } a(0), \frac{\partial}{\partial \Theta} J(0)$$

where $\underline{Y}'(t) = [y(k - 1) \cdots y(k - t)]'$

leading to the state-space convergence analysis of the previous section. We summarize the all-pole algorithm in Table 8.4–2.

8.5   LATTICE ADAPTIVE FILTERS

In this section, we develop the lattice filter for adaptive signal processing. The lattice, by its inherent orthogonality properties, discussed previously, offers many advantages over other techniques. The lattice filter is numerically well-behaved and enables the monitoring of stability during the calculation of each stage. The lattice model can be used to develop an estimator which minimizes either the forward prediction error or backward prediction error, or both. The "forward-backward" lattice method calculates the forward and backward reflection coefficients individually and then combines them to form the desired estimates. The method suffers from the problem that the reflection coefficient may not be less than unity. We choose the "Burg" approach, developed previously, which constrains the forward and backward reflection coefficients to be identical and less than unity. Here we will first briefly review the Burg block processing method developed in Sec. 7.4 and then extend it to the adaptive case.

The underlying model set of the signal is all-pole, given by

$$s(t) = -\sum_{i=1}^{N_a} a_i y(t - i) \tag{8.5-1}$$

with corresponding measurement characterized by

$$y(t) = s(t) + e(t)$$

which is an AR model, as before,

$$A(q^{-1})y(t) = e(t)$$

This signal model can be obtained by transforming from the estimated lattice parameters to the predictor coefficients, as before (see Table 4.6–1); therefore, we start with the lattice recursion

$$e_f(t, i) = e_f(t, i - 1) - k(t, i)e_b(t - 1, i - 1)$$
$$e_b(t, i) = e_b(t - 1, i - 1) - k(t, i)e_f(t, i - 1) \tag{8.5-2}$$

where $e_f(t, i), e_b(t, i)$, and $k(t, i)$ are the respective forward and backward prediction errors and the reflection coefficient of the ith section at time t.

For the adaptive algorithm, we start with the mean-squared error criterion and minimize it with respect to the reflection coefficient; that is,

$$\min_k J(t, i) = E\{e_f^2(t, i) + e_b^2(t, i)\} \tag{8.5-3}$$

Performing the required minimization as before (see Sec. 7.5 for details), we obtain an expression for the gradient

$$\frac{\partial J(t, i)}{\partial k(t, i)} = -2E\{e_f(t, i)e_b(t - 1, i - 1) + e_b(t, i)e_f(t, i - 1)\} \tag{8.5-4}$$

or, substituting the model recursion and combining, we obtain the gradient expression

$$\frac{\partial J(t, i)}{\partial k(t, i)} = -2E\{2e_f(t, i - 1)e_b(t - 1, i - 1) - k(t, i)(e_f^2(t, i - 1) + e_b^2(t - 1, i - 1))\} \tag{8.5-5}$$

Setting this equation to zero and solving for $k(t, i)$ we have, as before,

$$k(t, i) = \frac{2E\{e_f(t, i - 1)e_b(t - 1, i - 1)\}}{E\{e_f^2(t, i - 1) + e_b^2(t - 1, i - 1)\}} \tag{8.5-6}$$

Again, we must replace the ensemble averages with time averages. Anticipating the real-time adaptive algorithm, we can recursively estimate these statistics. Suppose we estimate the denominator statistic with the exponentially weighted estimator

$$V(t,i) = \frac{1}{t}\sum_{j=1}^{t}\lambda^{t-j}\left(e_f^2(j,i) + e_b^2(j-1,i)\right) \approx E\{e_f^2(t,i) + e_b^2(t-1,i)\}$$

which can be estimated recursively (see Example 9.2–2) by removing the tth term from the sum and identifying the previous iterate $V(t-1,i)$ to obtain

$$V(t,i) = \left(\frac{t-1}{t}\right)\lambda\, V(t-1,i) + \frac{1}{t}\left(e_f^2(t,i) + e_b^2(t-1,i)\right) \qquad (8.5-7)$$

Similarly, for the numerator in Eq. (8.5–6), we have

$$\gamma(t,i) = \frac{1}{t}\sum_{j=1}^{t}\lambda^{t-j}e_f(t,i)e_b(t-1,i) \approx E\{e_f(t,i-1)e_b(t-1,i-1)\}$$

which leads to the recursion

$$\gamma(t,i) = \left(\frac{t-1}{t}\right)\lambda\gamma(t-1,i) + \frac{1}{t}e_f(t,i)e_b(t-1,i) \qquad (8.5-8)$$

Note that using these recursions leads us to an alternative block-processing method rather than that in Table 7.5–1. We summarize the algorithm in Table 8.5–1 for convenience. Note that for *each* lattice section the *entire block* of data

TABLE 8.5–1
Block lattice filtering algorithm[†]

For $i = 1,\ldots,M$ (Order recursion)
For $t = 1,\ldots,N$ (Block recursion)

Statistics estimate

$$\gamma(t,i) = \left(\frac{t-1}{t}\right)\lambda\gamma(t-1,i) + \frac{1}{t}e_f(t,i)e_b(t-1,i)$$

$$V(t,i) = \left(\frac{t-1}{t}\right)\lambda V(t-1,i) + \frac{1}{t}\left(e_f^2(t,i) + e_b^2(t-1,i)\right)$$

Reflection coefficient estimate

$$k(t,i) = \frac{2\gamma(t,i)}{V(t,i)}$$

Prediction-error estimate

$$e_f(t,i) = e_f(t,i-1) - k(t,i)e_b(t-1,i-1)$$

$$e_b(t,i) = e_b(t-1,i-1) - k(t,i)e_f(t,i-1)$$

Initial conditions: $V(0,i),\ \gamma(0,i),\ \lambda$

[†] For signal estimation follow Steps 3 through 5 of Table 7.5–1, after the order recursions are completed.

is processed, and then a new section is added and the block processed again. This type of processing essentially deconvolves or removes the effects of each stage before the next one is added, owing to the orthogonality of the lattice processor.

With these recursions in mind, we can now begin the development of the adaptive algorithm based on the gradient approach. With the lattice model, we must update the reflection coefficient iteratively; that is,

$$k(m + 1, i) = k(m, i) + \Delta k(m, i)$$

With the choice of a gradient-based adaptation algorithm, the correction is given by

$$\Delta k(m, i) = \frac{-\Delta(m, i)}{2} \frac{\partial}{\partial k_i} J(k(m, i))$$

The gradient for this process, determined previously, leads to the corresponding parameter iteration

$$k(m + 1, i) = k(m, i) - \frac{\Delta(m, i)}{2} \frac{\partial}{\partial k_i} J(k(m, i))$$

or equivalently, using the gradient of Eq. (8.5–5),

$$k(m + 1, i) = k(m, i) + \Delta(m, i)E\{2e_f(m, i - 1)e_b(m - 1, i - 1)$$

$$- k(m, i)(e_f^2(m, i - 1) + e_b^2(m - 1, i - 1)\} \quad (8.5–9)$$

Replacing the ensemble statistics with their time-recursive counterparts of Eqs. (8.5–7,8), we obtain the parameter iterator with $m \rightarrow t$:

$$k(t + 1, i) = k(t, i) + \Delta(t, i)[2\gamma(t, i) - k(t, i)V(t, i)] \quad (8.5–10)$$

or, combining like terms, we have

$$k(t + 1, i) = (1 - \Delta(t, i)V(t, i))k(t, i) + 2\Delta(t, i)\gamma(t, i)$$

Note that this can be written in state-space form by expanding over i; that is,

$$\underline{k}(t + 1) = (I - \Delta(t, i)V(t))\underline{k}(t) + 2\Delta(t, i)\underline{\gamma}(t)$$

where $V(t) = \text{diag}[V(t, 1) \cdots V(t, t)]$ and $\underline{\gamma}(t) = [\gamma(t, 1) \cdots \gamma(t, t)]'$.

For analysis purposes, if we tacitly assume stationary properties and ignore t (as in Sec. 8.3), then with $\underline{k}_{\text{opt}} = 2V^{-1}\gamma$, we obtain

$$\tilde{\underline{k}}(t + 1) = (I - V)\tilde{\underline{k}}(t) \quad (8.5–11)$$

where $\tilde{\underline{k}}(t) := \underline{k}(t) - \underline{k}_{\text{opt}}$ and $V := \text{diag}[\Delta(1)V(1) \cdots \Delta(t)V(t)]$. Solving this equation as before, we have

$$\tilde{\underline{k}}(t) = (I - V)^t \tilde{\underline{k}}(0) \quad (8.5–12)$$

But here, because of the underlying lattice structure, the transition matrix is *already* diagonalized, with each reflection coefficient converging at the rate governed by

$$\tilde{k}(t, i) = (1 - \Delta(i)V(i))^t \tilde{k}(0)$$

which implies that the step size must satisfy

$$|1 - \Delta(i)V(i)| < 1$$

or the choice

$$\Delta(i) = \frac{\alpha}{V(i)} \qquad \text{for } 0 < \alpha < 1 \tag{8.5-13}$$

will satisfy this constraint. This relation implies that all of the reflection coefficients will converge at essentially the same rate. We summarize the stochastic gradient algorithm in Table 8.5–2. Note that, in implementing the algorithm, all of the data are still processed through the recursion while each coefficient converges.

TABLE 8.5–2
Stochastic gradient adaptive lattice algorithm

For $i = 1, \ldots, M$ (Order recursion)
For $t = 1, \ldots, N$ (Block recursion)

Statistics estimate

$$\gamma(t, i) = \left(\frac{t-1}{t}\right) \lambda \gamma(t-1, i) + \frac{1}{t} e_f(t, i) e_b(t-1, i)$$

$$V(t, i) = \left(\frac{t-1}{t}\right) \lambda V(t-1, i) + \frac{1}{t}\left(e_f^2(t, i) + e_b^2(t-1, i)\right)$$

Gradient estimate

$$\frac{\partial J}{\partial k_i} = 2\gamma(t, i) - k(t, i)V(t, i)$$

Reflection coefficient estimate

$$k(t+1, i) = k(t, i) + \Delta(i)\frac{\partial J}{\partial k_i}$$

Prediction-error estimate

$$e_f(t+1, i) = e_f(t+1, i-1) - k(t+1, i)e_b(t, i-1)$$

$$e_b(t+1, i) = e_b(t, i-1) - k(t+1, i)e_f(t+1, i-1)$$

Initial conditions: $V(0, i)$, $\gamma(0, i)$, λ

Lattice-to-predictor conversion

$$\underline{k}(t) \rightarrow \underline{a}(t)$$

Signal estimate

$$\hat{s}(t) = \underline{a}'(t)\underline{Y}(t)$$

For real-time application, we replace the stochastic gradient with the instantaneous gradient and develop the LMS algorithm for the lattice model. The instantaneous gradient is simply Eq. (8.5–4), ignoring the expectation and replacing $m \to t$:

$$\frac{\hat{\partial J}}{\partial k_i} = -2\{ e_f(t, i) e_b(t - 1, i - 1) + e_b(t, i) e_f(t, i - 1) \} \qquad (8.5-14)$$

Replacing the gradient in the parameter iteration of Eq. (8.5–9) with this relation gives the parameter iteration

$$k(t + 1, i) = k(t, i) + \Delta(t, i)[e_f(t, i) e_b(t - 1, i - 1) + e_b(t, i) e_f(t, i - 1)] \qquad (8.5-15)$$

Choosing the step-size adjustment to satisfy the convergence constraints of Eq. (8.5–13), the *gradient* (LMS) *adaptive lattice algorithm* evolves with

$$\Delta(t, i) = \frac{\alpha}{V(t, i)}$$

where $V(t, i)$ satisfies the time recursion of Eq. (8.5–7) [16,17]. We summarize this algorithm in Table 8.5–3. We note in passing that exact methods to estimate the gradient using least-squares techniques do exist; these computations are more intensive but possess superior performance [18,19]. The following example illustrates the performance of the lattice algorithms.

Example 8.5–1. Suppose we have data generated from a model given by the ARMAX(2,1,1,1) model of the previous example and that we apply the block lattice (Table 8.5–1) and gradient (LMS) adaptive lattice (Table 8.5–3). The results are shown in Fig. 8.5–1. The best we can hope for is that the adaptive results converge to the block solutions. The AIC selected a third-order model with corresponding parameters:

Parameter	True value	Block estimate	Adaptive estimate
a_1	−1.5	−1.47	−1.53
a_2	0.7	0.73	0.84
a_3	–	−0.10	−0.16
b_0	1.0	0.155	0.133
b_1	0.5	–	–
c_1	0.3	–	–
d_1	0.5	–	–

Using SIG [13], the simulation results are shown in Fig. 8.5–1; the estimated signal and filtered measurements are shown for both block and adaptive processors. Note that the adaptive results are not quite as good as the block.

TABLE 8.5–3
Gradient (LMS) adaptive lattice algorithm

For $t = 1, 2, \ldots$
For $i = 1, \ldots, M$ (Order recursion)

Statistics estimate

$$V(t, i) = \left(\frac{t-1}{t}\right) \lambda V(t-1, i) + \frac{1}{t} e_f^2(t, i) e_b^2(t-1, i)$$

Gradient estimate

$$\frac{\partial}{\partial k} J(t, i) = -2[e_f(t, i) e_b(t-1, i-1) + e_b(t, i) e_f(t, i-1)]$$

Step-size estimate

$$2\Delta(t, i) = \frac{\alpha}{V(t, i)}$$

Reflection coefficient estimate

$$k(t+1, i) = k(t, i) + \Delta(t, i) \frac{\partial}{\partial k} J(t, i)$$

Prediction-error estimate

$$e_f(t+1, i) = e_f(t+1, i-1) - k(t+1, i) e_b(t, i-1)$$

$$e_b(t+1, i) = e_b(t, i-1) - k(t+1, i) e_f(t+1, i-1)$$

Initial conditions: $V(0, i), \lambda, \alpha$

Lattice-to-predictor conversion

$$\underline{k}(t) \rightarrow \underline{a}(t)$$

Signal estimate

$$\hat{s}(t) = \underline{a}'(t) \underline{Y}(t)$$

Before we close this section, let us consider development of the adaptive lattice filter for the FIR (Wiener) or joint-process estimator case.

Basically, the idea behind the FIR adaptive lattice filters is to use the lattice model to transform correlated measurement data $\{x(t)\}$ to the uncorrelated backward prediction errors $\{e_b(t)\}$ of the lattice, and then to solve the standard Wiener problem. In the batch problem of Sec. 7.2, we transformed the correlated measurements to the uncorrelated innovations.

The underlying signal model is an all-zero (FIR) structure given by

$$s(t) = \sum_{i=0}^{L-1} l(i) e_b(t-i) = \underline{l}' \underline{e}_b \tag{8.5–16}$$

and, as before, we would like to minimize the criterion:

$$J(t) = E\{e^2(t)\} = E\{s_d^2(t)\} - 2\underline{l}' E\{s_d(t)\underline{e}_b\} + \underline{l}' E\{\underline{e}_b \underline{e}_b'\} \underline{l}$$

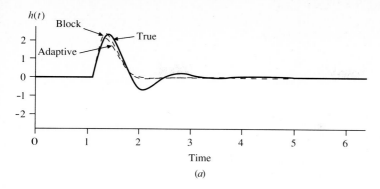

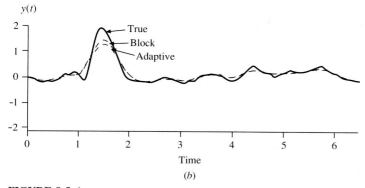

FIGURE 8.5–1
Block and adaptive lattice filter design for ARMAX(2,1,1,1) simulation (*a*) True and estimated impulse response (*b*) Actual and filtered measurements (signals)

where $e(t) = s_d(t) - \hat{s}(t)$. Replacing these operations with variances, we have

$$J(t) = E\{s_d^2(t)\} - 2\underline{l}'\underline{R}_{se_b} + \underline{l}'\mathbf{R}_{e_b e_b}\underline{l} \qquad (8.5\text{–}17)$$

Following the procedure of Sec. 8.3, we obtain the gradient

$$\frac{\partial}{\partial \underline{l}}J(\underline{l}) = E\{e(t)\underline{e}_b\} = -2\underline{R}_{se_b} + 2\mathbf{R}_{e_b e_b}\underline{l} \qquad (8.5\text{–}18)$$

which yields the equivalent Wiener solution

$$\underline{l}_{\text{opt}} = \mathbf{R}_{e_b e_b}^{-1}\underline{R}_{se_b} = (\mathbf{L}^{-1}\mathbf{R}_{xx}^{-1}\mathbf{L}^{-T})(\mathbf{L}^T\underline{R}_{sx}) \qquad (8.5\text{–}19)$$

for $\mathbf{L}$ the linear transformation matrix taking $\underline{X} \rightarrow \underline{e}_b$. Note that the $\mathbf{R}_{e_b e_b}$ is diagonal, with entries $r_{e_b}(i) = E\{e_b^2(i)\}$ (see [2] for more details). Clearly, the same analysis follows, as in the previous cases (see Sec. 8.3) and we obtain a stochastic gradient algorithm identical to that given in Table 8.3–1 with $X \rightarrow e_b$. The stochastic gradient parameter iteration simplifies considerably, because in the equation

$$\underline{l}(i+1) = \underline{l}(i) - \Delta(i)(-2\underline{R}_{se_b} + 2\mathbf{R}_{e_b e_b}\underline{l}(i)) \qquad (8.5\text{–}20)$$

$R_{e_b e_b} = \text{diag}[r_{e_b}(0) \cdots r_{e_b}(L-1)]$, which decouples the iteration into its components; that is, Eq. (8.5–20) becomes

$$l_j(i+1) = l_j(i) - \Delta_j(i)\big(-2r_{se}(j) + 2r_{e_b}(j)l_j(i)\big) \qquad \text{for } j = 0, \ldots, L-1 \quad (8.5\text{–}21)$$

Using the same analysis as in the previous case, we must select the step size as

$$2\Delta_j(i) = \frac{\alpha}{r_{e_b}(j)} \qquad 0 < \alpha < 1$$

which again shows that all lattice weights converge at the same rate.

The componential instantaneous gradient approximation of Eq. (8.5–18) is given by

$$\frac{\partial}{\partial l_i} \hat{J}(\underline{l}) = -2e(t)e_b(t, i) \qquad \text{for } i = 0, \ldots, L-1 \quad (8.5\text{–}22)$$

The parameter iteration becomes

$$l_i(t+1) = l_i(t) + \frac{\alpha}{\nu(t, i)} e(t)e_b(t, i) \quad (8.5\text{–}23)$$

where $\nu(t, i)$ is given by the time recursion

$$\nu(t, i) = \left(\frac{t-1}{t}\right) \lambda \nu(t-1, i) + \frac{1}{t} e_b^2(t, i) \quad (8.5\text{–}24)$$

The complete algorithm includes the lattice recursion as well as the reflection coefficient iteration. We summarize the FIR gradient adaptive lattice algorithm in Table 8.5–4.

8.6 APPLICATIONS

In this section, we briefly discuss some of the popular applications of adaptive signal-processing techniques after the selection of a particular adaptation method. We have seen in the previous sections that adaptation algorithms coupled with a signal model provide an approximate solution to the Wiener filter design problem for stationary as well as non-stationary processes. Here we will consider the construction of various signal-processing techniques which incorporate the adaptive algorithms as an integral part of their structure. Specifically, we will consider the adaptive noise canceller and its variants, the adaptive predictor and line enhancer.

In its simplest form, a noise canceller is an optimal Wiener filter which can be characterized by the usual measurement model,

$$y(t) = s(t) + n(t) \quad (8.6\text{–}1)$$

where $s(t)$ and $n(t)$ are the respective signal and noise sequences. The canceller assumes that a reference noise signal exists, $r(t)$, which is correlated to the measurement noise,

$$r(t) = h(t) * n(t) = H(q^{-1})n(t)$$

TABLE 8.5–4
FIR gradient (LMS) adaptive lattice algorithm

For $t = 1, 2, \ldots$
For $i = 0$ to $L-1$ (Order recursion)

Adaptive lattice

Statistics estimate

$$V(t, i) = \left(\frac{t-1}{t}\right) \lambda V(t-1, i) + \frac{1}{t} e_f^2(t, i) e_b^2(t-1, i)$$

Reflection coefficient estimate

$$k(t+1, i) = k(t, i) + \frac{\alpha}{V(t, i)} [e_f(t, i) e_b(t-1, i-1) + e_b(t, i) e_f(t, i-1)]$$

Prediction-error estimate (lattice recursion)

$$e_f(t+1, i) = e_f(t+1, i-1) - k(t+1, i) e_b(t, i-1)$$
$$e_b(t+1, i) = e_b(t, i-1) - k(t+1, i) e_f(t+1, i-1)$$

FIR recursion

Statistics estimate

$$\nu(t, i) = \left(\frac{t-1}{t}\right) \lambda \nu(t-1, i) + \frac{1}{t} e_b^2(t, i)$$

Error estimate

$$e(t) = s_d(t) - \hat{s}(t)$$

Parameter estimate

$$l_i(t+1) = l_i(t) + \frac{\alpha}{\nu(t, i)} e(t) e_b(t, i)$$

Signal estimate

$$\hat{s}(t+1) = \sum_{i=0}^{L-1} l_i(t+1) e_b(t+1-i, i)$$

Initial conditions: $V(0, i)$, $\nu(0, i)$, λ, α

If we further assume that $H(q^{-1})$ is invertible, then we have

$$n(t) = H_n(q^{-1}) r(t) = \sum_{i=0}^{L-1} h(i) r(t-i) = \underline{h}'_n \underline{r} \tag{8.6–2}$$

where $H_n(q^{-1}) = H^{-1}(q^{-1})$. Substituting this relation for $n(t)$ into the measurement model, we obtain

$$y(t) = s(t) + H_n(q^{-1}) r(t) \tag{8.6–3}$$

which is an input-output model relating the reference noise (input) to the measurement (output). The noise canceller is shown in Fig. 8.6–1 (ignoring dashed line). Here we see that the signal can be extracted from the noisy measurement, using the canceller, as

$$\hat{s}(t) = y(t) - \hat{n}(t) \qquad (8.6\text{--}4)$$

where $\hat{n}(t) = \hat{H}_n(q^{-1})r(t)$ or, substituting and using Eq. (8.6–3), we have

$$\hat{s}(t) = s(t) + [H_n(q^{-1}) - \hat{H}_n(q^{-1})]r(t) \qquad (8.6\text{--}5)$$

Clearly, when the canceller converges, we have

$$\hat{H}_n(q^{-1}) \rightarrow H_n(q^{-1}) \qquad \text{and} \qquad \hat{s}(t) \rightarrow s(t)$$

which shows that the canceller cancels that part of the measurement *correlated* with the reference input, much the same as the optimal solution of Sec. 7.2. The optimal solution to this problem is obtained by

$$\min_{\underline{h}} J(t) = E\{e^2(t)\}$$

where $e(t) = y(t) - \hat{n}(t) = \hat{s}(t)$. Performing this minimization, we obtain the gradient, as in Sec. 8.3,

$$\frac{\partial}{\partial h}J(t) = 2E\left\{ e(t)\frac{\partial}{\partial h}e(t) \right\} = -2E\{e(t)\underline{r}\} = -2\underline{R}_{yr} + 2\mathbf{R}_{rr}\underline{h}_n \qquad (8.6\text{--}6)$$

Setting this expression to zero and solving, we obtain the "optimal noise filter"

$$\hat{\underline{h}}_n = \mathbf{R}_{rr}^{-1}\underline{R}_{yr} \qquad (8.6\text{--}7)$$

which leads to an optimal estimate of the noise

$$\hat{n}_{\text{opt}}(t) = \hat{\underline{h}}_n'\underline{r} \qquad (8.6\text{--}8)$$

and therefore of the signal

$$\hat{s}_{\text{opt}}(t) = y(t) - \hat{n}_{\text{opt}}(t) \qquad (8.6\text{--}9)$$

Measurement
$y(t) = s(t) + n(t)$

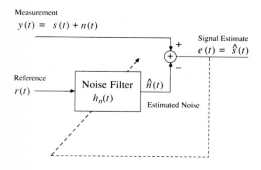

FIGURE 8.6–1
Noise canceller diagram (*a*) Stationary case (solid lines only) (*b*) Adaptive (quasi-stationary) case (solid and dashed lines)

Note that the noise canceller consists of a two-step operation: (1) obtaining the optimal estimate of the noise filter, $h_n(t)$, and corresponding noise signal, $\hat{n}(t)$; and (2) cancelling or subtracting the estimated noise to obtain the desired signal. So, we see from this construction that the function of the noise filter is to "pass" that noise-correlated part of the reference and remove it from the measurement to produce a signal estimate. The filter can be designed with any of the parametric (joint-process) methods of Chapter 7, if the processes under investigation are stationary, or with the adaptive techniques of this chapter with $y(t) \rightarrow s_d(t)$, $h_n(t) \rightarrow h(t)$, $r(t) \rightarrow x(t)$ (see Fig. 8.6–1).

Using the expression of Eq. (8.6–6), we easily have the stochastic gradient iteration

$$\underline{h}_n(i + 1) = \underline{h}_n(i) - \frac{\Delta_i}{2}\left(-2\underline{R}_{yr} + 2\mathbf{R}_{rr}\underline{h}_n(i)\right) \qquad (8.6\text{–}10)$$

or, using the instantaneous gradient estimate, we obtain the LMS iteration for adaptive noise cancelling:

$$\underline{h}_n(t + 1) = \underline{h}_n(t) + \Delta_t e(t)\underline{r}(t) \qquad (8.6\text{–}11)$$

where $\underline{r}'(t) = [r(t) \cdots r(t - L + 1)]$. Convergence results are identical with $y(t)$, $r(t)$, $h_n(t)$ and $\hat{n}(t)$ replacing $s_d(t)$, $x(t)$, $h(t)$ and $\hat{s}(t)$, respectively, of Table 8.3–2. We summarize the algorithm in Table 8.6–1. We also note in closing that the noise-cancelling filter design can be considered a system-identification problem, as noted by Ljung [5], since $\{r(t)\}$ and $\{y(t)\}$ are the respective input and output sequences, and $\{\hat{h}(t)\}$ is the "identified" system. This is the approach taken in [15,20].

Consider the following simple example of the noise cancelling.

Example 8.6–1. Suppose we have a triangular pulse contaminated with a harmonic disturbance at 10 Hz and random noise,

$$y(t) = s(t) + n(t)$$

where $n(t) = \sin 2\pi(10)t + e(t)$ for uniform e with variance 0.001. We would like to design a noise-cancelling filter to pass the desired signal (triangular pulse) and eliminate the noise. We have a reference measurement of the harmonic disturbance. We design the stationary (optimal Wiener) filter; the results are shown in Fig. 8.6–2. Here we see the signal and noise and corresponding spectrum. Note the low-pass Fourier spectrum (triangular pulse), harmonic disturbance at 10 Hz, and random noise. The optimal cancelling FIR filter design of Eq. (8.6–13) is shown in Fig. 8.6–2(b). Here we see from the corresponding spectrum that the filter will pass the sinusoid unattenuated to produce the estimated noise. The cancelled signal and spectrum are shown in (c). Note that the triangular pulse evolves including the random noise, and that the sinusoidal disturbance at 10 Hz has been removed, as shown in the spectrum. An adaptive (LMS) processor also achieved essentially the same performance.

TABLE 8.6–1
Normalized LMS noise-cancelling algorithm

For $t = 1, 2, \ldots$

Noise estimate

$$\hat{n}(t) = \sum_{i=0}^{L-1} h_n(i) r(t - i) = \underline{h}_n' \underline{r}(t)$$

Prediction-error estimate

$$e(t) = y(t) - \hat{n}(t)$$

Step-size update

$$\hat{V}_{rr}(t) = \gamma \hat{V}_{rr}(t - 1) + (1 - \gamma) r^2(t)$$

$$\Delta_t = \frac{\alpha}{\hat{V}_{rr}(t) + \beta}$$

Parameter estimate

$$\underline{h}_n(t + 1) = \underline{h}_n(t) + \Delta_t e(t) \underline{r}(t)$$

Signal estimate†

$$\hat{s}(t) = y(t) - \hat{n}(t) = e(t)$$

Initial conditions: $\underline{h}_n(0), \hat{V}_{rr}(0), \alpha, \beta, \gamma$

† Note that this is redundant, since $\hat{s}(t) = e(t)$, but we include this step for consistency.

Next let us consider a special case of the noise-cancelling filter—the predictor. Suppose we have a reference signal that is a delayed version of the primary signal,

$$r(t) = s(t - d)$$

or equivalently $\qquad\qquad r(t + d) = s(t) \qquad\qquad\qquad (8.6\text{–}12)$

Substituting into the gradient expression of Eq. (8.6–6) for the canceller, with $y(t) = s(t)$, we obtain

$$\frac{\partial}{\partial h} J(t) = -2E\{e(t)\underline{r}(t + d)\} = -2\underline{R}_{yr}(d) + 2\mathbf{R}_{rr}\underline{h}$$

which leads to the optimal *d-step predictor*

$$\hat{\underline{h}}_{\text{opt}} = \mathbf{R}_{rr}^{-1}\underline{R}_{yr}(d) \qquad\qquad\qquad (8.6\text{–}13)$$

The *d*-step predictor is shown in Fig. 8.6–3.

The predictor implementation has two outputs which are used in various applications, (1) the uncorrelated channel, and (2) the correlated channel. By selecting the prediction distance, we control that part of the signal available at the error or uncorrelated output channel, $e(t)$, since the predictor is in fact

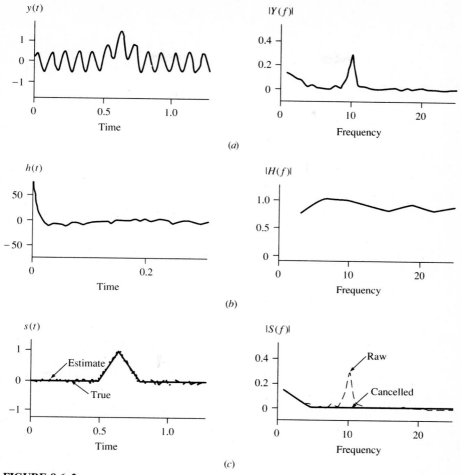

FIGURE 8.6–2
Noise-cancelling filter design for sinusoid signal (*a*) Measured signal and spectrum (*b*) Cancelling filter impulse response and spectrum (*c*) Signal estimates (stationary) and spectrum

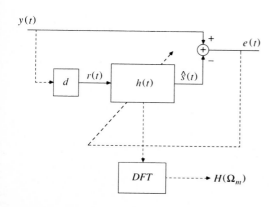

FIGURE 8.6–3
D-step predictor (noise-canceller) diagram (*a*) Stationary case (solid lines only) (*b*) Adaptive (quasi-stationary) case (solid and dashed lines)

a "correlation canceller" (see Sec. 7.2 for details). The correlated channel $\hat{s}(t)$ consists of the portion of the signal that is strongly autocorrelated at lag "d" or greater and is therefore the output of the predictor. This configuration can also be used to enhance spectral lines, as indicated in the dotted lines of Fig. 8.6–3. Here we see that taking the DFT of the cancelling filter impulse response leads to an enhanced line spectrum (see [1] for more details).

Clearly, the predictor can be implemented adaptively using the LMS of Table 8.6–1. The predictor can be used to separate a signal which consists of both narrowband and wideband components, depending on the choice of the *delay* or *prediction distance*, d. The following example illustrates the operation of the processor.

Example 8.6–2. Suppose we have a signal that is composed of two uncorrelated exponentials such that

$$y(t) = \alpha^t + \beta^t \qquad 0 < \alpha < \beta < 1$$

We would like to design a predictor to separate the signals for both stationary as well as quasi-stationary cases, using the Wiener solution of Eq. (8.6–13) and the adaptive LMS algorithm. The correlation function of this process can be approximated by the components as

$$R_{yy}(k) = \alpha^{|k|} + \beta^{|k|} \qquad 0 < \alpha < \beta < 1$$

where the broadband signal correlation time,[†] τ_α, decays much more rapidly than the narrowband, τ_β. If we select the prediction distance according to

$$\tau_\alpha < d \leq \tau_\beta$$

then d will be larger than τ_α of the broadband. Therefore, the reference signal will be completely uncorrelated with the broadband component; that is (assuming α is uncorrelated with β),

$$y(t) = s_\alpha(t) + s_\beta(t)$$

and $\qquad\qquad r(t) = s_\alpha(t - d) + s_\beta(t - d) \qquad$ for $\tau_\alpha < d < \tau_\beta$

$$E\{y(t)r(t)\} = E\{s_\alpha(t)s_\alpha(t - d)\} + E\{s_\beta(t)s_\beta(t - d)\}$$

or $\qquad\qquad E\{y(t)r(t)\} = R_{s_\alpha s_\alpha}(d) + R_{s_\beta s_\beta}(d) = R_{s_\beta s_\beta}(d) \qquad$ for $\tau_\alpha < d$

which implies that the correlated part of the measurement is $s_\beta(t)$, and therefore the predictor will produce an estimate of $\hat{s}_\beta(t)$ and remove it from the output:

$$\hat{n}(t) \approx s_\beta(t)$$

and of course $\qquad\qquad \hat{s}(t) = e(t) = y(t) - \hat{n}(t) = s_\alpha(t)$

Suppose we select $\alpha = 0.5$ and $\beta = 0.9$, which implies that $\tau_\alpha = 1.5$ and $1.5 < d$. We simulated the data using SIG [13]; the results are shown in Fig. 8.6–4. Here

---

[†]Correlation time is simply the time it takes for the correlation function to decay to zero.

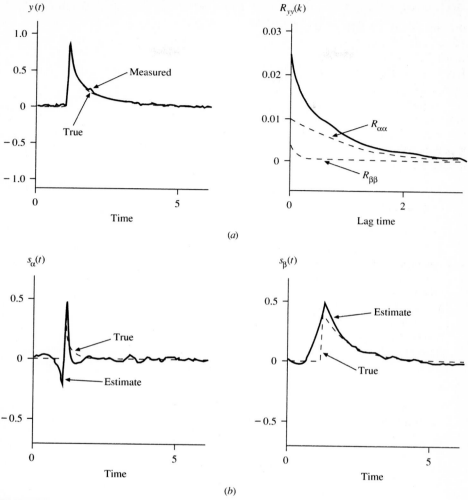

FIGURE 8.6–4

D-step predictor simulation (a) Simulated data and composite (solid) and signal covariances (b) Estimated and true (dashed) broadband and narrowband signals

we see the signal and individual functions and the outputs of the predictor, both narrow and broadband signals. Clearly the predictor has been able to separate and estimate the signals.

There are many applications of adaptive signal-processing, ranging from adaptive identification, to deconvolution, array beamforming, and adaptive digital filter design [1,2,3,4,5,6].

8.7 CASE STUDY:
PLASMA-PULSE ESTIMATION

In this section, we consider the design of a noise-cancelling filter to extract a plasma pulse from contaminated measurement data [20]. First, we discuss the background of the problem and then the design of the processor.

Controlled *fusion* of heavy isotopes of hydrogen (deuterium and tritium) would enable virtually limitless energy and therefore provide a solution to the dwindling supply of conventional energy sources. Fusion reactions occur when ions of the hydrogen isotopes, heated to sufficient temperatures, collide and overcome the electrical forces of separation. The basic requirement for controlled fusion is to heat a plasma (or ionized gas) to temperatures in excess of 10^8 degrees and *confine* it for times long enough that a significant number of fusion events occur. One method of accomplishing this is called *magnetic confinement*. The magnetic-confinement method presently used at Lawrence Livermore National Laboratory (LLNL) is the Tandem Mirror Experiment-Upgrade (TMX-U) experiment. Results from this experiment are leading to design principles for a commercial fusion reactor.

A parameter of significant importance to magnetically confined plasmas is the diamagnetism of the plasma. The diamagnetism is a measure of energy density stored in hot particles which is used to determine the efficiency of the applied magnetic fields. In the TMX-U, a single-turn loop transformer, the diamagnetic loop (DML) is used as the sensor for the plasma diamagnetism. The DML sensor is subjected to various noise sources, which makes the plasma-estimation problem difficult. Variations of the magnetic field used to contain the plasma are present because of feedback circuits and ripple currents in the main power system. In many cases, the signal that is used to determine the plasma diamagnetism is so badly corrupted with coherent frequency noise (ripple) that the plasma perturbation due to diamagnetism is not even visible. When the signals are approaching the noise level, or when the feedback-control system has introduced a trend to the data, a sophisticated technique must be used for the processing of the measured signals. It must incorporate trend removal with the capability of removing the coherent noise, without affecting the frequency content of the plasma perturbation itself. We choose the noise-cancelling processor of (Eq. 8.6–9) in the previous section.

First, we analyze the acquired diamagnetic loop (DML) sensor measurements and show how the data can be processed to retain the essential information required for post-experimental analysis. The measured DML data is analog (anti-alias) filtered and digitized at a 25 kHz sampling rate (40 μsec sample interval). A typical experiment generates a transient signal (plasma) which is recorded for approximately 650 msec. Pre-processed data (decimated, etc.) and the frequency spectrum are shown in Fig. 8.7–1, along with an expanded section of the transient pulse and noise. We note that the raw data are contaminated with a sinusoidal drift, linear trend, and random noise, as well as sinusoidal disturbances at harmonics of 60 Hz; the largest, at 360 Hz, is caused by the feedback circuits and

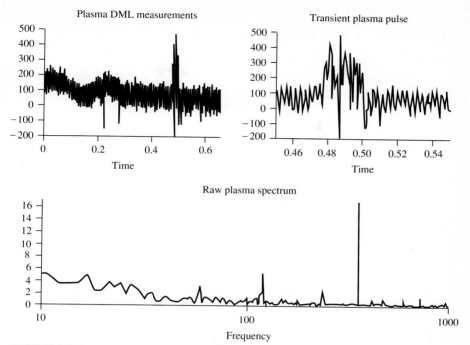

FIGURE 8.7–1
Preprocessed diamagnetic loop measurement data and spectrum

ripple currents in the main power system. The pulse is also contaminated by these disturbances. We also note that some of the plasma information appears as high energy spikes (pulses) *riding* on the slower plasma build-up pulse.

A processor must be developed to eliminate these disturbances yet preserve all of the essential features of the transient plasma pulse and associated energy spikes. This application is ideally suited for noise cancelling, since the signal and noise should not be correlated. This condition is satisfied by the DML measurement data, since the onset of the measurement consists only of the disturbances (trend and sinusoids), and the signal is available at the time of the transient plasma pulse.

The model of the TMX measurement is given by

$$y(t) = s(t) + n(t) + v(t) \tag{8.7–1}$$

$$n(t) = H(q^{-1})r(t) \tag{8.7–2}$$

where
    y is the measured data
    s is the plasma pulse signal
    n is the disturbance (drift, trend, sinusoids)

v is the random noise

H is the noise dynamics (unknown)

r is the reference noise

The reference noise data are obtained by selecting only that portion of the data prior to the onset of the plasma pulse, thus assuring no signal information is present. The noise canceller produces a pulse estimate given by

$$\hat{s}(t) = y(t) - \hat{n}(t) = s(t) + [H(q^{-1}) - \hat{H}_n(q^{-1})]r(t) + v(t) \qquad (8.7\text{–}3)$$

where the estimated noise is obtained from

$$\hat{n}(t) = \hat{H}_n(q^{-1})r(t) \qquad (8.7\text{–}4)$$

We see that as $\hat{H}_n \to H$, then $\hat{s} \to s + v$, as discussed in Sec. 8.6, indicating the importance of designing a processor capable of "matching" the dominant noise characteristics. As before, the minimum variance solution is given by

$$\hat{h}_n = \mathbf{R}_{rr}^{-1}\underline{R}_{yr} \qquad (8.7\text{–}5)$$

which can be implemented in the stationary case using the LWR algorithm (see Table 7.4–1) or in the nonstationary/real-time case using any of the adaptive all-zero processors discussed in Sec. 8.3. We implemented both the LWR and adaptive LMS (see Table 8.3–2) cancellers. We were able to achieve almost identical performance from both processors, so we limit our discussion to the results of the stationary design.

An acceptable design of a processor is also dependent on the timing requirements and computational capability available. During the operation of the TMX experiment, a "shot" (injection of a plasma into the reactor) terminates after a few seconds; during this time, data are collected and displayed, so that the experimenter can adjust process parameters and criteria and perform another shot within a five-minute time period. Even though the processor need not be on-line, it still must function in a real-time environment. Clearly, post-experimental analysis creates no restrictions on the processor design and allotted computational time. So we analyze the performance of the processor to function for both real-time and post-experimental modes of operation. We study the performance of the processor by varying its order L. The real-time processor must perform well enough to enable the experimenter to make the necessary decisions regarding the selection of process parameters for the next shot.

After some preliminary runs of the processor over various data sets, we decided to use $L = 512$ weights for the post-experimental design, since it produced excellent results. Using the post-experimental design as a standard, we then evaluated various designs for weights in the range of $8 < L < 512$. Before we discuss the comparisions, let us consider the heuristic operation of the processor. The crucial step in the design of the canceller is the estimation of the optimal noise filter $\hat{h}$, which is required to produce the minimum-variance estimate of the noise, $\hat{n}$ of Eq. (8.7–4). In essence, we expect the filter to match the corresponding

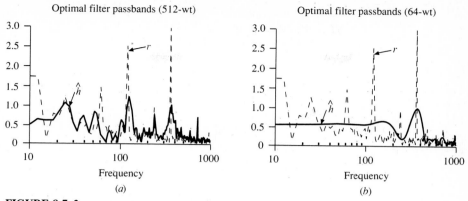

FIGURE 8.7–2

Optimal noise filter and noise spectra: (*a*) Post-filter (*b*) Real-time filter

noise spectrum in magnitude and phase. This means that we expect the optimal filter to pass the spectral peaks of the noise and attenuate any signal information not contained in the reference.

These results are confirmed as shown by the performance of the 512-weight filter shown in Fig. 8.7–2(*a*). Here we see that the filter passbands enable most of the noise resonances to pass while signal energy is attenuated. The real-time design is shown in Fig. 8.7–2(*b*). We see that the 64-weight filter still passes much of the noise energy but does not spectrally match the noise as well as the 512-weight filter, since there are fewer weights. These results are again confirmed in Fig. 8.7–3, where the estimated and actual noise spectra are shown. Again we see that the 512-weight produces a much better spectral match than the 64-weight design, due to its increased resolution. Note that the highest energy noise spectral

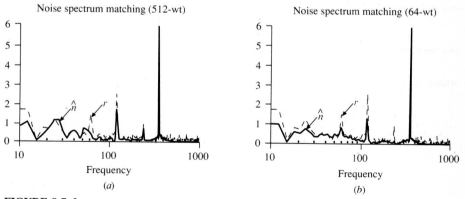

FIGURE 8.7–3

Optimal noise spectral matching: (*a*) Post-filter (*b*) Real-time filter

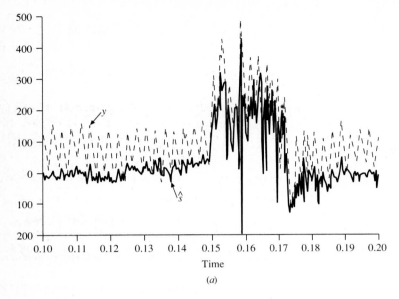

Estimated plasma signal (512-wt)

(a)

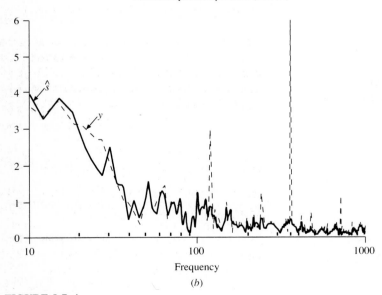

Estimated plasma spectrum (512-wt)

(b)

FIGURE 8.7–4
Post-experiment noise-canceller design: (*a*) Plasma pulse (*b*) Spectra

peaks were matched by both processors reasonably well, thereby eliminating these disturbances in the cancelling operation. Intermediate designs for the real-time processor fall in between these results, where selecting a higher number of weights results in better processor performance.

The noise-canceller algorithm is constructed using various commands in SIG [13], as discussed previously. Both the post-experimental and real-time designs were run on the data set described in Fig. 8.7–1; the results of the 512-weight design are shown in Fig. 8.7–4. Here we see the raw and processed data and corresponding spectra. A closer examination of the estimated transient pulse shows that not only have the disturbances been removed, but the integrity of the pulse has been maintained, and all of the high-frequency energy spikes have been preserved. Note that the 512-weight processor has clearly eliminated the trends and sinusoidal disturbances (spectrum) and has retained the transient plasma information. The real-time (64-weight) processor performs satisfactorily, but does not attenuate the disturbances as well as the post-experimental processor, as discussed previously. Once these disturbances have been removed, the resulting signal can be processed further to remove the random noise and provide an estimate of the stored energy build-up in the machine.

8.8   SUMMARY

In this chapter, we have introduced the adaptive signal-processing method and discussed various gradient-based adaptation algorithms: (i) the stochastic gradient, (ii) the Newton, and (iii) the instantaneous gradient. We then applied this approach to all-zero adaptive filter designs and investigated the stochastic and instantaneous gradient or least mean-squared (LMS) approaches, as well as the Newton-like recursive least squares (RLS) design. Next, we investigated pole-zero adaptive filter designs and showed how the RPEM algorithms of the previous chapter can be used to solve the adaptive problem as well. We then developed an LMS algorithm. All-pole and all-zero adaptive lattice filters were developed, and an alternative approach to the block processor of the previous chapter also evolved. We investigated noise-cancelling filter designs and various applications of the adaptive algorithm and its variants, the adaptive predictor and line enhancer. Finally, we discussed a case study involving the design of noise-cancelling filters for the extraction of a transient plasma pulse.

REFERENCES

1. B. Widrow and S. Stearns, *Adaptive Signal Processing* (Englewood Cliffs, N. J.: Prentice-Hall, 1984).
2. S. Haykin, *Introduction to Adaptive Filters* (New York: Macmillan, 1984).
3. R. Monzingo and T. Miller, *Introduction to Adaptive Arrays* (New York: Wiley, 1980).
4. G. Goodwin and K. Sin, *Adaptive Filtering, Prediction and Control* (Englewood Cliffs, N. J.: Prentice-Hall, 1984).

5. L. Ljung and T. Soderstrom, *Theory and Practice of Recursive Identification* (Boston: MIT Press, 1983).
6. P. Eykhoff, *System Identification—Parameter and State Estimation* (New York: Wiley, 1974).
7. D. Luenberger, *Introduction to Linear and Nonlinear Programming* (Reading, Pa.: Addison-Wesley, 1973).
8. D. Luenberger, *Optimization by Vector Space Methods* (New York: Wiley, 1969).
9. G. Noble, *Applied Linear Algebra* (Englewood Cliffs, N. J.: Prentice-Hall, 1967).
10. S. Orfanidis, *Optimum Signal Processing* (New York: Macmillan, 1985).
11. R. Bitmead and B. Anderson, "Adaptive Frequency Sampling Filters," *IEEE Trans. Circuits Systems*, Vol. CAS-28, 1981.
12. R. Bitmead and B. Anderson, "Performance of Adaptive Estimation Algorithms in Dependent Random Environments," *IEEE Trans. Auto. Control*, Vol. AC-25, 1980.
13. D. Lager and S. Azevedo, "SIG—A General Purpose Signal Processing Code," *Proc. IEEE*, 1987.
14. C. Cowen and P. Grant, *Adaptive Filters* (Englewood Cliffs, N. J.: Prentice-Hall, 1985).
15. B. Friedlander, "System Identification Techniques for Adaptive Signal Processing," *Circuits, Systems, Signal Process*, Vol. 1, 1982.
16. L. Griffiths "A Continuously Adaptive Filter Implemented as a Lattice Structure," *Proc. ICASSP*, 1977.
17. L. Griffiths, "An Adaptive Lattice Structure for Noise-Cancelling Applications," *Proc. ICASSP*, 1978.
18. M. Morf and D. Lee, "Recursive Least-Squares Ladder Forms for Fast Parameter Tracking," *Proc. IEEE CDC Conf.*, 1979.
19. D. Lee, M. Morf, and B. Friedlander, "Recursive Least-Squares Ladder Estimation Algorithms," *IEEE Trans. Acoustics, Speech and Signal Proc.* Vol. ASSP-29, 1981.
20. J. Candy, T. Casper, and R. Kane, "Plasma Estimation: A Noise-Cancelling Application," *Automatica*, Vol. 22, 1986.

SIG NOTES

SIG can be used to develop adaptive signal processors. The primary command is the ANOISECANCEL command, which is an implementation of the LMS all-zero algorithm (see Sec. 8.6). From this command we can construct many applications simply by conditioning the input signal (signal plus noise; e.g., predictor, line enhancer, deconvolver, etc.). The stationary case of the noise canceller can be implemented using the set of commands IMPULSERESPONSE, CONVOLVE, and SUBTRACT. The LSIABC can be used as an adaptive processor to implement the RLS all-zero, all-pole, and pole-zero structures. Finally, the ALATTICE and GRADAPTLATTICE (SHARE) commands implement the block and LMS Burg algorithms which can then be transformed to an all-pole or all-zero processor.

EXERCISES

8.1. Suppose we are given a set of measurements $\{y(1), y(2)\}$ and asked to design a two-weighted adaptive signal estimator with $\underline{w}' = [w(1)\, w(2)]$, such that

$$\hat{s}(t) = \sum_{i=1}^{2} w(i)y(i)$$

(a) Find the optimal Wiener estimator.
(b) Find the gradient estimator.
(c) Find the Newton estimator.

8.2. We are asked to design and analyze the performance of an all-zero processor for a desired signal given by an AR model

$$s_d(t) = \tfrac{3}{4} s_d(t-1) + n(t)$$

and measurement given by

$$x(t) = \tfrac{1}{8} s_d(t) + v(t)$$

for n, v zero-mean, white noise, with variance 0.1 and 1 respectively. Design *both* a single-weight and two-weight processor using the following:

(a) optimal Wiener filter
(b) stochastic gradient processor
(c) stochastic Newton processor
(d) LMS processor
(e) RLS processor

8.3. Suppose that for the previous problem the measurement is given by

$$x(t) = s_d(t-1) + s_d(t-2) + v(t)$$

Repeat the designs.

8.4. We are given a two-weight, adaptive signal processor characterized by

$$\underline{h}(i+1) = \underline{h}(i) - \Delta \frac{\partial J(\underline{h})}{\partial h}$$

where $e(t) = s_d(t) - \underline{h}' \underline{X}$.

(a) Derive the stochastic gradient, Newton, and LMS solutions, and explicitly show the entries of the associated matrices.
(b) Suppose $R_{xx}(k) = \alpha^{|k|}$ and $R_{s_d x}(k) = \beta^{|k|}$; repeat part (a). What are reasonable step sizes for convergence? Calculate the misadjustment for a selected step size.
(c) Repeat part (a) with $\alpha = 0.5$ and $\beta = 0.8$.
(d) Simulate this process and apply the adaptive algorithms. Are the results as predicted?

8.5. We are asked to compare the performance of two processors, each evolving from the output covariance matrix

$$R_{xx} = \begin{bmatrix} 1 & 1/2 \\ 1/2 & 1 \end{bmatrix}$$

The corresponding cross-covariance vectors are given by

$$\gamma_{s_d x}^1 = \begin{bmatrix} 1 \\ 1/2 \end{bmatrix}, \qquad \gamma_{s_d x}^2 = \begin{bmatrix} 2 \\ 1/4 \end{bmatrix}$$

(a) Find the corresponding optimal weight vectors, h_{opt}^i, $i = 1, 2$.
(b) Calculate the corresponding minimum mean-squared error, J_{min}, $i = 1, 2$. Which estimator can be expected to perform better?
(c) Find expressions for each learning curve in terms of J_{min}^i.
(d) Assume you are going to implement a stochastic gradient algorithm. What is a reasonable step size for stability? For speed?
(e) Perform the eigenvalue-eigenvector transformation and sketch the error surface.

8.6. Suppose that we have a desired signal characterized by a second-order AR process; that is,

$$s_d(t) + \alpha_1 s_d(t - 1) + \alpha_2 s_d(t - 2) = e(t)$$

with $e(t)$ zero-mean, white, of variance R_{ee}. Assume we are to design a two-weight, all-pole stochastic gradient processor characterized by

$$\hat{s}_d(t) = \underline{h}'\underline{X} = [h_1\ h_2]\begin{bmatrix} x(t) \\ x(t-1) \end{bmatrix}$$

(a) In terms of the AR process parameters, calculate the covariance matrix, R_{xx}.
(b) Determine the eigenvalue spread of R_{xx} and the corresponding eigenvectors (normalized to unit length).
(c) Calculate the minimum mean-squared error and sketch a plot as a function of eigenvalue spread. What is the overall conclusion?

8.7. Calculate the optimal all-zero Wiener solutions for the following process statistics, find $\underline{h}_{\text{opt}}$, J_{min}, and obtain the corresponding eigensolutions

(a) $R_{xx} = \begin{bmatrix} 6 & 3 \\ 3 & 6 \end{bmatrix}$, $R_{s_d x} = \begin{bmatrix} 2 & 1 \\ 2 & 4 \end{bmatrix}$, $R_{s_d s_d} = 126$

(b) $R_{xx} = \begin{bmatrix} 1 & 1/2 \\ 1/2 & 1 \end{bmatrix}$, $R_{s_d x} = \begin{bmatrix} 1/4 \\ 1/2 \end{bmatrix}$, $R_{s_d s_d} = 64$

8.8. An unknown plant can be "identified" with the stochastic gradient algorithm by using the connections shown below:

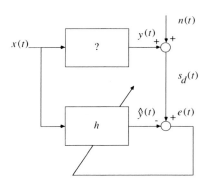

(a) Develop the stationary solution to this problem.
(b) Develop the stochastic gradient and LMS iterators.
(c) Suppose the "unknown" model is given by an FIR model.

$$y(t) = (1 + 1.8q^{-1} + 0.9q^{-2})x(t)$$

and assume x and n are white with $R_{xx} = 1/12$, $R_{nn} = 1/12$. Simulate this system and obtain the optimal stationary and LMS solutions. Calculate the minimum mean-squared error in this case. Compare the solutions.

(*d*) FIR filters can also be designed in this manner by specifying the unknown system coefficients and allowing the adaptive algorithm to learn the response. Start with a second-order Butterworth low-pass response ($f_c = 10$ Hz) and excite the filter with pseudorandom noise. Design a 16, 32, 64-weight filter and "teach" the adaptive filter. Discuss your results.

8.9. The unknown input can be estimated or "deconvolved" adaptively as well, as shown in the figure below:

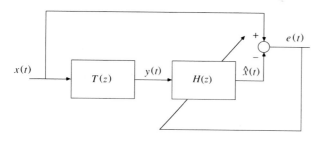

(*a*) Develop the stationary solution to this problem.
(*b*) Develop the stochastic gradient and LMS iterators.
(*c*) Suppose we use ARMAX(2,1,1,1) model of Example 7.3–2 and design an adaptive deconvolver using the LMS algorithm. Choose weights as in Example 7.4–2 and compare with the stationary solution. This is also called a "channel equalizer" if $T(z)$ is the model of a channel.

8.10. Suppose we are asked to develop a single-weight noise canceller with signal

$$y = s + n$$

$$n = Fr$$

where r is the reference noise source.
(*a*) Show that the optimal weight $h = F$ and that complete noise cancellation occurs with this choice.
(*b*) Suppose that the noise model is

$$n = Fr + v$$

with v random, white. What is the optimal weight now? Does complete cancellation still occur?
(*c*) Suppose that the signal model is

$$y = s + n + v$$

Are the results different from those in (*b*)?
(*d*) Suppose the noise model also contains signal information; that is

$$\hat{n} = Fr + \Delta s$$

Calculate the output of the canceller $e(t)$, and show that when $\Delta = 0$ the results are identical to (*a*).

8.11. In the adaptive LMS noise canceller, the *complex* algorithm is given by

$$\underline{h}(t + 1) = \underline{h}(t) + \Delta e(t)\underline{x}^*(t)$$

Suppose $x(t) = Ae^{j\omega_0 t}$, then obtain a scalar weight update equation and, using Z-transforms, show that the transfer function is that of a "notch filter" which can be controlled by the step size. That is, a pole at $z = e^{j\omega_0}(1 - \Delta|A|^2(L + 1))$.

8.12. Simulate a 10 Hz sinusoidal signal contaminated with random noise of variance $R_{nn} = 0.01$. Now assume it is related to a reference noise signal that is AR:

$$r(t) = 1.5r(t - 1) + 0.1562r(t - 2) = n(t)$$

Design a stationary and adaptive solution to this problem.

8.13. Suppose we are asked to develop the optimal d-step prediction solution for a stationary signal. Using an all-zero filter,

$$\underline{r}(t + d) = \underline{h}'\underline{y}$$

(a) Derive the optimal estimate of Eq. (8.6–13).

(b) Derive the adaptive stochastic gradient and corresponding LMS solutions to this problem.

8.14. An antenna array can be considered a "linear combiner," as shown below:

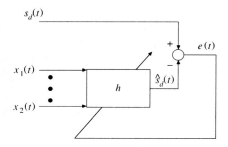

(a) Derive the optimal Wiener "stationary" solution to this problem.

(b) Derive the adaptive stochastic gradient and LMS solutions.

(c) Suppose this represents a two-element antenna array and we are asked to design an adaptive nulling scheme, $s_d(t) = 0$. What are the optimal weights for this problem?

8.15. An adaptive predictor can be used as a "self-tuning filter." Suppose we would like to extract a periodic signal in broadband noise using the predictor shown below:

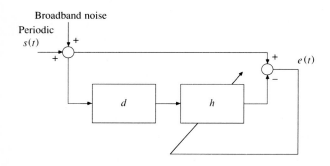

(a) Discuss the operation of this device and suggest an appropriate choice for d, the prediction distance.

(b) Simulate a 50 Hz sinusoid in bandpass Gaussian noise (cutoffs: 10, 90 Hz) of variance 0.01. Based on estimated covariances, select d and obtain the adaptive (LMS) solution.

CHAPTER
9

MODEL-BASED SIGNAL PROCESSING

In this chapter, we discuss the concepts behind model-based processing. We develop the state-space model-based processor—the Kalman filter—and show how it can be used to solve various problems: state estimation, identification, and deconvolution. Finally, we show the equivalence of Wiener and Kalman filters.

9.1 INTRODUCTION

This chapter is concerned with the development of signal-processing techniques to extract pertinent information from random signals, using any a priori information available. Recall that we call these techniques *signal estimation*, and we call a particular algorithm a *signal estimator* or just an *estimator*. Sometimes estimators are called filters (e.g., the Wiener filter), because they perform the same function as a deterministic filter except for random signals; i.e., they remove unwanted disturbances. Noisy measurements are processed by the estimator to produce "filtered" data. Estimation can be thought of as a procedure made up of three primary parts:

1. The criterion function
2. Models
3. The algorithm

314

The criterion function can take many forms and can also be classified as deterministic or stochastic. Models represent a broad class of information formalizing the a priori knowledge about the process generating the signal, about measurement instrumentation, about noise characterization, about underlying probabilistic structure, etc. Finally, the algorithm or technique chosen to minimize (or maximize) the criterion can take many different forms, depending on (1) the models, (2) the criterion, and (3) the choice of solution. For example, one may choose to solve the well-known least-squares problem recursively or with a numerical-optimization algorithm. Another important aspect of most estimation algorithms is that they provide a "measure of quality" of the estimator. Usually what this means is that the estimator also predicts vital statistical information about its own performance.

Intuitively, we can think of the estimation procedure as these processes:

1. The specification of a criterion
2. The selection of models from a priori knowledge
3. The development and implementation of an algorithm

Criterion functions are usually selected on the basis of information that is meaningful about the process, or the ease with which an estimator can be developed. Criterion functions that are useful in estimation can be classified as deterministic and probabilistic. Some typical functions follow:

Deterministic:
    Squared error
    Absolute error
    Integral absolute error
    Integral squared error
Probabilistic:
    Maximum likelihood
    Maximum a posteriori (bayesian)
    Maximum entropy
    Minimum (error) variance

Models can also be deterministic as well as probabilistic; however, here we prefer to limit their basis to knowledge of the process phenomenology (physics) and the underlying probability mass functions with the associated statistics. Phenomenological models fall into the usual classes defined by the type of underlying mathematical equations and their structure; i.e., linear or nonlinear, differential or difference, ordinary or partial, time-invariant or varying. Usually these models evolve to a stochastic model by the inclusion of uncertainty or noise processes.

Finally, the estimation algorithm can evolve from various influences. A preconceived notion of the structure of the estimator heavily influences the resulting algorithm. We may choose, based on computational considerations, to calculate

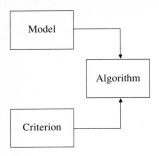

FIGURE 9.1–1
Interaction of model and criterion in an estimation algorithm

an estimate recursively, rather than as a result of a batch process, because we require an on-line, pseudo-real-time estimate. Also, each algorithm must provide a measure of estimation quality, usually in terms of the expected estimation error. This measure provides a means for comparing estimators. Thus the estimation procedure is a combination of these three major ingredients: criterion, models, and algorithm. The development of a particular algorithm is an interaction of selecting the appropriate criterion and models, as depicted in Fig. 9.1–1.

This completes the discussion of the general estimation procedure. Many estimation techniques have been developed independently from various viewpoints (optimization, probabilistic) and have been shown to be equivalent. In most cases, it is easy to show that they can be formulated in this signal-processing framework. Thus, it is more appropriate to call the processing discussed in this chapter "model-based" to differentiate it from pure statistical techniques. In the next section we give some examples of model-based processors.

9.2 MODEL-BASED SIGNAL PROCESSING

In this section, we discuss the levels of models that can be used for estimation purposes. It is important to investigate what, if anything, can be gained by filtering the data. The amount of information available in the data is related to the precision (variance) of the particular measurement instrumentation used, as well as to any signal-processing devices or algorithms employed to improve the estimates. As we utilize more and more information about the physical principles underlying the given data, we expect to improve our estimates (decrease estimation error) significantly.

A typical measurement y is depicted in Fig. 9.2–1. The ideal measurement instrument is presented in Fig. 9.2–2(a), and a more realistic model of the measurement process is depicted in Fig. 9.2–2(b). If we were to use y to estimate s (that is, $\hat{s}$), we have the noise lying within the Δ confidence limits superimposed on the signal (see Fig. 9.2–1). The best estimate of s we can hope for lies within the accuracy (bias) and precision (variance) of the instrument. If we include a model of the measurement instrument (see Fig. 9.2–3) as well as its associated uncertainties, then we can improve on the preceding noisy measurements. This technique is the filtering of noise or signal processing based on our model of

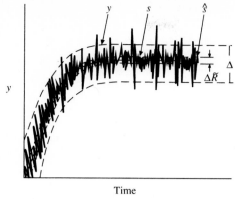

Time

$y = s + v$

$\hat{s} \approx s$

FIGURE 9.2–1

Model-based processing of a noisy measurement

instrument and noise. We can also specify the estimation error variance $\tilde{R}$, or quality, of this estimator (e.g., least-squares estimator) in estimating s. Finally, if we incorporate not only instrumentation knowledge but also knowledge of the physical process, then we expect to do even better; i.e., we expect the estimation error ($\tilde{s} = s - \hat{s}$) variance $\tilde{P}$ to be small (see Fig. 9.2–1 for $\Delta\tilde{R}$ confidence limits) as we incorporate more and more knowledge (see Fig. 9.2–4). In fact, this is the case, because it can be shown that

$$\text{Instrument variance} > \tilde{R} > \tilde{P}$$

This is the basic idea in model-based signal processing: *The more a priori information we can incorporate into the algorithm, the smaller the resulting error variance.* The following example illustrates these ideas.

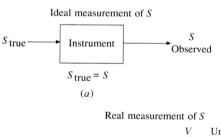

Ideal measurement of S

S_{true} — Instrument — S Observed

$S_{\text{true}} = S$

(a)

Real measurement of S

V Uncertainty

S_{true} — Instrument — S Observed + Y Measured

(b)

FIGURE 9.2–2

Simple measurement systems

Incorporate Knowledge of Instrument and Uncertainty

$y = c(s)$ (Instrument model)

$V \sim N(0, R_{vv})$ (Measurement uncertainty model)

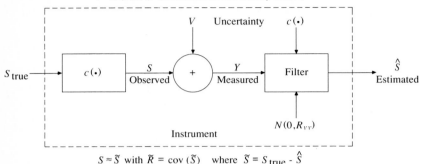

$$S \approx \tilde{S} \text{ with } \tilde{R} = \text{cov}(\tilde{S}) \quad \text{where } \tilde{S} = S_{\text{true}} - \hat{S}$$

FIGURE 9.2–3
Model-based signal processing with measurement instrument modeling

Example 9.2–1. The voltage at the output of an RC circuit is to be measured using a high impedance voltmeter, shown in Fig. 9.2–5. The measurement is contaminated with random instrumentation noise, which can be modeled as

$$e_{\text{out}} = K_e e + n$$

where

e_{out} = measured voltage
K_e = instrument amplification factor
e = true voltage
n = zero-mean random noise of variance R_{nn}

Incorporate Knowledge of Instrument and Uncertainty

$s = a(s, w)$ (Process model)
$W \sim N(0, R_{ww})$ (Process uncertainty model)
$y = c(s)$ (Instrument model)
$V \sim N(0, R_{vv})$ (Measurement uncertainty model)

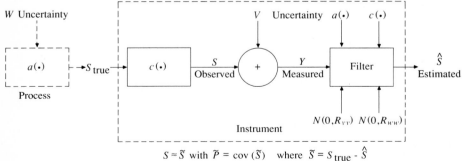

$$S \approx \tilde{S} \text{ with } \tilde{P} = \text{cov}(\tilde{S}) \quad \text{where } \tilde{S} = S_{\text{true}} - \hat{S}$$

FIGURE 9.2–4
Model-based signal processing with measurement uncertainty and process modeling

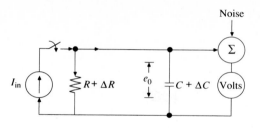

FIGURE 9.2–5
Model-based processing of an RC circuit

This model corresponds to those described in Fig. 9.2–3. A processor is to be designed to improve the precision of the instrument. Then, as in the figure, we have

Measurement:

$$s \rightarrow e$$

$$c(\cdot) \rightarrow K_e$$

$$v \rightarrow n$$

$$R_{vv} \rightarrow R_{nn}$$

and for the filter,

$$\hat{s} \rightarrow \hat{e}$$

$$\tilde{R}_{ss} \rightarrow \tilde{R}_{ee}$$

The precision of the instrument can be improved even further by including a model of the process (circuit). Writing the Kirchoff node equations, we have

$$\dot{e} = \frac{1}{C} I_{in} - \frac{e}{RC} + q$$

where

R = resistance
C = capacitance
I_{in} = excitation current
q = zero-mean random noise of variance R_{qq}

The improved model-based processor employs both measurement and process models, as in Fig. 9.2–4. Thus, we have

Process:

$$\dot{s} \rightarrow \dot{e}$$

$$a(\cdot) \rightarrow -\frac{e}{RC} + \frac{1}{C} I_{in}$$

$$w \rightarrow q$$

$$R_{ww} \rightarrow R_{qq}$$

Measurement:

$$s \rightarrow e$$

$$c(\cdot) \rightarrow K_e$$

$$v \to n$$

$$R_{vv} \to R_{nn}$$

Therefore, the filter becomes

$$\hat{s} \to \hat{e}$$

$$\tilde{P}_{ss} \to \tilde{P}_{ee}$$

such that

$$R_{vv} > \tilde{R}_{ee} > \tilde{P}_{ee}$$

There are many different forms of model-based processors, depending on the models used and the manner in which the estimates are calculated. For example, there are process model-based processors (Kalman filters [1,2]), statistical model-based processors (Box-Jenkins filters [3], bayesian filters [4]), statistical model-based processors (covariance filters [5]), or even optimization-based processors (gradient filters [6]). In any case, many processors can be placed in a recursive form, with various subtleties emerging in the calculation of the current estimate ($\hat{S}_{\text{old}}$). The standard technique employed is based on correcting or updating the current estimate as a new piece of measurement data becomes available. The estimates generally take the *recursive form*:

$$\hat{S}_{\text{new}} = \hat{S}_{\text{old}} + K E_{\text{new}}$$

where

$$E_{\text{new}} = Y - \hat{Y}_{\text{old}} = Y - C\hat{S}_{\text{old}}$$

Here we see that the new estimate is obtained by correcting the old estimate by a K-weighted amount. The error term E_{new} is the new information or innovation; that is, the error term is the difference between the actual measurement and the predicted measurement ($\hat{Y}_{\text{old}}$), based on the old estimate ($\hat{S}_{\text{old}}$). The computation of the weight K depends on the criterion used (e.g., mean-squared error, absolute error, etc.).

The following example shows how to estimate the sample mean recursively.

Example 9.2–2. The sample-mean estimator can easily be put in recursive form. The estimator is given by

$$\hat{X}(N) = \frac{1}{N} \sum_{t=1}^{N} y(t)$$

Extracting the Nth term from the sum, we obtain

$$\hat{X}(N) = \frac{1}{N} y(N) + \frac{1}{N} \sum_{t=1}^{N-1} y(t)$$

Identifying $\hat{X}(N-1)$ from the last term,

$$\hat{X}(N) = \frac{1}{N} y(N) + \frac{N-1}{N} \hat{X}(N-1)$$

The recursive form is given by

$$\hat{X}(N) = \hat{X}(N-1) + \frac{1}{N}[y(N) - \hat{X}(N-1)]$$

9.3 MODEL-BASED PROCESSORS: STATE-SPACE (KALMAN) FILTERS

In this section, we develop model-based processors for the dynamic estimation problem, that is, the estimation of processes that vary with time. First, we develop the Kalman filter for estimating states from noisy data [7]. Here the state-space representation is used as the basic model. We discuss the operation of the filter as a predictor-corrector algorithm. Then we briefly outline the estimator theoretically, using the innovations approach. The practice of filtering is mentioned and an example given.

In this section we discuss the Kalman filter from the model-based processing viewpoint using the state-space representation. First, we describe the algorithm and attempt to give some insight into the operations. Then, we briefly develop the algorithm. The Kalman filter can be thought of as a processor that produces three types of outputs, given a noisy measurement sequence and the associated models. The filter is depicted in Fig. 9.3–1. It can be thought of as a *state estimator* or *reconstructor*; i.e., it reconstructs estimates of the state $x(t)$ from noisy measurements $y(t)$. Second, the Kalman estimator can be thought of as a *measurement filter* which, on input, accepts the noisy sequence $\{y(t)\}$ and, on output, produces a filtered measurement sequence $\{\hat{y}(t|t)\}$. Finally, the estimator can be thought of as a *whitening filter* that accepts noisy correlated measurements $\{y(t)\}$ and produces uncorrelated or white equivalent measurements $\{e(t)\}$, the innovation sequence. All these properties of the filter have been exploited in many different applications [1,2,3,4]. Next we present the algorithm in the predictor-corrector form. We use this form because it provides much insight into the operation of the Kalman filter.

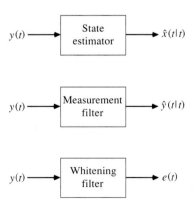

FIGURE 9.3–1
Various representations of the Kalman filter estimator

TABLE 9.3–1
Kalman filter algorithm (predictor-corrector form)

Prediction

$$\hat{x}(t \mid t - 1) = A(t - 1)\hat{x}(t - 1 \mid t - 1) + B(t - 1)u(t - 1) \quad \text{(State prediction)}$$

$$\tilde{P}(t \mid t - 1) = A(t - 1)\tilde{P}(t - 1 \mid t - 1)A'(t - 1) + W(t - 1)R_{ww}(t - 1)W'(t - 1)$$

(Covariance prediction)

Innovation

$$e(t) = y(t) - \hat{y}(t \mid t - 1) = y(t) - C(t)\hat{x}(t \mid t - 1) \quad \text{(Innovation)}$$

$$R_{ee}(t) = C(t)\tilde{P}(t \mid t - 1)C'(t) + R_{vv}(t) \quad \text{(Innovation covariance)}$$

Gain

$$K(t) = \tilde{P}(t \mid t - 1)C'(t)R_{ee}^{-1}(t) \quad \text{(Kalman gain or weight)}$$

Correction

$$\hat{x}(t \mid t) = \hat{x}(t \mid t - 1) + K(t)e(t) \quad \text{(State correction)}$$

$$\tilde{P}(t \mid t) = [I - K(t)C(t)]\tilde{P}(t \mid t - 1) \quad \text{(Covariance correction)}$$

Initial conditions: $\hat{x}(0 \mid 0)$ $\tilde{P}(0 \mid 0)$

The operation of the Kalman filter algorithm can be viewed as a predictor-corrector algorithm, as in standard numerical integration. Referring to the algorithm in Table 9.3–1[†] and Fig. 9.3–2, we see the inherent timing in the algorithm. First, suppose we are currently at time t and have not received a measurement $y(t)$ as yet. We have available to us the previous filtered estimate $\hat{x}(t - 1 \mid t - 1)$ and covariance $\tilde{P}(t - 1 \mid t - 1)$ we would like to obtain the best estimate of the state, based on $[t - 1]$ data samples. We are in the "prediction phase" of the algorithm. We use the state-space model to predict the state estimate $\hat{x}(t \mid t-1)$ and associated error covariance $\tilde{P}(t \mid t-1)$. Once the prediction based on the model is completed, we then calculate the innovation covariance $R_{ee}(t)$ and Kalman gain $K(t)$. As soon as the measurement at time t becomes available, that is, $y(t)$, then we determine the innovation $e(t)$. Now we enter the "correction phase" of the algorithm. Here we correct or update the state based on the new information in the measurement— the innovation. The old, or predicted, state estimate $\hat{x}(t \mid t - 1)$ is used to form the filtered, or corrected, state estimate $\hat{x}(t \mid t)$ and $\tilde{P}(t \mid t)$. Here we see that the error, or innovation, is the difference between the actual measurement and the predicted measurement $\hat{y}(t \mid t - 1)$. The innovation is weighted by the gain

---

[†]Note that the preferred numerical technique of implementing this algorithm is the U-D factorization method (see [18] for details).

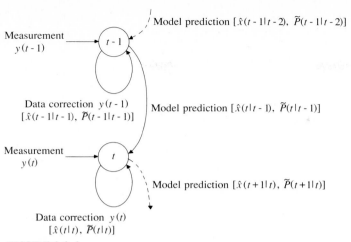

FIGURE 9.3–2
Predictor-corrector form of the Kalman filter

$K(t)$ to correct the old state estimate (predicted) $\hat{x}(t \mid t - 1)$; the associated error covariance is corrected as well. The algorithm then awaits the next measurement at time $(t + 1)$.

Note that in the absence of a measurement, the state-space model is used to perform the prediction, since it provides the best estimate of the state. Note also that the covariance equations can be interpreted in terms of the various signal (state) and noise models (see Table 9.3–1). The first term of the predicted covariance, $\tilde{P}(t \mid t - 1)$, relates to the uncertainty in predicting the state using the model A. The second term indicates the increase in error variance due to the contribution of the process noise (R_{ww}). The corrected covariance equation indicates the predicted error covariance or uncertainty due to the prediction, decreased by the effect of the update (KC), thereby producing the corrected error covariance $\tilde{P}(t \mid t)$.

We outline the derivation of the Kalman filter algorithm from the innovations viewpoint, following the approach by Kailath [8,9]. First, recall from Sec. 4.7 that a model of a stochastic process is given by the state-space representation

$$x(t) = A(t - 1)x(t - 1) + B(t - 1)u(t - 1) + W(t - 1)w(t - 1) \quad (9.3\text{–}1)$$

where w is assumed zero-mean and white, with covariance R_{ww}, and x and w are uncorrelated. The measurement model is given by

$$y(t) = C(t)x(t) + v(t) \quad (9.3\text{–}2)$$

where v is a zero-mean, white sequence, with covariance R_{vv}, and v is uncorrelated with x and w.

The model-based processing problem can be stated in terms of the preceding state-space model. The linear *state estimation problem* is as follows:

Given a set of noisy measurements $\{y(i)\}$, for $i = 1, \ldots, t$, characterized by the measurement model of Eq. (9.3–2), find the linear minimum (error) variance estimate of the state characterized by the state-space model of Eq. (9.3–1). That is, find the best estimate of $x(t)$, given the measurement data up to time t, $Y(t) := \{y(1), \ldots, y(t)\}$.

First, we develop the batch minimum-variance estimator for this problem, using the results derived in Sec. 5.2. Then, we develop an alternative solution, using the innovations sequence. The recursive solution follows almost immediately from the innovations and, therefore, we derive equations for the predicted state, gain, and innovations. The corrected, or filtered, state equation is then developed.

We constrain the estimator to be linear. Therefore, we see that for a batch of N data, the minimum variance estimator is given by[†]

$$\hat{X}_{MV} = K_{MV}\underline{Y} = R_{xy}R_{yy}^{-1}\underline{Y} \tag{9.3–3}$$

where $\hat{X}_{MV} \in R^{N_x N}$, R_{xy} and $K \in R^{(N_x \times N_y)N}$, $R_{yy} \in R^{(N_y \times N_y)N}$, and $\underline{Y} \in R^{N_y N}$. Similarly, in the dynamic case the linear estimator is

$$\hat{X}_{MV}(N) = R_{xy}(\underline{N})R_{yy}^{-1}(\underline{N}) = K_{MV}(\underline{N})\underline{Y} \tag{9.3–4}$$

where $\hat{X}_{MV}(N) = [\hat{x}'(1) \ldots \hat{x}'(N)]'$, $\underline{Y} = [y'(1) \ldots y'(N)]'$, $\hat{x} \in R^{N_x}$, $y \in R^{N_y}$, and we define the notation $\underline{N} := 1, 2, \ldots, N$.

Note that we are investigating a "batch" solution to the state-estimation problem, since all the N_y-vector data $\{y(1) \ldots y(N)\}$ are processed in one batch. However, we require a recursive solution to this problem, of the form

$$\hat{X}_{new} = \hat{X}_{old} + KE_{new}$$

In order to achieve the recursive solution, it is necessary to transform the covariance matrix R_{yy} to be block diagonal, since

$$R_{yy}(N) = \begin{bmatrix} E\{y(1)y'(1)\} & \cdots & E\{y(1)y'(N)\} \\ \vdots & & \vdots \\ E\{y(N)y'(1)\} & \cdots & E\{y(N)y'(N)\} \end{bmatrix} = \begin{bmatrix} R_{yy}(1,1) & \cdots & R_{yy}(1,N) \\ \vdots & & \vdots \\ R_{yy}(1,N) & \cdots & R_{yy}(N,N) \end{bmatrix}$$

R_{yy} block diagonal implies that all the off-diagonal block matrices $R_{yy}(t, j) = 0$, for i not equal to j, which in turn implies that $\{y(t)\}$ must be uncorrelated or orthogonal. Therefore, we must construct a sequence of independent N_y vectors, say $\{e(t)\}$, such that

$$E\{e(t)e'(k)\} = 0 \qquad \text{for } t \neq k$$

The sequence $\{e(t)\}$ can be constructed using the orthogonality property of the minimum-variance estimator:

[†]Note that the complete form of the linear minimum-variance estimator for nonzero mean is given by $\hat{X}_{MV} = \underline{m}_x + R_{xy}R_{yy}^{-1}(\underline{Y} - \underline{m}_y)$

$$[y(t) - E\{y(t) \mid Y(t-1)\}] \perp Y(t-1)$$

We define the *innovation* or new information [8] as

$$e(t) = y(t) - \hat{y}(t \mid t-1) \tag{9.3-5}$$

with the orthogonality property that

$$\text{cov}\,[y(T), e(t)] = 0 \qquad \text{for } T \le t-1 \tag{9.3-6}$$

Since $\{e(t)\}$ is a time-uncorrelated N_y-vector sequence, we have

$$R_{ee}(\underline{N}) = \begin{bmatrix} R_{ee}(1) & & 0 \\ & \ddots & \\ 0 & & R_{ee}(N) \end{bmatrix} \qquad \text{for each } R_{ee}(i) \in R^{N_y \times N_y}$$

The correlated measurement vector can be transformed to an uncorrelated innovation vector through a linear transformation [9], say L, given by

$$\underline{Y} = L\underline{e} \tag{9.3-7}$$

where $L \in R^{(N_y \times N_y)N}$ is a nonsingular transformation matrix and $\underline{e} := [e'(1) \cdots e'(N)]'$. Multiplying Eq. (9.3–7) by its transpose and taking expected values, we obtain

$$R_{yy}(\underline{N}) = L R_{ee}(\underline{N}) L'$$

Inverting, we have

$$R_{yy}^{-1}(\underline{N}) = (L')^{-1} R_{ee}^{-1}(\underline{N}) L^{-1}$$

Similarly, we obtain

$$R_{xy}(\underline{N}) = R_{xe}(\underline{N}) L'$$

Substituting these results into Eq. (9.3–4) gives

$$\underline{\hat{X}}_{\text{MV}}(N) = R_{xy}(\underline{N}) R_{yy}^{-1}(\underline{N}) \underline{Y} = [R_{xe}(\underline{N}) L'][(L)'^{-1} R_{ee}^{-1}(\underline{N}) L^{-1}](L\underline{e})$$

or $\qquad \underline{\hat{X}}_{\text{MV}}(N) = R_{xe}(\underline{N}) R_{ee}^{-1}(\underline{N}) \underline{e}$ $\qquad\qquad$ (9.3–8)

Since the values for $\{e(t)\}$ are time-uncorrelated, $R_{ee}(\underline{N})$ is block diagonal. From the orthogonality properties of e, it can be shown that $R_{xe}(\underline{N})$ is lower-block triangular; that is,

$$R_{xe}(\underline{N}) = \begin{cases} R_{xe}(t, i) & t > i \\ R_{xe}(t, t) & t = i \\ 0 & t < i \end{cases}$$

Substituting into Eq. (9.3–8), we obtain

$$\underline{\hat{X}}_{\text{MV}}(N) = \begin{bmatrix} \hat{x}(1 \mid 1) \\ \hat{x}(2 \mid 2) \\ \vdots \\ \hat{x}(N \mid N) \end{bmatrix}$$

$$= \begin{bmatrix} R_{xe}(1,1) & & & 0 \\ R_{xe}(2,1) & R_{xe}(2,2) & & \\ \vdots & & \ddots & \\ R_{xe}(N,1) & \cdots & & R_{xe}(N,N) \end{bmatrix} \begin{bmatrix} R_{ee}^{-1}(1) & & & 0 \\ & R_{ee}^{-1}(2) & & \\ & & \ddots & \\ 0 & & & R_{ee}^{-1}(N) \end{bmatrix} \begin{bmatrix} e(1) \\ e(2) \\ \vdots \\ e(N) \end{bmatrix}$$

$$(9.3\text{--}9)$$

where $e(t) \in R^{N_y}$, $R_{xe}(t,t) \in R^{N_x \times N_y}$, and $R_{ee}(t) \in R^{N_y \times N_y}$. The recursive filtered solution follows easily, if we realize that we want the best estimate of $x(t)$, given $Y(t)$; therefore, any block row of Eq. (9.3–9) can be written (for $N = t$) as

$$\hat{x}(t \mid t) = \sum_{i=1}^{t} R_{xe}(t,i) R_{ee}^{-1}(i) e(i)$$

If we extract the last (tth) term out of the sum (recall Example 9.2–2), we get

$$\hat{X}_{\text{new}} = \hat{x}(t \mid t) = \sum_{i=1}^{t-1} R_{xe}(t,i) R_{ee}^{-1}(i) e(i) + R_{xe}(t,t) R_{ee}^{-1}(t) e(t) \qquad (9.3\text{--}10)$$

$$\overset{\hat{X}_{\text{old}}}{} \qquad\qquad \overset{K}{}$$

or $\quad \hat{X}_{\text{new}} = \hat{x}(t \mid t) = \hat{x}(t \mid t-1) + K(t) e(t) \qquad (9.3\text{--}11)$

where $K(t) = R_{xe}(t) R_{ee}^{-1}(t)$ and $R_{xe}(t,t) = R_{xe}(t)$. So we see that the recursive solution using the innovations sequence, instead of the measurement sequence, has reduced the computations to inverting a $N_y \times N_y$ matrix $R_{ee}(t)$ instead of a $(N_y \times N_y)N$ matrix $R_{yy}(\underline{N})$. Before we develop the expression for the filtered estimate of Eq. (9.3–11), let us investigate the innovations sequence more closely. Recall that the minimum-variance estimate of $y(t)$ is just a linear transformation of the minimum-variance estimate of $x(t)$; that is,

$$\hat{y}(t \mid t-1) = C(t) \hat{x}(t \mid t-1) \qquad (9.3\text{--}12)$$

Thus, the innovation can be decomposed using Eqs. (9.3–2,12) as

$$e(t) = y(t) - C(t) \hat{x}(t \mid t-1) = C(t)[x(t) - \hat{x}(t \mid t-1)] + v(t)$$

or $\quad e(t) = C(t) \tilde{x}(t \mid t-1) + v(t) \qquad (9.3\text{--}13)$

Consider the innovation covariance $R_{ee}(t)$ using this equation:

$$R_{ee}(t) = C(t) E\{\tilde{x}(t \mid t-1) \tilde{x}'(t \mid t-1)\} C'(t) + E\{v(t) \tilde{x}'(t \mid t-1)\} C'(t)$$
$$+ C(t) E\{\tilde{x}(t \mid t-1) v'(t)\} + E\{v(t) v'(t)\}$$

This gives the following, since v and $\tilde{x}$ are uncorrelated:

$$R_{ee}(t) = C(t) \tilde{P}(t \mid t-1) C'(t) + R_{vv}(t) \qquad (9.3\text{--}14)$$

The cross-covariance R_{xe} is obtained, using Eq. (9.3–13), by

$$R_{xe}(t) = E\{x(t)e'(t)\} = E\{x(t)[C(t)\tilde{x}(t \mid t - 1) + v(t)]'\}$$

or $\qquad R_{xe}(t) = E\{x(t)\tilde{x}'(t \mid t - 1)\}C'(t)$

Using the definition of the estimation error $\tilde{x}$ and substituting for x, we obtain

$$R_{xe}(t) = E\{[\hat{x}(t \mid t - 1) + \tilde{x}(t \mid t - 1)]\tilde{x}'(t \mid t - 1)\}C'(t)$$

or

$$R_{xe}(t) = E\{\hat{x}(t \mid t - 1)\tilde{x}'(t \mid t - 1)\}C'(t) + E\{\tilde{x}(t \mid t - 1)\tilde{x}'(t \mid t - 1)\}C'(t) \tag{9.3-15}$$

From the orthogonality property of the estimation error for dynamic variables [9], that is,

$$E\{f(Y(T))\tilde{x}'(t \mid t - 1)\} = 0 \qquad \text{for } T \le t - 1 \tag{9.3-16}$$

the first term of Eq. (9.3–15) is zero, because $\tilde{x}(t \mid t - 1) = f(Y(t - 1))$, giving

$$R_{xe}(t) = \tilde{P}(t \mid t - 1)C'(t) \tag{9.3-17}$$

Thus, we see that the weight, or Kalman gain matrix is given by

$$K(t) = R_{xe}(t)R_{ee}^{-1}(t) = \tilde{P}(t \mid t - 1)C'(t)R_{ee}^{-1}(t) \tag{9.3-18}$$

Before we can calculate the corrected state estimate, we require the predicted, or old, estimate; that is,

$$X_{\text{old}} = \hat{x}(t \mid t - 1) = E\{x(t) \mid Y(t - 1)\}$$

If we employ the state-space model of Eq. (9.3–1), then from the linearity properties of the conditional expectation, we have

$$\hat{x}(t \mid t - 1) = E\{A(t - 1)x(t - 1) + B(t - 1)u(t - 1) + W(t - 1)w(t - 1) \mid Y(t - 1)\}$$

or

$$\hat{x}(t \mid t - 1) = A(t - 1)\hat{x}(t - 1 \mid t - 1) + B(t - 1)u(t - 1) + W(t - 1)\hat{w}(t - 1 \mid t - 1)$$

However, we have

$$\hat{w}(t \mid T) = E\{w(t) \mid Y(T)\} = 0 \qquad \text{for } t \ge T$$

which is not surprising, since the best estimate of zero-mean, white noise is zero (unpredictable). Thus, the prediction is given by

$$\hat{X}_{\text{old}} = \hat{x}(t \mid t - 1) = A(t - 1)\hat{x}(t - 1 \mid t - 1) + B(t - 1)u(t - 1) \tag{9.3-19}$$

To complete the algorithm, the expressions for the predicted and corrected error covariances must be determined; we refer the interested reader to Candy [2] for details.

We summarize the *Kalman filter estimator* as follows:

Criterion: $J(t \mid t) = \text{trace } \tilde{P}(t \mid t)$
Models:
  Signal: $x(t) = A(t-1)x(t-1) + B(t-1)u(t-1)$
                            $+ W(t-1)w(t-1)$
  Measurement: $y(t) = C(t)x(t) + v(t)$
  Noise: w and v are zero-mean and white with covariances
                    R_{ww} and R_{vv}
  Initial state: $x(0)$ has mean $\hat{x}(0 \mid 0)$ and covariance $\tilde{P}(0 \mid 0)$
Algorithm (Kalman filter): $\hat{x}(t \mid t) = \hat{x}(t \mid t-1) + K(t)e(t)$
Quality: $\tilde{P}(t \mid t) = [I - K(t)C(t)]\tilde{P}(t \mid t-1)$

Kalman filters can be designed using the *tuning analysis procedure*, which is developed on sound theoretical grounds:

1. Innovations sequence is *zero-mean*.
2. Innovations sequence is *white*.
3. Innovations sequence is *uncorrelated* in time and with input u.
4. Innovations sequence *lies within* the confidence limits constructed from R_{ee} of the Kalman filter algorithm.
5. Innovations variance is *reasonably close* to the sample variance estimate $\hat{R}_{ee}$.
6. Tracking or estimation error *lies within* the confidence limits constructed from $\tilde{P}$ of the Kalman filter algorithm.
7. Error variance is *reasonably close* to the sample variance estimate $\hat{P}$.

Consider an application of model-based processing to an RLC circuit design problem.

Example 9.3–1. Consider the design of an estimator for a series RLC circuit (second-order system) excited by a pulse train [2]. The circuit diagram is shown in Fig. 9.3–3. Using Kirchhoff's voltage law, we can obtain the circuit equations with $i = C(de/dt)$:

$$\frac{d^2e}{dt^2} + \frac{R}{L}\frac{de}{dt} + \frac{1}{LC}e = \frac{1}{LC}e_{\text{in}}$$

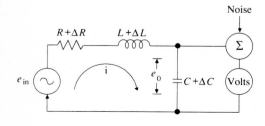

FIGURE 9.3–3
RLC circuit problem

where e_{in} is a unit pulse train. This equation is that of a second-order system which characterizes an electrical RLC circuit, or a mechanical vibration system, or a hydraulic flow system, etc. The dynamic equations can be placed in state-space form by choosing $x := [e \mid de/dt]'$ and $u = e_{in}$:

$$\frac{dx}{dt} = \begin{bmatrix} 0 & 1 \\ -\dfrac{1}{LC} & -\dfrac{R}{L} \end{bmatrix} x + \begin{bmatrix} 0 \\ -\dfrac{1}{LC} \end{bmatrix} u + \begin{bmatrix} 0 \\ -\dfrac{1}{LC} \end{bmatrix} w$$

where $w \sim N(0, R_{ww})$ is used to model component inaccuracies.

A high-impedance voltmeter is placed in the circuit to measure the capacitor voltage e. We assume that it is a digital (sampled) device contaminated with noise of variance R_{vv}; that is,

$$y(t) = e(t) + v(t)$$

where $v \sim N(0, R_{vv})$. For our problem we have the following parameters: $R = 5$ kΩ, $L = 2.5$ H, $C = 0.1$ μF, and $T = 0.1$ ms (the problem will be scaled in milliseconds). We assume that the component inaccuracies can be modeled using $R_{ww} = 0.01$, characterizing a deviation of ± 0.1 V uncertainty in the circuit representation. Finally, we assume that the precision of the voltmeter measurements are $(e \pm 0.2$ V), the two standard deviation value, so that $R_{vv} = 0.01$ (V)^2. Summarizing the circuit model, we have the continuous-time representation

$$\frac{dx}{dt} = \begin{bmatrix} 0 & 1 \\ -4 & -2 \end{bmatrix} x + \begin{bmatrix} 0 \\ -4 \end{bmatrix} u + \begin{bmatrix} 0 \\ -4 \end{bmatrix} w$$

and the discrete-time measurements

$$y(t) = [1 \quad 0]x(t) + v(t)$$

where $$R_{ww} = \frac{(0.1)^2}{T} = 0.1(\text{V})^2 \quad \text{and} \quad R_{vv} = 0.01(\text{V})^2$$

Before we design the discrete Kalman estimator, we must convert the system or process model to a sampled-data (discrete) representation. Using SSPACK (see Appendix E, [10]), this is accomplished automatically with the Taylor-series approach to approximating the matrix exponential. For an error tolerance of 1×10^{-12}, a 15-term series expansion yields the following discrete-time Gauss-Markov model:

$$x(t) = \begin{bmatrix} 0.98 & 0.09 \\ -0.36 & 0.801 \end{bmatrix} x(t-1) + \begin{bmatrix} -0.019 \\ -0.36 \end{bmatrix} u(t-1) + \begin{bmatrix} -0.019 \\ -0.36 \end{bmatrix} w(t-1)$$

$$y(t) = [1 \quad 0]x(t) + v(t)$$

where $$R_{ww} = 0.1(\text{V})^2 \quad \text{and} \quad R_{vv} = 0.01(\text{V})^2$$

Using SSPACK with initial conditions $x(0) = 0$ and $P = \text{diag}(0.01, 0.04)$, the simulated system is depicted in Fig. 9.3–4. In Fig. 9.3–4(a) through (c) we see the simulated states and measurements with corresponding confidence limits about the mean (true) values. In each case, the simulation satisfies the statistical properties of the model. The corresponding true (mean) trajectories are also shown, along with the pulse-train excitation. Note that the measurements are merely a noisier version (process and measurement noise) of the voltage x_1.

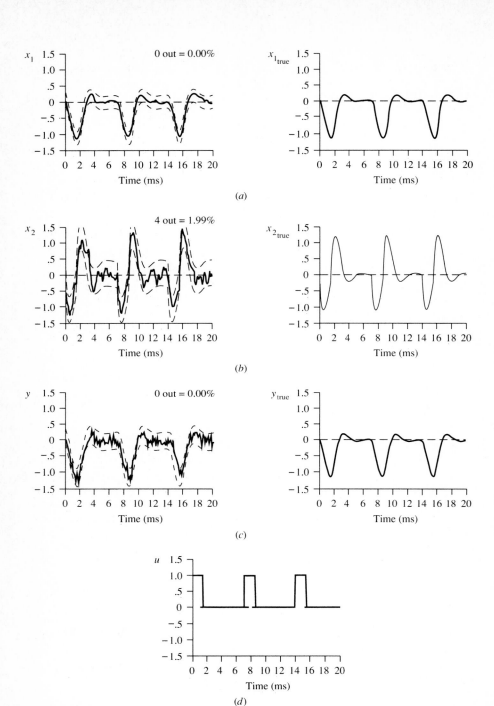

FIGURE 9.3–4
Gauss-Markov simulation of RLC circuit problem (*a*) Simulated and true state (voltage) (*b*) Simulated and true state (current) (*c*) Simulated and true measurement (*d*) Pulse-train excitation

A discrete Kalman estimator was designed using SSPACK (Appendix E) to improve the estimated voltage $\hat{x}_1$. The results are shown in Fig. 9.3–5. In Fig. 9.3–5(a) through (c), we see the filtered states and measurements, as well as the corresponding estimation errors. The true (mean) states are superimposed as well, to indicate the tracking capability of the estimator. The estimation errors lie within the bounds (3 percent out), for the second state, but the error covariance is slightly underestimated for the first state (14 percent out). The predicted and sample variances are close ($0.002 \approx 0.004$ and $0.028 \approx 0.015$) in both cases. The innovations lie within the bounds (3 percent out), with the predicted sample variances close ($0.011 \approx 0.013$). The innovations are statistically zero-mean ($0.0046 \ll 0.014$) and white (5 percent out, WSSR below threshold),[†] indicating a well-tuned estimator. We summarize the results of this problem as follows:

Criterion: $J(t \mid t) = \text{trace } \tilde{P}(t \mid t)$
Models:
   Signal:

$$x(t) = \begin{bmatrix} 0.98 & 0.09 \\ -0.36 & 0.801 \end{bmatrix} x(t-1) + \begin{bmatrix} -0.019 \\ -0.36 \end{bmatrix} u(t-1) + \begin{bmatrix} -0.019 \\ -0.36 \end{bmatrix} w(t-1)$$

   Measurement: $y(t) = [1 \quad 0]x(t) + v(t)$
   Noise: w and v are zero-mean and white with covariances
                 R_{ww} and R_{vv}
Algorithm: $\hat{x}(t \mid t) = \hat{x}(t \mid t-1) + K(t)e(t)$
Quality: $\tilde{P}(t \mid t) = [I - K(t)c(t)]\tilde{P}(t \mid t-1)$

9.4 KALMAN FILTER IDENTIFIER

The Kalman filter can easily be formulated to solve the well-known *system identification problem* [11]:

Given a set of noisy measurements $\{y(t)\}$ and inputs $\{u(t)\}$, find the minimum (error) variance estimate $\hat{\Theta}(t)$ of $\Theta(t)$, a vector of unknown parameters of a linear system.

The identification problem is depicted in Fig. 9.4–1. The scalar linear system is given by the autoregressive model with exogenous inputs (ARX) of Sec. 4.5 as

$$A(q^{-1})y(t) = B(q^{-1})u(t) + e(t) \qquad (9.4\text{--}1)$$

where A and B are polynomials in the backward-shift operator q^{-i} and $\{e(t)\}$ is white.

We must convert this model to the state-space/measurement-system framework required by the Kalman filter. The measurement model can be expressed as

---

[†]WSSR is the weighted-sum squared residual statistic which aggregates the innovation vector information over a window to perform a vector-type whiteness test (see [2] for details).

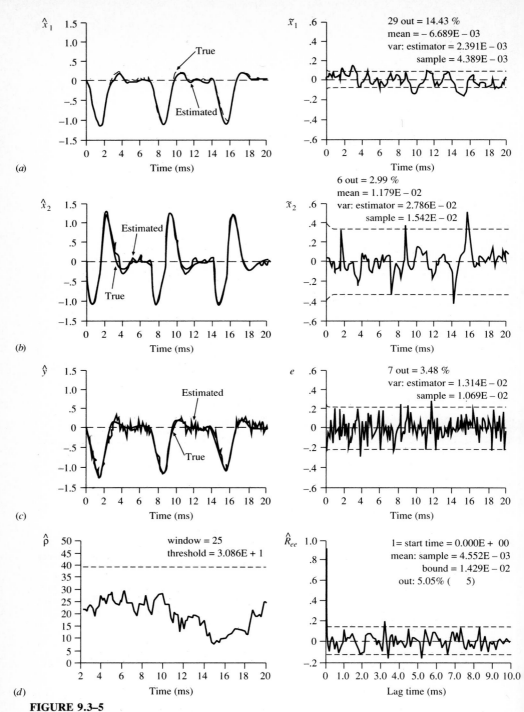

FIGURE 9.3–5
Kalman filter design for RLC circuit problem (*a*) Estimated state (voltage) and error (*b*) Estimated state (current) and error (*c*) Filtered and true measurement (voltage) and error (innovation) (*d*) WSSR and whiteness test

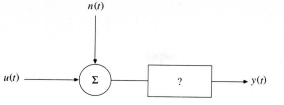

FIGURE 9.4–1
The identification problem

$$y(t) = c'(t)\Theta(t) + n(t) \tag{9.4–2}$$

where $c'(t) = [-y(t-1) \cdots -y(t-N_a) \mid u(t) \cdots u(t-N_b)]$

for $n \sim N(0, R_{nn})$, and

$$\Theta(t) = [a_1 \cdots a_{N_a} \mid b_0 \cdots b_{N_b}]'$$

which represents the measurement equation in the Kalman filter formulation. The parameters can be modeled as constants with uncertainty $w(t)$; i.e., the state-space model used in this formulation takes the form

$$\Theta(t) = \Theta(t-1) + w(t) \tag{9.4–3}$$

where $w \sim N(0, R_{ww})$. We summarize the Kalman filter used as an identifier in Table 9.4–1. Thus, the Kalman filter provides the minimum-variance estimates of the parameters of the ARX model.

TABLE 9.4–1
Kalman filter identifier

<table>
<tr><td colspan="2" align="center">**Prediction**</td></tr>
<tr><td align="center">$\hat{\Theta}(t \mid t-1) = \hat{\Theta}(t-1 \mid t-1)$</td><td align="right">(Parameter prediction)</td></tr>
<tr><td align="center">$\tilde{P}(t \mid t-1) = \tilde{P}(t-1 \mid t-1) + R_{ww}(t-1)$</td><td align="right">(Covariance prediction)</td></tr>
<tr><td colspan="2" align="center">**Innovation**</td></tr>
<tr><td align="center">$e(t) = y(t) - \hat{y}(t \mid t-1) = y(t) - c'(t)\hat{\Theta}(t \mid t-1)$</td><td align="right">(Innovation)</td></tr>
<tr><td align="center">$R_{ee}(t) = c'(t)\tilde{P}(t \mid t-1)c(t) + R_{vv}(t)$</td><td align="right">(Innovation covariance)</td></tr>
<tr><td colspan="2" align="center">**Gain**</td></tr>
<tr><td align="center">$K(t) = \tilde{P}(t \mid t-1)c(t)R_{ee}^{-1}(t)$</td><td align="right">(Kalman gain or weight)</td></tr>
<tr><td colspan="2" align="center">**Correction**</td></tr>
<tr><td align="center">$\hat{\Theta}(t \mid t) = \hat{\Theta}(t \mid t-1) + K(t)e(t)$</td><td align="right">(Parameter correction)</td></tr>
<tr><td align="center">$\tilde{P}(t \mid t) = [I - K(t)c'(t)]\tilde{P}(t \mid t-1)$</td><td align="right">(Covariance correction)</td></tr>
</table>

Initial conditions: $\hat{\Theta}(0 \mid 0)$ and $\tilde{P}(0 \mid 0)$

where $c'(t) = [-y(t-1) \cdots -y(t-N_a) \mid u(t) \cdots u(t-N_b)]$

and $\Theta(t) = [a_1 \cdots a_{N_a} \mid b_0 \cdots b_{N_b}]'$

Example 9.4–1. Suppose we have the RC circuit model of Chapter 1, that is,

$$x(t) = 0.97x(t-1) + 100u(t-1) + w(t-1)$$

and

$$y(t) = 2.0x(t) + v(t)$$

where $x(0) = 2.5$, $R_{ww} = 10^{-6}$, and $R_{vv} = 10^{-12}$. The "identification" model is given by

$$\Theta(t) = \Theta(t-1) + w^*(t)$$

$$y(t) = c'(t)\Theta(t) + v(t)$$

where $c'(t) = [-y(t-1) \mid u(t)u(t-1)]$, $\Theta'(t) = [a_1 \mid b_0 \quad b_1]$, $v \sim N(0, 10^{-12})$, and $w^* \sim N(0, 10^{-6})$.

Using the Gauss-Markov model and SSPACK (see Appendix E, [10]), we simulated the data with the specified variances. The performance of the Kalman identifier is shown in Fig. 9.4–2. The parameter estimates in Fig. 9.4–2(a) and (c) show that the estimator identifies the a parameter in about 10 samples, but b_0 does not converge at all in this time interval. The innovations are clearly zero-mean ($2.3 \times 10^{-4} < 2.8 \times 10^{-4}$) and white (0 percent lie outside), as depicted in Fig. 9.4–2(c). The measurement filtering property of the filter is shown in Fig. 9.4–2(b). We summarize the results as follows:

Criterion:	$J(t \mid t) = \text{trace } \tilde{P}(t \mid t)$
Models:	
Signal:	$\Theta(t) = \Theta(t-1) + w^*(t-1)$
Measurement:	$y(t) = c'(t)\Theta(t) + v(t)$
Noise:	$w^* \sim N(0, 10^{-6})$, $v \sim N(0, 10^{-12})$
Initial state:	$\hat{\Theta}(0 \mid 0) = [0.0 \quad 0.0 \quad 0.0]$, $\tilde{P}(0 \mid 0)$
Algorithm:	$\hat{\Theta}(t \mid t) = \hat{\Theta}(t \mid t-1) + K(t)e(t)$
Quality:	$\tilde{P}(t \mid t) = [I - K(t)c'(t)]\tilde{P}(t \mid t-1)$

Note that the Kalman identifier is identical to the recursive least-squares algorithm of Chapter 7, with the exception that the parameters can include process noise $w^*(t)$. Although this difference appears minimal, avoiding estimator divergence actually becomes very significant and necessary (see [11] or [5] for details).

9.5 KALMAN FILTER DECONVOLVER

In this section, we consider extending the Kalman filter algorithm to solve the problem of estimating an unknown input from data that have been "filtered." This problem is called *deconvolution* in signal-processing literature and occurs commonly in seismic and speech processing [12,13] as well as transient problems [14].

In many measurement systems, it is necessary to deconvolve or estimate the input to an instrument, given that the data are noisy. The basic deconvolution problem is depicted in Fig. 9.5–1(a) for deterministic inputs $\{u(t)\}$ and outputs $\{y(t)\}$. The problem can be simply stated as follows:

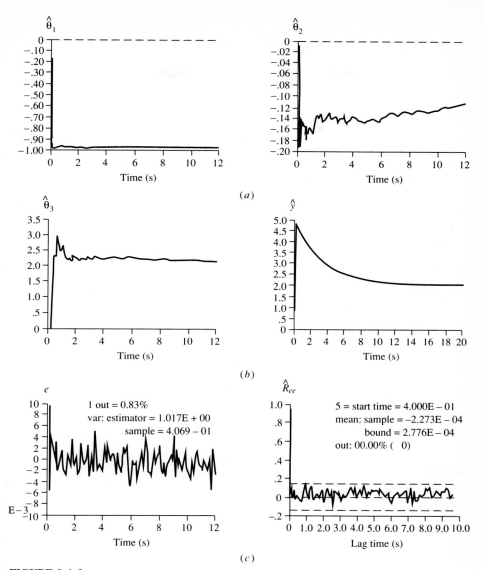

FIGURE 9.4–2
Kalman filter identifier output for tuning example (*a*) Parameter estimates for a_1 and b_0 (*b*) Parameter b_1 and filtered measurement (*c*) Prediction error (innovation) and whiteness test

Given the impulse response $H(t)$ of a linear system and outputs $\{y(t)\}$, find the unknown input $\{u(t)\}$ over some time interval.

In practice, this problem is complicated by the fact that the data are noisy, and impulse-response models are uncertain. Therefore, a more pragmatic view of the problem would account for these uncertainties. The uncertainties lead us

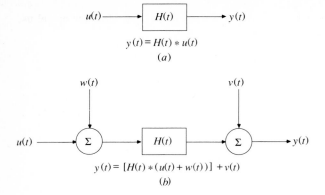

$$y(t) = H(t) * u(t)$$

(a)

$$y(t) = [H(t) * (u(t) + w(t))] + v(t)$$

(b)

FIGURE 9.5–1
The deconvolution problem
(a) Deterministic case
(b) Stochastic case

to define the *stochastic deconvolution problem*, shown in Fig. 9.5–1(b). This problem can be stated as follows:

> Given a model of the linear system $H(t)$ and discrete noisy measurements $\{y(t)\}$, find the minimum (error) variance estimate of the input sequence $\{u(t)\}$ over some time interval.

The solution to this problem, using the Kalman filter algorithm, again involves developing a model for the input and augmenting the state vector.

Suppose we use the standard discrete Gauss-Markov model of Sec. 4.7 and define the following Gauss-Markov model for the input signal:

$$u(t) = F(t-1)u(t-1) + n(t-1) \tag{9.5-1}$$

where $n \sim N(0, R_{nn}(t))$. The augmented Gauss-Markov model is given by $X_u :=$ $[x' \mid u']'$:

$$X_u(t) = A_u(t-1)X_u(t-1) + W_u(t-1)w_u(t-1)$$

and $$y(t) = C_u(t)X_u(t) + v(t)$$

The matrices in the augmented model are given by

$$A_u(t-1) = \begin{bmatrix} A(t-1) & B(t-1) \\ 0 & F(t-1) \end{bmatrix} \qquad R_{w_u} = \begin{bmatrix} R_{ww}(t-1) & R_{wn}(t-1) \\ R_{nw}(t-1) & R_{nn}(t-1) \end{bmatrix}$$

and $$W_u(t-1) := [W'(t-1) \mid I]' \qquad C_u(t) = [C(t) \mid 0]$$

This model can be simplified by choosing $F = I$; that is, u is a piecewise constant. This model becomes valid if the system is oversampled (see [14] for details). The Kalman filter for this problem is the standard algorithm of Sec. 9.2, with the augmented matrices. The augmented Kalman filter is sometimes called the *Schmidt-Kalman filter*. Schmidt (see [4]) developed a suboptimal method to estimate parameters and decrease the computational burden of the augmented system.

Example 9.5–1. For the simulated measurements we again use the tuning example with a higher signal-to-noise ratio, as shown in Fig. 9.5–2(c). Here the model uses $Bu \to u$, since B is assumed unknown:

$$x(t) = 0.97x(t-1) + 100u(t-1) + w(t-1)$$

$$y(t) = 2.0x(t) + v(t)$$

where $x(0) = 2.5$, $R_{ww} = 10^{-4}$, and $R_{vv} = 4 \times 10^{-3}$. The Schmidt-Kalman filter for this problem, using the "augmented" model and a piecewise constant approximation for the unknown input, is given by

$$X_u(t) = \begin{bmatrix} 0.97 & 1.0 \\ 0 & 1.0 \end{bmatrix} X_u(t-1) + \begin{bmatrix} w(t-1) \\ n(t-1) \end{bmatrix}$$

and

$$y(t) = [2 \quad 0]X_u(t) + v(t)$$

The results of the estimator simulation using SSPACK (see Appendix E, [10]), are shown in Fig. 9.5–2. Here the state estimates are shown in Fig. 9.5–2(a), along with the filtered measurement and innovations in Fig. 9.5–2(b). The filter tracks the input quite reasonably after an initial transient, as indicated. The optimal results of the estimator are confirmed by the corresponding zero-mean ($0.0045 < 0.01$) and white (5 percent lie outside) innovations sequence. We summarize the results as follows:

Criterion:	$J(t \mid t) = \text{trace } \tilde{P}(t \mid t)$
Models:	
Signal:	$x(t) = 0.97x(t-1) + 100u(t-1) + w(t-1)$
Measurement:	$y(t) = 2.0x(t) + v(t)$
Noise:	$w^* \sim N(0, 10^{-4})$, $n \sim N(0, 5 \times 10^{-7})$, $v \sim N(0, 4 \times 10^{-3})$
Initial state:	$\hat{x}(0 \mid 0) = [2.55 \quad 0.0]'$, $\tilde{P}(0 \mid 0) = \text{diag}[5 \times 10^{-2}, 5 \times 10^{-2}]$
Algorithm:	$\hat{X}_u(t \mid t) = \hat{X}_u(t \mid t-1) + K(t)e(t)$
Quality:	$\tilde{P}(t \mid t) = [I - K(t)C(t)]\tilde{P}(t \mid t-1)$

9.6 THE KALMAN/WIENER FILTER EQUIVALENCE

In this section, we show the relationship between the Wiener filter and its state-space counterpart, the Kalman filter. Detailed proofs of these relations are available for both the continuous and discrete cases [15]. Our approach is to state the Wiener solution and then show that the steady-state Kalman filter provides a solution with all the necessary properties. We use frequency-domain techniques to show this equivalence. The time-domain approach would be to use the batch-innovations solution of Sec. 9.3.

Recall that the Wiener filter solution in the frequency domain can be solved by spectral factorization, since

$$K(z) = [S_{sy}(z)S_{yy}^{-1}(z^-)]_{cp} S_{yy}^{-1}(z^+) \tag{9.6-1}$$

where $K(z)$ has all its poles and zeros within the unit circle. The classical approach to Wiener filtering can be accomplished in the frequency domain by factoring the

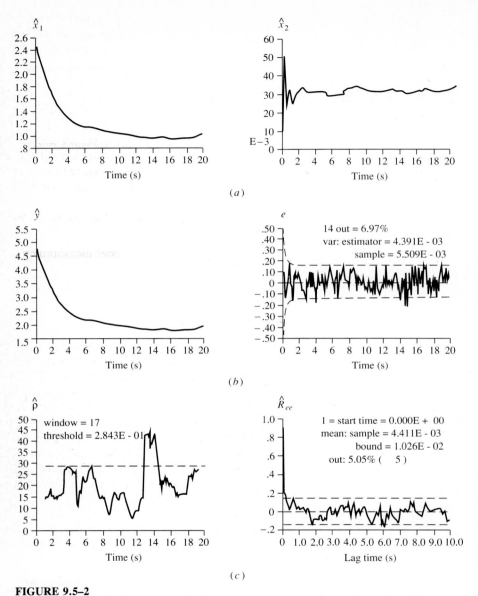

FIGURE 9.5–2
The Schmidt-Kalman filter deconvolver (*a*) Estimated state and input (*b*) Filtered measurement and innovation (*c*) WSSR and whiteness test

power spectral density (PSD) of the measurement sequence; that is,

$$S_{yy}(z) = K(z)K'(z^{-1}) \qquad (9.6-2)$$

The factorization is unique, stable, and minimum-phase (see [1] for proof).

Next, we must show that the *steady-state Kalman filter* or the *innovations model*, given by the following equation, is stable and minimum-phase and therefore, in fact, the Wiener solution:

$$\hat{x}(t) = A\hat{x}(t-1) + Ke(t-1)$$

$$y(t) = C\hat{x}(t) + e(t) \qquad (9.6-3)$$

where e is the zero-mean, white innovation with covariance R_{ee}. The "transfer function" of the innovations model is defined as

$$T(z) = C(zI - A)^{-1}K \qquad (9.6-4)$$

Let us calculate the measurement covariance of Eq. (9.6–3):

$$R_{yy}(k) = \text{cov}\,[y(t+k)y(t)] = R_{\hat{y}\hat{y}}(k) + R_{\hat{y}e}(k) + R_{e\hat{y}}(k) + R_{ee}(k) \quad (9.6-5)$$

where $\hat{y}(t) := C\hat{x}(t)$. Taking Z-transforms, we obtain the measurement PSD as

$$S_{yy}(z) = S_{\hat{y}\hat{y}}(z) + S_{\hat{y}e}(z) + S_{e\hat{y}}(z) + S_{ee}(z) \qquad (9.6-6)$$

Using the linear system relations of Chapter 4, we see that

$$S_{\hat{y}\hat{y}}(z) = CS_{\hat{x}\hat{x}}(z)C' = T(z)S_{ee}(z)T'(z^{-1})$$

$$S_{ee}(z) = R_{ee}$$

$$S_{\hat{y}e}(z) = CS_{\hat{x}e}(z) = T(z)S_{ee}(z)$$

and $\qquad S_{e\hat{y}}(z) = S_{ee}(z)T'(z^{-1}) \qquad (9.6-7)$

Thus, the measurement PSD is given by

$$S_{yy}(z) = T(z)S_{ee}(z)T'(z^{-1}) + T(z)S_{ee}(z) + S_{ee}(z)T'(z^{-1}) + S_{ee}(z) \quad (9.6-8)$$

Since $S_{ee}(z) = R_{ee}$ and $R_{ee} \geq 0$, the following factorization always exists as

$$R_{ee} = R_{ee}^{1/2}(R'_{ee})^{1/2} \qquad (9.6-9)$$

Thus, using Eq. (9.6–9), $S_{yy}(z)$ of Eq. (9.6–8) can be written as

$$S_{yy}(z) = [T(z)R_{ee}^{1/2} \quad R_{ee}^{1/2}]\begin{bmatrix} (R'_{ee})^{1/2}T'(z^{-1}) \\ (R'_{ee})^{1/2} \end{bmatrix} := T_e(z)T'_e(z^{-1}) \qquad (9.6-10)$$

which shows that the innovations model indeed admits a spectral factorization of the type desired. To show that $T_e(z)$ is the unique, stable, minimum-phase spectral factor, it is necessary to show that $|T_e(z)|$ has all its poles within the unit circle (stable). It has been shown that $T_e(z)$ does satisfy these constraints [1,7,16] and therefore

$$T_e(z) = K(z)$$

is the Wiener solution.

It is interesting to note that there is an entire class of solutions given by

$$S_{yy}(z) = [C(zI - A)^{-1}LN \quad N]\begin{bmatrix} N'L'(z^{-1}I - A')^{-1}C' \\ N' \end{bmatrix}$$

The unique Wiener solution is given by the relations

$$L = PC'(NN')^{-1} \quad \text{and} \quad (NN') = CPC' + R_{vv} \qquad (9.6\text{--}11)$$

only when P is either the minimum P_* or maximum P^* solution of the Riccati equation (see [17] for details); that is, $P_* < P < P^*$, $L = K$ (Kalman gain), and $NN' = R_{ee}$ (innovations covariance) for $P = P^*$.

9.7 SUMMARY

We have developed the concept of a model-based processor, both heuristically and theoretically, using the popular state-space or Kalman filter. We showed how the estimator can be tuned according to the procedure developed in Candy [2], and we applied it to an RLC circuit problem. We then showed how the processor can be used to solve the identification and deconvolution problems. Finally, we showed the equivalence (in steady-state) of the Kalman filter to the Wiener filter.

REFERENCES

1. B. D. Anderson and J. B. Moore, *Optimal Filtering* (Englewood Cliffs, N.J.: Prentice-Hall, 1979).
2. J. V. Candy, *Signal Processing: The Model-Based Approach* (New York: McGraw-Hill, 1986).
3. G. E. Box and G. M. Jenkins, *Time Series Analysis: Forecasting and Control* (San Francisco: Holden-Day, 1976).
4. A. Jazwinski, *Stochastic Processes and Filtering Theory* (New York: Academic Press, 1970).
5. G. C. Goodwin and R. L. Payne, *Dynamic System Identification* (New York: Academic Press, 1976).
6. A. P. Sage and J. L. Melsa, *System Identification* (New York: Academic Press, 1971).
7. R. E. Kalman, "A New Approach to Linear Filtering and Prediction Problems," *J. Basic Eng.*, Vol. 82, 1960.
8. T. Kailath, "The Innovations Approach to Detection and Estimation Theory," *Proc. IEEE*, Vol. 58, 1970.
9. T. Kailath, *Lectures on Kalman and Wiener Filtering Theory* (New York: Springer-Verlag, 1981).
10. S. Azevedo, J. Candy, and D. Lager, "SSPACK: An Interactive Multi-channel, Model-Based Signal Processing Package," *Proc. IEEE Confr. Circuits Systems*, 1986.
11. L. Ljung and T. Soderstrom, *Theory and Practice of Recursive Identification* (Boston: MIT Press, 1983).
12. M. T. Silvia and E. A. Robinson, *Deconvolution of Geophysical Time Series in the Exploration of Oil and Natural Gas* (New York: Elsevier, 1979).
13. J. Mendel, J. Kormylo, J. Lee, and F. Ashirafi, "A Novel Approach to Seismic Signal Processing and Modeling," *Geophysics*, Vol. 46, 1981.

14. J. V. Candy and J. E. Zicker, "Deconvolution of Noisy Transient Signals: A Kalman Filtering Application," *LLNL Rep.*, UCID-87432, and *Proc. of CDC Confr.*, Orlando, 1982.
15. P. Maybeck, *Stochastic Models, Estimation, and Control* (New York: Academic Press, 1979).
16. A. P. Sage and J. L. Melsa, *Estimation Theory with Applications to Communications and Control* (New York: McGraw-Hill, 1971).
17. P. Faurre, "Stochastic Realization Algorithms," *System Identification: Advances and Case Studies*, edited by R. Mehra and C. Laineotis (New York: Academic Press, 1976).
18. G. J. Bierman, *Factorization Methods of Discrete Sequential Estimation* (New York: Academic Press, 1977).
19. I. Rhodes, "A Tutorial Introduction to Estimation and Filtering," *IEEE Trans. Autom. Contr.*, Vol. AC-16, 1971.

SSPACK NOTES

SSPACK can be used to design model-based signal processors. The development environment (see Appendix E for details) consists of a Supervisor, which enables the choice of model set (linear or nonlinear) as well as techniques (Kalman filter, extended Kalman [nonlinear] filter, Kalman identifier, etc.) and of processors to implement the various models and perform statistical analysis of the resulting designs. The package enables the signal processor to simulate data using the Gauss-Markov model (see Sec. 4.7), and to approximate Gauss-Markov (nonlinear) models while designing signal (state-space) and/or parameter estimators. Steady-state (Wiener filter) designs can also be easily achieved from the package. The versatility of the state-space, model-based signal processing approach allows for many equivalent applications, after performing the required transformation to state-space form (see Sec. 4.8).

EXERCISES

9.1. We are asked to estimate the displacement of large vehicles (semi-trailers) when they are parked on the shoulder of a freeway and subjected to wind gusts created by passing vehicles. We measure the displacement of the vehicle by placing an accelerometer on the trailer. The accelerometer has inherent inaccuracies, which are modeled as

$$y = K_a x + n$$

with y and x the measured and actual displacement, n the white measurement noise of variance R_{nn}, and K_a the instrument gain. The dynamics of the vehicle can be modeled by a simple mass-spring-damper.
(*a*) Construct and identify the measurement model of Fig. 9.2–3.
(*b*) Construct and identify the process model and model-based estimator of Fig. 9.2–4.

9.2. Think of measuring the temperature of a liquid in a beaker heated by a burner. Suppose we use a thermometer immersed in the liquid and periodically observe and record the temperature.
(*a*) Construct a measurement model, as in Fig. 9.2–3, assuming that the thermometer is linearly related to the temperature; that is, $y(t) = k\Delta T(t)$. Also, model the uncertainty of the visual measurement as a random sequence $v(t)$ with variance R_{vv}.
(*b*) Suppose we model the heat transferred to the liquid from the burner as

$$Q(t) = CA\Delta T(t)$$

where C is the coefficient of thermal conductivity, A is the cross-sectional area, and $\Delta T(t)$ is the temperature gradient, with assumed random uncertainty $w(t)$ and variance R_{ww}. Using this process model and the models developed above, identify the representation of Fig. 9.2–4.

9.3. The variance of a zero-mean sequence of N data values $\{y(t)\}$ can be estimated using

$$R_{yy}(N) = \frac{1}{N} \sum_{i=1}^{N} y^2(i)$$

(a) Develop a recursive form for this estimator.
(b) Repeat part (a) if $\{y(t)\}$ has mean m_y.

9.4. The data set $\{y(t)\} = \{2.5, -0.5, 2.2, 1.7\}$ was generated by the following Gauss-Markov model:

$$x(t) = -0.5x(t - 1) + w(t - 1)$$

$$y(t) = x(t) + v(t)$$

where $R_{ww} = 1$, $R_{vv} = 4$, and $x(0) \sim N(0, 1.33)$.
(a) Using the Kalman algorithm of Table 9.3–1, calculate $\hat{x}(2 \mid 2)$.
(b) Assume that $\tilde{P}(t) = \cdots = \tilde{P}$ is a constant (steady-state value). Calculate $\tilde{P}$ using the prediction and correction covariance equations.
(c) Calculate the corresponding (steady-state) gain $\overline{K}$.
(d) Under the assumptions of steady-state operation of the algorithm ($P = \tilde{P}$, $K = \overline{K}$), recalculate $\hat{x}(2 \mid 2)$.
(e) What calculations have been eliminated by using $\overline{K}(\tilde{P})$ instead of $K(t)(\tilde{P}(t \mid t))$?
(f) What are the tradeoffs between the steady-state filter and a transient or dynamic filter?

9.5. Show that for a batch of N measurements the $(N_x \times N_y)N$ matrix $R_{xe}(\underline{N})$ is block lower triangular and that $R_{ee}(\underline{N})$ is block diagonal, as in Eq. (4.2–11).

9.6. Using the predictor-corrector Kalman filter equations of Table 9.3–1, develop the *prediction form* of the filter by rewriting the filter equations solely in terms of predicted estimates and covariances.
(a) What is the equation for the *prediction gain* $K_p(t)$ in this algorithm?
(b) How is K_p related to K in the predictor-corrector form?
(c) Develop the predicted form of the error covariance equation (Riccati equation).

9.7. Suppose $\{y(t)\}$ is a zero-mean scalar process with covariance

$$R_{yy}(i,j) = E\{y(i)y(j)\} = \alpha^{|i-j|} \qquad 0 < \alpha < 1$$

Find the corresponding innovations sequence $\{e(t)\}$.

9.8. Suppose we have a *scalar* process given by

$$x(t) = \frac{1}{2}x(t - 1) + w(t - 1) \qquad w \sim N(0, R_{ww})$$

$$y(t) = cx(t) + v(t) \qquad v \sim N(0, R_{vv})$$

(a) What is the steady-state Kalman gain for this system?
(b) What is the effect of increasing R_{ww} on K_{ss} for fixed R_{vv}?
(c) What is the effect of decreasing R_{vv} on K_{ss} for fixed R_{ww}?

(d) Suppose $R_{vv} = 0$; what is the effect of increasing R_{ww} on K_{ss}?
(e) Suppose $R_{ww} = 0$; what is the effect of decreasing R_{vv} on K_{ss}?

9.9. In a simple radar system, a large-amplitude, narrow-width pulsed sinusoid is transmitted by an antenna. The pulse propagates through space at light speed ($c = 3 \times 10^8$ m/s) until it hits the object being tracked. The object absorbs and reflects the energy; the radar antenna receives a portion of the reflected energy. By locating the time difference between the leading edge of the transmitted and received pulse Δt, the distance or range from the radar r can be determined; that is,

$$r = \frac{c\Delta t}{2}$$

The pulse is transmitted periodically; that is,

$$r(t) = \frac{c\Delta(t)}{2}$$

Assume that the object is traveling at a constant velocity ρ with random disturbance $w \sim N(0, 100 \text{ m/s})$. Therefore, the range of the object at $(t + 1)$ is

$$r(t + 1) = r(t) + T\rho(t) \qquad T = 1 \text{ m/s}$$

where T is the sampling interval between received range values. The measured range is contaminated by noise

$$y(t) = r(t) + n(t) \qquad \text{for } n \sim N(0, 10)$$

(a) Develop a model to simulate this radar system.
(b) Design a range estimator, using the Kalman filter and assuming the initial target range is $r(0) \approx 20 \pm 2$ km.

9.10. Suppose we are given the ARMAX model

$$y(t) + a_1 y(t - 1) = b_1 u(t - 1) + c_1 e(t - 1)$$

(a) Assuming c_1 is known, develop the Kalman filter/identifier for this problem.
(b) Assuming c_1 is unknown, "extend" the Kalman filter/identifier to solve this problem.

9.11. Suppose we are given the ARMAX model

$$y(t) - 1.5y(t - 1) + 0.7y(t - 2) = u(t - 1) + 0.5u(t - 2) + e(t)$$

where u is a pseudo-random binary sequence and e is white gaussian noise with variance 0.01.
(a) Simulate this process for 256 sample points.
(b) Apply the Kalman filter/identifier algorithm to the data. What are the final parameter estimates? Show plots of results.

9.12. Suppose we have the following system Gauss-Markov model:

$$x(t) = -ax(t - 1) + bu(t - 1) + w(t - 1)$$
$$y(t) = cx(t) + v(t)$$

with $w \sim N(0, R_{ww})$. Develop the Kalman filter/deconvolver if
(a) u is assumed piecewise constant
(b) u is assumed piecewise linear

(c) *u* is random and exponentially correlated

(d) *u* is modeled by the Taylor series

$$u(t + T) \approx \sum_{i=0}^{N} \frac{\partial^i u(t)}{\partial t^i} \frac{T^i}{i!}$$

9.13. Suppose we are given the following innovations model (in steady state):

$$\hat{x}(t) = a\hat{x}(t - 1) + ke(t - 1)$$

$$y(t) = c\hat{x}(t) + e(t)$$

where *e* is the zero-mean, white innovations sequence with covariance R_{ee}. Derive the equivalent Wiener filter.

APPENDIX

A

In this appendix, we develop the fast Fourier Transform (FFT) algorithm, an efficient technique to calculate the discrete Fourier transform (DFT) [1]. The Fourier transform is an integral part of many different schemes for real-time signal processing. The frequency domain approach is popular primarily because of the efficient FFT algorithms to implement it.

The set of algorithms known as FFT consists of a variety of tricks for reducing the computation time of DFT by a factor of 100.

The DFT is given by

$$X(m) = \sum_{t=0}^{N-1} x(t)e^{-j(2\pi/N)mt} = \sum_{t=0}^{N-1} x(t)W_N^{mt} \qquad m = 0, 1, \ldots, N-1 \quad \text{(A–1)}$$

where $X(m) := X(\Omega_m)$, $W_N = e^{-j(2\pi/N)}$. The IDFT can be written

$$x(t) = \frac{1}{N} \sum_{m=0}^{N-1} X(m)W_N^{-mt} \qquad t = 0, 1, \ldots, N-1 \qquad \text{(A–2)}$$

Note that W_N^{mt} is periodic with period N, since

$$W^{(m+lN)t} = W^{mt}W^{lNt} = W^{mt}(e^{-j(2\pi/N)lN})^t = W^{mt} \qquad t, l = 0, \pm 1, \pm 2, \ldots$$

Note also that Eqs. (A–1) and (A–2) differ only in the sign of exponent of W_N, and in the scale factor (both x, X complex).

Consider *direct* evaluation of the DFT; with

$$x(t) = \text{Re } x(t) + j \text{ Im } x(t) \qquad \text{and} \qquad W_N^{mt} = \text{Re } W_N^{mt} + j \text{ Im } W_N^{mt}$$

substituting into Eq. (A–1), we have

$$X(m) = \sum_{t=0}^{N-1} \left(\text{Re } x(t) + j \text{ Im } x(t) \right)\left(\text{Re } W_N^{mt} + j \text{ Im } W_N^{mt} \right)$$

$$= \sum_{t=0}^{N-1} \left(\text{Re } x(t) \text{ Re } W_N^{mt} - \text{Im } x(t) \text{ Im } W_N^{mt} \right) +$$

$$j \left(\text{Re } x(t) \text{ Im } W_N^{mt} + \text{Im } x(t) \text{ Re } W_N^{mt} \right) \quad \text{(A–3)}$$

Investigating Eq. (A–3), we see that for an N-point sequence of DFT coefficients, that is, $\{X(m)\}$, $m = 0, \ldots, N-1$, we have the number of operations indicated in Table A–1.

The idea behind the FFT is to break the original N-point sequence into two shorter sequences (decimation-in-time), the DFT of which can be combined to give the DFT of the original N-point sequence. For example,

1. Break N-point sequence into two $(N/2)$-point sequences:

$$N\text{-point sequence} \Rightarrow 2 \cdot \left(\frac{N}{2}\right)\text{-point sequence}$$

and the savings for N^2 computations is

$$N^2 \Rightarrow 2 \cdot \left(\frac{N}{2}\right)^2 = \frac{N^2}{2} \sim \text{factor of two savings}$$

2. Breaking the $(N/2)$-point sequence into two $(N/4)$-point sequences gives a savings of

$$\left(\frac{N}{2}\right)^2 \Rightarrow 2\left(\frac{N}{4}\right)^2 = \frac{N^2}{8} \sim \text{factor of two savings}$$

Next, we show how this can be accomplished mathematically. Suppose $\{x(t)\}$ is an N-point sequence, where N is assumed a power of 2. Define two $(N/2)$-point

TABLE A–1
DFT operations

	m	t	Operations
Multiplies:	N	$4N$	$4N^2$ real multiplies (N^2 complex)
Adds:	N	$4N - 2$	$N(4N - 2)$ real adds ($N(N - 1)$ complex)

sequences, g and h, consisting of only the even-indexed samples and the odd-indexed samples of $x(t)$ respectively; that is,

$$g(t) = x(2t) \qquad t = 0, 1, \ldots, N/2 - 1$$

$$h(t) = x(2t + 1) \qquad t = 0, 1, \ldots, N/2 - 1$$

The N-point DFT can be written as

$$X(m) = \sum_{t=0(\text{even})}^{N-1} x(t) W_N^{mt} + \sum_{t=0(\text{odd})}^{N-1} x(t) W_N^{mt}$$

$$= \sum_{t=0}^{N/2-1} x(2t) W_N^{2mt} + \sum_{t=0}^{N/2-1} x(2t + 1) W_N^{(2t+1)m}$$

but $W_N^2 = [e^{j(2\pi/N)}]^2 = [e^{j(2\pi/(N/2))}] = W_{N/2}$; therefore, substituting, we have

$$X(m) = \sum_{t=0}^{N/2-1} g(t) W_{N/2}^{mt} + W_N^m \sum_{t=0}^{N/2-1} h(t) W_{N/2}^{mt} \qquad (A\text{--}4)$$

or, using the definition of the DFT, we have

$$X(m) = G(m) + W_N^m H(m) \qquad (N^2/2 + N) \text{ complex multiplications}$$

$$\underset{\frac{N}{2}\text{DFT}}{} \qquad \underset{\frac{N}{2}\text{DFT}}{}$$

where $G(m)$ and $H(m)$ are the respective $(N/2)$-point DFT of the even- and odd-indexed points of $x(t)$. Now, since $W_N^{m+N/2} = -W_N^m$, we can rewrite the expression for an N-point DFT as

$$X(m) = \begin{cases} G(m) + W_N^m H(m) & 0 \leq m \leq N/2 - 1 \\ G(m - N/2) + W_N^m H(m - N/2) & N/2 \leq m \leq N - 1 \end{cases} \qquad (A\text{--}5)$$

This is the underlying principle behind the FFT calculations using the "decimation-in-time" approach. The $(N/2)$-point DFT calculation can be further decimated to two $(N/4)$-point calculations, and so on. For instance, we can proceed a step further and break the $(N/2)$-point DFT calculation into two $(N/4)$-point calculations; consider the computation of $G(m)$:

$$G(m) = \sum_{t=0}^{N/2-1} g(t) W_{N/2}^{mt} = \sum_{t=0}^{N/4-1} g(2t) W_{N/2}^{2mt} + \sum_{t=0}^{N/4-1} g(2t + 1) W_{N/2}^{m(2t+1)} \qquad (A\text{--}6)$$

or, using the properties of the discrete exponential,

$$W_N^2 = W_{N/2}$$

which implies that

$$W_{N/2}^{2mt} = W_{N/4}^{mt}$$

Thus, using these relations in Eq. (A–6), we obtain

$$G(m) = \sum_{t=0}^{N/4-1} g(2t)W_{N/4}^{mt} + W_{N/2}^{t} \sum_{t=0}^{N/4-1} g(2t + 1)W_{N/4}^{mt} \qquad (A-7)$$

and a similar relation for $H(m)$ as

$$H(m) = \sum_{t=0}^{N/4-1} h(2t)W_{N/4}^{mt} + W_{N/2}^{t} \sum_{t=0}^{N/4-1} h(2t + 1)W_{N/4}^{mt} \qquad (A-8)$$

Similar to the $(N/2)$-point calculations for $X(m)$ of Eq. (A–5), we have the decomposition and $(N/4)$-point calculation for $G(m)$ or $H(m)$ as

$$G(m) = \hat{G}(m) + W_{N/2}^{t}\hat{H}(m) \qquad (A-9)$$

Proceeding, we can decompose $\hat{G}$ and $\hat{H}$ into $(N/8)$-point computations, and so on.

This technique is a "power-of-two" or radix-two transform, since $N = 2^{\nu}$ and we require $\nu = \log_2 N$ decompositions, which implies that the FFT requires $N \log_2 N$ computations to calculate an N-point DFT. We illustrate this idea with the following example and refer the interested reader to [3] for more details.

Example A–1. Consider the computation of a four-point DFT using the FFT approach:

$$X(m) = \begin{cases} G(m) + W_4^m H(m) & 0 \le m \le 1 \\ G(m-2) + W_4^m H(m-2) & 2 \le m \le 3 \end{cases}$$

Then we have

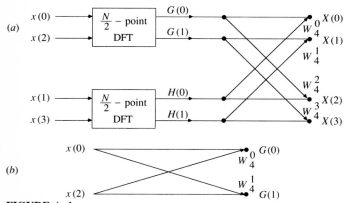

FIGURE A–1
Decimation-in-time decomposition of N-point DFT computation ($N = 4$) (a) $(N/2)$-point decomposition (b) Butterfly computation

m	$X(m)$
0	$G(0) + W_4^0 H(0)$
1	$G(1) + W_4^1 H(1)$
2	$G(0) + W_4^2 H(0)$
3	$G(1) + W_4^3 H(1)$

We illustrate this computation with the corresponding flow graph in Fig. A–1. Using the decimation-in-time approach, we continue to decimate the (N/k)-point sequences until we reach a two-point DFT calculation represented by the "butterfly" computation of Eq. (A–5) and Fig. A–1(b). For more details see [2] or [3].

REFERENCES

1. J. Cooley and J. Tukey, "An Algorithm for the Machine Calculation of Complex Fourier Series," *Math. Comput.*, 1965.
2. E. Brigham, *The Fast Fourier Transform* (Englewood Cliffs, N.J.: Prentice-Hall, 1974).
3. A. Oppenheim and R. Shafer, *Digital Signal Processing* (Englewood Cliffs, N.J.: Prentice-Hall, 1975).

APPENDIX
B

MAXIMUM ENTROPY SPECTRAL ESTIMATION

In this appendix, we develop the Maximum-Entropy Method (MEM) of spectral estimation. The estimation of the PSD from finite-length data and covariance sequences is a classical problem. Early work was based on the use of window functions to weight the sequences and evolved to the FFT-based correlation and periodogram methods discussed in Sec. 5.4. The shortcoming of these classical approaches is that the estimated spectrum agrees with these weighted values, rather than with the actual covariance sequence. The windowing also introduces the inherent assumption that the covariance is zero outside the window length. Burg [1] introduced the concept of entropy to resolve this problem. Maximum-entropy spectral estimation is based on choosing a spectrum corresponding to the most random sequence, with covariance corresponding to the given set of values. We shall show that the MEM leads to an all-pole or autoregressive solution of the spectral estimation problem; therefore, it can also be considered a model-based technique.

From the definition (see Chapter 4) and information theory it can be shown that the entropy of a time series is proportional to the integral of its PSD (see [2]). The *maximum-entropy principle* demands that for a given time series $x(t)$, $t =$

$0, \ldots, N$ taken from a gaussian process, bandlimited to $|\Omega| \le \Omega_B$, with entropy rate

$$H = \frac{1}{4\Omega_B} \int_{-\Omega_B}^{\Omega_B} \ln S_{xx}(\Omega) d\Omega \qquad (B-1)$$

we find the PSD function that maximizes H, subject to the constraint that the *Wiener-Khintchine* theorem must be satisfied. Since the corresponding covariance is known only for values of lag, $|k| \le |N|$, the MEM reduces to finding the PSD that *maximizes* H, subject to

$$R_{xx}(k) = \frac{1}{2\pi} \int_{-\Omega_B}^{\Omega_B} \ln S_{xx}(\Omega) e^{j\Omega k} d\Omega \qquad |k| \le N$$

Using the *Wiener-Khintchine* theorem, we can transform this problem to the equivalent problem of maximizing the entropy with respect to the covariance; that is, using

$$S_{xx}(\Omega) = \sum_{k=-\infty}^{\infty} R_{xx}(k) e^{-j\Omega k} \qquad (B-2)$$

we have from Eq. (B-1) that

$$H = \frac{1}{4\Omega_B} \int_{-\Omega_B}^{\Omega_B} \ln \left(\sum_{k=-\infty}^{\infty} R_{xx}(k) e^{-j\Omega k} \right) d\Omega \qquad (B-3)$$

Now the maximum entropy principle becomes simply $\max_{R_{xx}(k)} H$. Clearly, partial derivatives can be calculated as

$$\frac{\partial S_{xx}(\Omega)}{\partial R_{xx}(k)} = \frac{\partial}{\partial R_{xx}(k)} \left(\sum_{l=-\infty}^{\infty} R_{xx}(l) e^{-j\Omega l} \right) = e^{-j\Omega k}$$

and therefore $\quad \dfrac{\partial \ln S_{xx}(\Omega)}{\partial R_{xx}(k)} = \left[\dfrac{1}{S_{xx}(\Omega)} \right] \dfrac{\partial S_{xx}(\Omega)}{\partial R_{xx}(k)} = S_{xx}^{-1}(\Omega) e^{-j\Omega k} \qquad (B-4)$

Since we know $R_{xx}(k)$ for $|k| \le N$, we can maximize the entropy function of Eq. (B-1) for the *unknown* values of $R_{xx}(k)$ for $|k| > N$:

$$\max_{R_{xx}(k)} H = \frac{\partial H}{\partial R_{xx}(k)} = \frac{1}{4\Omega_B} \int_{-\Omega_B}^{\Omega_B} \frac{\partial}{\partial R_{xx}(k)} \ln S_{xx}(\Omega) d\Omega \qquad |k| > N$$

Substituting Eq. (B-4), we have

$$\max_{R_{xx}(k)} H = \frac{1}{4\Omega_B} \int_{-\Omega_B}^{\Omega_B} S_{xx}^{-1}(\Omega) e^{-j\Omega k} d\Omega = 0 \qquad |k| > N \qquad (B-5)$$

Since the PSD of a stationary process must be minimum-phase and positive on the unit circle (see Table 4.3-2), then its inverse spectrum also satisfies these properties; that is,

$$S_{xx}^{-1}(\Omega) = \frac{1}{S_{xx}(\Omega)} \geq 0$$

Thus, the Wiener-Khintchine relations hold for the inverse spectrum as well:

$$S_{xx}^{-1}(\Omega) = \sum_{k=-\infty}^{\infty} C_{xx}(k)e^{-j\Omega k}$$

and
$$C_{xx}(k) = \frac{1}{2\pi} \int_{2\pi} S_{xx}^{-1}(\Omega)e^{+j\Omega k}d\Omega \qquad (B\text{--}6)$$

where we have used the symmetry property of the covariance, $C_{xx}(k) = C_{xx}(-k)$. Now, for the maximum-entropy process, Eq. (B–5) must hold, implying that

$$C_{xx}(k) = 0 \qquad \text{for } |k| > N$$

after manipulating and equating with Eq. (B–6). Using this result, we see that the *inverse* spectrum of a maximum-entropy process must satisfy the finite series

$$S_{xx}^{-1}(\Omega) = \sum_{k=-N}^{N} C_{xx}(k)e^{-j\Omega k} \qquad (B\text{--}7)$$

Equivalently, this implies that the PSD of a maximum-entropy process is given by

$$S_{xx}(\Omega) = \frac{1}{\sum_{k=-N}^{N} C_{xx}(k)e^{-j\Omega k}} \qquad (B\text{--}8)$$

or, in terms of Z-transforms,

$$S_{xx}(z) = \frac{1}{\sum_{k=-N}^{N} C_{xx}(k)z^{-k}} \qquad (B\text{--}9)$$

but since Eq. (B–7) is a legitimate PSD, the spectral factorization theorem holds (see Sec. 4.4) and, therefore, it follows that

$$S_{xx}^{-1}(z) = A(z)A^*(z)$$

or equivalently
$$S_{\text{MEM}}(z) = S_{xx}(z) = \frac{R_{ee}\Delta T}{|A(z)|^2} \qquad (B\text{--}10)$$

That is, the maximum entropy spectral estimate is an autoregressive process.

From the representation theorem (see Sec. 4.4), we know that $A(z)$ is a finite N_ath order polynomial; therefore, using the sum decomposition of Eq. (4.4–6), and equating coefficients of

$$S_{xx}^{-1}(z) = A(z)A^*(z)$$

we obtain the normal equations of Eq. (6.2–14), which can be solved as shown previously. So we see that spectral estimation techniques utilizing an all-pole or AR model can be considered MEM spectral estimators. The appeal of the MEM is based on the fact that since the covariance function satisfies the AR recursion,

$$R_{xx}(k) = -\sum_{j=1}^{N_a} a_j R_{xx}(k-j) \qquad \text{for } k > N$$

then it can be extended indefinitely using this recursion rather than windowing it to zero; thus, the resulting spectra do not exhibit sidelobes due to windowing.

REFERENCES

1. J. Burg, "Maximum Entropy Spectral Analysis," Ph.D. Dissertation, Stanford Univ., 1975.
2. C. Shannon, *The Mathematical Theory of Communications* (Urbana, Ill.: Univ. of Illinois Press, 1948).

APPENDIX

C

RECURSIVE PREDICTION-ERROR METHOD (RPEM)

In this appendix, we derive the RPEM following Ljung and Soderstrom [1]. We first note that the ARMAX model can be rewritten in vector form as

$$y(t) = \phi'(t)\Theta + e(t) \tag{C–1}$$

where

$$\phi' := [-y(t-1) \cdots -y(t-N_a) \,|\, u(t) \cdots u(t-N_b) \,|\, e(t-1) \cdots e(t-N_c)]$$

and
$$\Theta' := [a_1 \cdots a_{N_a} \,|\, b_0 \cdots b_{N_b} \,|\, c_1 \cdots c_{N_c}]$$

We define the *prediction error criterion*[†] as

$$\min_{\Theta} J_t(\Theta) = \frac{1}{2} \sum_{k=1}^{t} e^2(k, \Theta) \tag{C–2}$$

where the *prediction error* is given (as before) by

$$e(t, \Theta) := y(t) - \hat{y}(t \,|\, t-1) \tag{C–3}$$

and
$$\hat{y}(t \,|\, t-1) = \hat{s}(t \,|\, t-1) = \left(1 - \hat{A}(q^{-1})\right)y(t) + \hat{B}(q^{-1})u(t) \tag{C–4}$$

[†]We explicitly show the dependence of the prediction error on the unknown parameter vector Θ.

The parameter estimator evolves in the standard manner by performing the indicated minimization in Eq. (C–2) after performing a Taylor-series expansion about $\Theta = \hat{\Theta}(t-1)$:

$$J_t(\Theta) = J_t(\hat{\Theta}) + \frac{\partial}{\partial\Theta}J'_t(\hat{\Theta})(\Theta - \hat{\Theta}) + \frac{1}{2}(\Theta - \hat{\Theta})'\frac{\partial^2}{\partial\Theta^2}J_t(\hat{\Theta})(\Theta - \hat{\Theta}) + h.o.t$$

(C–5)

where $\qquad\qquad\qquad\Theta$ is the $N_\Theta \times 1$ parameter vector

$\qquad\qquad\dfrac{\partial^2 J}{\partial\Theta^2}$ is an $N_\Theta \times N_\Theta$ Hessian matrix

$\qquad\qquad\dfrac{\partial J}{\partial\Theta}$ is an $N_\Theta \times 1$ gradient vector

$\qquad\qquad h.o.t$ is higher order terms

If we perform the minimization, then differentiating we obtain

$$\frac{\partial}{\partial\Theta}J_t(\Theta) = \frac{\partial}{\partial\Theta}J_t(\hat{\Theta}) + \frac{\partial}{\partial\Theta}\left[\frac{\partial'}{\partial\Theta}J_t(\hat{\Theta})(\Theta - \hat{\Theta})\right]$$

$$+ \frac{\partial}{\partial\Theta}\left[\frac{1}{2}(\Theta - \hat{\Theta})'\frac{\partial^2}{\partial\Theta^2}J_t(\hat{\Theta})(\Theta - \hat{\Theta})\right] + h.o.t.$$

Using the chain rule of vector calculus (see Eq. (5.1–4)), performing the indicated operations, and neglecting the *h.o.t.*, we have

$$\frac{\partial}{\partial\Theta}J_t(\Theta) \approx \frac{\partial}{\partial\Theta}J_t(\hat{\Theta}) + \frac{\partial^2}{\partial\Theta^2}J_t(\hat{\Theta})(\Theta - \hat{\Theta}(t-1))\bigg|_{\Theta = \hat{\Theta}(t)} = 0 \qquad (C–6)$$

Solving for $\hat{\Theta}(t)$, we obtain the *Gauss-Newton* parameter estimator

$$\hat{\Theta}(t) = \hat{\Theta}(t-1) - \left[\frac{\partial^2}{\partial\Theta^2}J_t(\hat{\Theta})\right]^{-1}\frac{\partial}{\partial\Theta}J_t(\hat{\Theta}) \qquad (C–7)$$

Thus, we must construct the gradient and Hessian in order to construct the RPEM estimator. The gradient is given by

$$\frac{\partial}{\partial\Theta}J_t(\Theta) = \frac{\partial}{\partial\Theta}\frac{1}{2}\sum_{k=1}^{t}e^2(k, \Theta) = \sum_{k=1}^{t}e(k, \Theta)\frac{\partial}{\partial\Theta}e(k, \Theta) \qquad (C–8)$$

We define the negative innovation gradient vector as

$$\psi(t, \Theta) := -\frac{\partial}{\partial\Theta}e(t, \Theta) \qquad (C–9)$$

and therefore, substituting into Eq. (C–8), we obtain

$$\frac{\partial}{\partial\Theta}J_t(\Theta) = -\sum_{k=1}^{t}\psi(k, \Theta)e(k, \Theta) \qquad (C–10)$$

Since we are interested in a *recursive-in-time* technique, we can factor out the tth term

$$\frac{\partial}{\partial\Theta}J_t(\Theta) = -\sum_{k=1}^{t-1}\psi(k,\Theta)e(k,\Theta) - \psi(t,\Theta)e(t,\Theta)$$

Recognizing the first term as the gradient of the previous time step, we have the *gradient recursion*

$$\frac{\partial}{\partial\Theta}J_t(\Theta) = \frac{\partial}{\partial\Theta}J_{t-1}(\Theta) - \psi(t,\Theta)e(t,\Theta) \tag{C-11}$$

The Hessian can also be determined in a similar manner; using this recursion, we have

$$\frac{\partial^2}{\partial\Theta^2}J_t(\Theta) = \frac{\partial^2}{\partial\Theta^2}J_{t-1}(\Theta) - \psi(t,\Theta)\frac{\partial}{\partial\Theta}e'(t,\Theta) - e(t,\Theta)\frac{\partial}{\partial\Theta}\psi'(t,\Theta) \tag{C-12}$$

The expressions for the gradient and Hessian can be simplified even further, if we assume that when the algorithm converges the following hold true:

(i) $\hat{\Theta} \approx \Theta_{\text{TRUE}}$

(ii) $\dfrac{\partial^2}{\partial\Theta^2}J_t(\hat{\Theta}(t)) \approx \dfrac{\partial^2}{\partial\Theta^2}J_t(\hat{\Theta}(t-1))$

(iii) $\dfrac{\partial}{\partial\Theta}J_{t-1}(\hat{\Theta}(t-1)) \approx 0$

(iv) $e(t,\hat{\Theta})\dfrac{\partial^2}{\partial\Theta^2}e(t,\hat{\Theta}) = 0$

Under these assumptions, when $\hat{\Theta} \to \Theta_{\text{TRUE}}$, the gradient recursion simplifies to

$$\frac{\partial}{\partial\Theta}J_t(\hat{\Theta}(t-1)) = -\psi(t,\hat{\Theta}(t-1))e(t,\hat{\Theta}(t-1)) \tag{C-13}$$

Under the convergence assumptions, the Hessian recursion also simplifies to

$$\frac{\partial^2}{\partial\Theta^2}J_t(\hat{\Theta}(t-1)) = \frac{\partial^2}{\partial\Theta^2}J_{t-1}(\hat{\Theta}(t-1)) + \psi(t,\hat{\Theta}(t-1))\psi'(t,\hat{\Theta}(t-1)) \tag{C-14}$$

If we define the Hessian matrix as

$$R(t) = \frac{\partial^2}{\partial\Theta^2}J_t(\Theta)$$

then Eq. (C–14) becomes simply

$$R(t) = R(t-1) + \psi(t,\hat{\Theta}(t-1))\psi'(t,\hat{\Theta}(t-1)) \tag{C-15}$$

and the Gauss-Newton parameter estimator becomes

$$\hat{\Theta}(t) = \hat{\Theta}(t-1) + R^{-1}(t)\psi(t,\hat{\Theta})e(t,\hat{\Theta}) \tag{C-16}$$

These equations make up the RPEM algorithm, all that is required is to determine $e(t, \Theta)$ and $\psi(t, \Theta)$. Using the ARMAX model, we have[†]

$$\frac{\partial}{\partial \Theta} C(q^{-1}) e(t) = \frac{\partial}{\partial \Theta} [A(q^{-1}) y(t) - B(q^{-1}) u(t)]$$

or

$$
\begin{bmatrix} \dfrac{\partial}{\partial a_i} \\ \hline \dfrac{\partial}{\partial b_i} \\ \hline \dfrac{\partial}{\partial c_i} \end{bmatrix}
\begin{bmatrix} \displaystyle\sum_{j=0}^{N_c} c_j e(t-j) \end{bmatrix}
=
\begin{bmatrix} \dfrac{\partial}{\partial a_i} \\ \hline \dfrac{\partial}{\partial b_i} \\ \hline \dfrac{\partial}{\partial c_i} \end{bmatrix}
\begin{bmatrix} \displaystyle\sum_{j=0}^{N_a} a_j y(t-j) - \sum_{j=0}^{N_b} b_j u(t-j) \end{bmatrix}
$$

Performing the gradient operation, we obtain

$$
\begin{bmatrix} \displaystyle\sum_{j=0}^{N_c} c_j q^{-j} \dfrac{\partial e(t)}{\partial a_i} \\ \hline \displaystyle\sum_{j=0}^{N_b} c_j q^{-j} \dfrac{\partial e(t)}{\partial b_i} \\ \hline e(t-i) + \displaystyle\sum_{j=0}^{N_c} c_j q^{-j} \dfrac{\partial e(t)}{\partial c_i} \end{bmatrix}
=
\begin{bmatrix} +y(t-i) \\ \hline -u(t-i) \\ \hline 0 \end{bmatrix}
\qquad \text{(C–17)}
$$

which can be written

$$
- \begin{bmatrix} C(q^{-1}) \dfrac{\partial e(t)}{\partial a_i} \\ \hline C(q^{-1}) \dfrac{\partial e(t)}{\partial b_i} \\ \hline C(q^{-1}) \dfrac{\partial e(t)}{\partial c_i} \end{bmatrix}
=
\begin{bmatrix} -y(t-i) \\ \hline u(t-i) \\ \hline e(t-i) \end{bmatrix}
$$

or, more succinctly,

$$C(q^{-1}) \psi(t) = \phi(t) \qquad \text{(C–18)}$$

This is equivalent to an inverse filtering operation characterized by

$$
\begin{bmatrix} y_f(t) + \displaystyle\sum_{i=1}^{N_c} c_i y_f(t-i) \\ \hline u_f(t) + \displaystyle\sum_{i=1}^{N_b} c_j u_f(t-i) \\ \hline e_f(t) + \displaystyle\sum_{i=1}^{N_c} c_i e_f(t-i) \end{bmatrix}
=
\begin{bmatrix} -y(t-i) \\ \hline u(t-i) \\ \hline e(t-i) \end{bmatrix}
\qquad \text{(C–19)}
$$

[†]We drop the Θ-dependent notation, $e(t, \Theta) = e(t)$, $\psi(t, \Theta) = \psi(t)$.

to give
$$\psi(t) = [\underline{y}_f(t) \mid \underline{u}_f(t) \mid \underline{e}_f(t)]'$$

after the "filtering" operation of Eq. (C–19) has been performed.

Finally, the prediction error is calculated using the most current available parameter estimate,

$$e(t) = y(t) - \phi'(t)\hat{\Theta}(t-1) \tag{C–20}$$

The RPEM algorithm can be implemented at this point; however, it requires the inversion of the $N_\Theta \times N_\Theta$ Hessian matrix at each time step. This calculation can be reduced by applying the matrix inversion lemma† to $R^{-1}(t)$; that is, taking the inverse of both sides of Eq. (C–15) and letting $P(t) := R^{-1}(t)$, we obtain

$$P(t) = [P^{-1}(t-1) + \psi(t)\psi'(t)]^{-1}$$

Applying the inversion lemma with $A = P^{-1}$, $B = \psi$, $C = 1$, $D = \psi'$, we obtain

$$P(t) = P(t-1) - \frac{P(t-1)\psi(t)\psi'(t)P(t-1)}{1 + \psi'(t)P(t-1)\psi(t)} \tag{C–21}$$

which completes the algorithm. We summarize the RPEM in Table 7.6–1.

REFERENCE

1. L. Ljung and T. Soderstrom, *Theory and Practice of Recursive Identification* (Boston: MIT Press, 1983).

---

†The matrix inversion lemma is given by $[A + BCD]^{-1} = A^{-1} - A^{-1}B[DA^{-1}B + C^{-1}]^{-1}DA^{-1}$

APPENDIX
D

SIG: A
GENERAL-PURPOSE
SIGNAL-PROCESSING
PACKAGE

D.1   INTRODUCTION

SIG is a general-purpose signal-processing, analysis, and display program. Its main purpose is to perform manipulations on time- and frequency-domain signals. However, it has been designed to ultimately accommodate other representations for data, such as multiplexed signals and complex matrices. Many of the basic operations one would perform on digitized data are contained in the core SIG package. Out of these core commands, more powerful processing algorithms may be constructed.

Two user interfaces are provided in SIG: a menu mode for the occasional user and a command mode for more experienced daily users. In both modes, errors are detected as early as possible in the processing and are indicated by clear, friendly messages. Command arguments not entered by the user will cause SIG to prompt for them or default to them at the user's option. Other options exist, such as multiple commands per line, command files with arguments, commenting lines, defining commands, automatic execution, etc.

Many different operations on time- and frequency-domain signals can be performed by SIG. They include operations on the samples of a signal, such as adding a scalar to each sample; operations on the entire signal, such as digital filtering; and operations on two or more signals, such as adding two signals.

Signals such as a pulse train or a random waveform may be simulated. Many of the signal-processing algorithms (e.g., FFT, correlation, coherence, Parks-McClellan filter, and convolution) have been taken from the IEEE text "Programs for Digital Signal Processing" [1], considered by many as the standard.

Graphics operations display signals and spectra. When spectra are displayed, the user may select the scaling according to the continuous or discrete domains. Internally, all SIG algorithms scale according to the continuous domain (i.e., multiply by the sample interval). Examples of the plot capability in SIG are shown throughout this text.

The commands are summarized in Table D–1. A user's manual [2] is also available, which contains a complete description of the software from the user's standpoint.

SIG has been implemented using the techniques of Structured Analysis and Design popularized by Yourdon and others [3]. In particular, SIG has been designed so users can work independently on developing signal-processing algorithms and can easily install them in the software.

D.2 COMMAND PROCESSOR

The command processor in SIG can be thought of as a miniature operating system. A user may execute commands, start other programs, run from command files, define (or redefine) new commands, keep a log of all interactions, or do many other activities normally reserved for operating systems. Even though SIG was written with signal processing in mind, the SIG Command Processor (CP) is a very general tool that can be used in a variety of applications. In addition, the CP runs on several different machines under very different software environments.

The SIG command processor is designed for computers with either a large address space (or virtual memory) or a small, fixed address space. On small-memory machines, the command processor acquires a command from the user, a file, or its own stack; then it decides which of many programs (called "tasks") can perform that command. The command processor then chains to that program, sending any supplied arguments on its execute line, and exits. The task then performs the command and chains back to the command processor. On large-memory machines it is inefficient to chain, so the command processor simply calls the task subroutines directly.

D.3 COMMUNICATION BY FILES

Two major design goals that make it possible to customize SIG are that it *remembers what you said*, and it has *no hard-coded parameters*. These goals are satisfied by having SIG read almost everything from disk files. This includes menus, help, command definitions, default arguments for commands, plot layouts (viewports), plot scalings, plot axis types, plot device types, terminal device types, the state of the command processor, and, of course, the data being processed.

The parameter file serves as a general database for much of SIG. Data are stored in the parameter file as strings accessed by "keys." A library of subroutines provides the capability to store and extract the strings and convert them to integers, reads, etc., as needed. SIG commands allow the user to store strings in the parameter file and list their values. The command processor will extract strings and use them as arguments for commands.

Menu files are read by the MENU command. They end in the extension ".MNU" and consist of two partitions, one for commands and one for text. The command partition begins with the first line of the file. Each line contains the SIG commands to be executed when the user picks the menu item corresponding to that line.

Help files are read by the HELP command. They end in the extension ".HLP" and like the menu files (discussed above) consist of two partitions. The help processor behaves differently from the menu processor when the command partition is empty (i.e., the file begins with a blank line). In that case, the text portion of the file is displayed on the terminal and control returns to the SIG command processor.

Command files have the default extension ".SCF" (for Sig Command File). They consist of lines of SIG commands and are executed by the EXECUTE command.

The graphics commands in SIG are implemented with Device Independent Graphics LIBrary (DIGLIB) [4]. The default graphics device is a VT-640 terminal, a DEC VT-100 with RETROGRAPHICS. SIG is "tuned" to work best with the VT-640; however, only slight differences appear when other graphics devices are used. The types of devices which are supported range from low- and high-resolution graphics terminals to printer-plotter, laser printers, and sophisticated frame buffers. Both color and black and white devices are supported.

D.4 SIG COMMANDS

In this section, we briefly discuss the commands available in SIG. First, we discuss each of the categories available and then each of the commands. Besides being developed to solve many signal-processing problems, SIG is also a general-purpose analysis and display package with its own internal database and command processor—somewhat like a miniature operating system. So, as we shall see, there are various categories in which the over 150 SIG commands can be grouped. Referring to the list of commands in Table D–1, we see the following major categories:

1. Database manipulation
2. Input/output
3. Graphics
4. Signal edit
5. Simulation

COMMAND SUMMARY

Data Base Manipulation Commands

For a data base:

DBCREATE	Create
DBDELETE	Delete
DBLIST	List contents
DBSUBTITLE	Set its sub-title
DBTITLE	Set its title

Reading/Writing Data In/Out of SIG

For a Frequency Spectrum (FS) data store:

FSREAD	Read
FSWRITE	Write

For a Real Coefficient (RC) data store:

RCREAD	Read
RCWRITE	Write

For a Time Signal (TS) data store:

TSREAD	Read
TSWRITE	Write

Data Store Commands

For a data store:

DSCOLUMNS	Set the number of columns
DSCOPY	Copy
DSDELETE	Delete
DSLIST	Print the header information
DSNOPROTECT	Make read-write
DSPROTECT	Make read-only
DSROWS	Set the number of rows
DSSUBTITLE	Set the sub-title
DSTITLE	Set the title
DSTYPE	Set the type

For a Frequency Spectrum data store:

FSINITIAL	Set the initial value
FSINTERVAL	Set the sample interval
FSSAMPLES	Set the number of points

For a Real Coefficient data store:

RCINTERVAL	Set the sample interval
RCLONGEST	Set the maximum number of coefficients
RCSETS	Set the number of sets of coefficients

For a Time Signal (TS) data store:

TSCATENATE	Catenate data stores
TSINITIAL	Set the initial time value
TSINTERVAL	Set the sample interval
TSSAMPLES	Set the number of points

Simulation Commands

Fill a Time Signal data store with a:

CHIRP	Chirping (changing frequency) sine wave
CONSTANT	Constant value
GAUSSIANRANDOM	Gaussian random sequence
PSEUDORANDOM	Pseudorandom binary sequence
PULSETRAIN	Pulse train
RAMP	Ramp
UNIFORMRANDOM	Uniform random sequence

Graphics Commands

CURSORVIEWPORT	Use the cursor to define a set of viewports
ERASE	Erase the graphics screen

For Frequency Spectrum data stores:

FSPLOT	Plot one or more
FSQUICKPLOT	Quick and dirty plot

For Time Signal data stores:

TSPLOT	Plot one or more
TSQUICKPLOT	Quick and dirty plot
TSTSQUICKPLOT	Quick plot of one versus another

Arithmetic Commands

For each element of a data store:

ABSOLUTE	Take the absolute value
ADD	Add one to another
ARCCOSINE	Compute the arccosine
ARCSINE	Compute the arcsine
ARCTANGENT	Compute the arctangent
CADD	Add a constant
CDIVIDE	Divide by a constant
CMULTIPLY	Multiply by a constant
COSINE	Take the cosine
CSUBTRACT	Subtract a constant
DEGREESTORADIA	Convert from degrees to radians
DIVIDE	Divide one by another
EXPONENTIAL	Take the exponential
LOGE	Take the base E logarithm
LOG10	Take the base 10 logarithm
MULTIPLY	Multiply one by another
POWER	Raise to a power
RADIANSTODEGRE	Convert from radians to degrees
RECIPROCAL	Take the reciprocal
SINE	Take the sine
SQRT	Take the square root
SQUARE	Square
SUBTRACT	Subtract one from another
TANGENT	Take the tangent

Time Signal Manipulation Commands

For a Time Signal data store:

ALIGN	Align the slave to the master
CONVOLVE	Convolve one with another
DECIMATEINTERP	Decimate and/or interpolate while filtering
DIFFERENTIATE	Compute derivative
ENSEMBLE	Compute the mean and variance of an ensemble
IMPULSERESPONSE	Compute the impulse response of a system
INTEGRATE	Compute integral
LEASTSQUARES	Find polynomial coefficients giving least squares fit
NORMALIZE	Divide by largest value
POLYNOMIAL	Expand polynomial given coefficients
RESAMPLE	Resample and Interpolate using Wiggins method
STATISTICS	Compute and print statistical moments

Time Signal Edit Commands

For a Time Signal data store:

DEINITIAL	Subtract the initial value
DEMEAN	Subtract the mean value
DETREND	Remove a trend
IICUT	Cut data given start and stop indexes
INCUT	Cut data given start index and number of points
ITCUT	Cut data given start index and time duration
TICUT	Cut data given start time and stop index
TNCUT	Cut data given start time and number of points
TTCUT	Cut data given start time and stop time

Filter Commands

For a Time Signal data store, perform a:

ANOISECANCEL	Adaptive noise canceler
MEDIAN	Median filter
PMFILTER	Parks-McClellan filter design
BPBESSEL	Band-pass Bessel filter
BPBUTTERWORTH	Band-pass Butterworth filter
BP1CHEBYCHEV	Band-pass type 1 Chebychev filter
BP2CHEBYCHEV	Band-pass type 2 Chebychev filter
BRBESSEL	Band-reject Bessel filter
BRBUTTERWORTH	Band-reject Butterworth filter
BR1CHEBYCHEV	Band-reject type 1 Chebychev filter
BR2CHEBYCHEV	Band-reject type 2 Chebychev filter
HPBESSEL	High-pass Bessel filter
HPBUTTERWORTH	High-pass Butterworth filter
HP1CHEBYCHEV	High-pass type 1 Chebychev filter
HP2CHEBYCHEV	High-pass type 2 Chebychev filter
LPBESSEL	Low-pass Bessel filter
LPBUTTERWORTH	Low-pass Butterworth filter
LP1CHEBYCHEV	Low-pass type 1 Chebychev filter
LP2CHEBYCHEV	Low-pass type 2 Chebychev filter

Fourier Transforms, Spectral Estimation, Windowing

COHERENCE	Compute coherence function
CORRELATE	Compute auto/cross correlation function
ECOHERENCE	Coherence for an ensemble of signals
FFT	Fast Fourier Transform
HAMMING	Hamming window
HANNING	Hanning window
IFFT	Inverse Fourier Transform
KAISER	Kaiser window
SDAONLY	Spectral density with A-polynomial
SDCORRELATION	Spectral density by correlation method
SDLATTICE	Spectral density using lattice algorithm
SDWELCH	Spectral density by Welch's method
TRIANGLE	Triangle window

Parameter File Manipulation Commands

PFCOPY	Copy a parameter file
PFDELETE	Delete a parameter
PFINSERT	Insert a parameter
PFKEYLIST	List the parameter keys
PFLIST	List the parameter keys and their values
PFREPLACE	Replace a parameter
PFWRITE	Write a parameter

SIG Command Processor Commands Input Stack Commands

CLEAR	Clear the stack
CONTINUE	Resume reading commands
EXECUTE	Read commands from a file
HELP	Print a description of a command
MENU	Display a menu
STACK	List the current items on the stack
TERMINAL	Read commands from the terminal

Interaction Commands

DEFAULT	Turn on the default-from-right flag
ECHO	Echo commands on terminal
LFCLOSE	Close command Log File
LFCREATE	Create a command Log File
MESSAGE	Write to the user's terminal
NEWS	List items of interest
NODEFAULT	Turn off the default-from-right flag
NOECHO	Turn off the echo flag
PAUSE	Stop execution and send out the message: Hit (RETURN) to continue...

Commands to Operate on Commands

PFLIST.	List command definitions
RSCREATE	Create a Repeat Sequence
SCDEFINE	Define a command
SCDELETE	Delete a command
SCLIST	List the commands

External Commands

EXIT, END	Terminate execution of SIG
NULLTASK	Start an empty task, used for timing tests
PROGRAM	Initiate a program
SUBPROGRAM	Initiate a subroutine
SYSTEM	Execute operating system commands

Linear System Commands

LSD = Linear System Deconvolution using:

LSDAONLY	A-polynomial only (All-pole)
LSDBONLY	B-polynomial only (All-zero)
LSDLATTICE	Lattice algorithm

LSI = Linear System Identification using:

LSIABC	A-, B-, C-polynomials (ARMAX)
LSIAONLY	A-polynomial only (All-pole)
LSILATTICE	Lattice algorithm

LSS = Linear System Simulation using:

LSSABCD	A, B, C, D-polynomials (ARMAX)
LSSABONLY	A, B-polynomials
LSSIMPULSE	Simulate impulse response

363

6. Linear systems
7. Spectral estimation
8. Digital filtering
9. Signal manipulation
10. Signal arithmetic
11. Operations
12. Processor
13. Parameter manipulation
14. Interaction
15. External

The SIG database consists of "files" called data stores, with headers and parameters describing the particular data "stored" in the data base. Thus, the *database-manipulation commands* enable the SIG user to list, copy, delete, protect, and set parameters such as titles, subtitles, type, sampling intervals, initial values, and the number of samples.

The SIG internal database files are in a packed binary format for efficient storage, but standard ASCII files may be "read into" or "written out" of the SIG database, using the *input/output commands*. Thus, the user can easily transform data to and from SIG database.

SIG has a self-contained graphics processor which is based on DIGLIB [4], thus enabling it to execute on small as well as large computers and to support many independent terminal and graphics devices. The SIG *graphics commands* include cursor viewporting, erase, frequency and time displays, as well as *xy*-plot capability.

Time signals can be edited using the SIG *signal edit commands*, which include such operations as deinitial, demean, detrend, and catenate, as well as a full suite of "cut" commands.

Signals can be simulated using the SIG *simulation commands*, which consist of basic signals such as chirping sinusoids, constant (step), gaussian, uniform, and pseudo random sequences, as well as pulsetrains and ramps.

SIG *linear system commands* consist of a suite of algorithms to simulate, identify, and deconvolve signals and discrete-time systems. Deterministic difference equation models, as well as the random ARMAX model, may be simulated; deconvolution is performed using the all-pole, all-zero, and lattice techniques (see Chapter 7). Linear-system identification/parameter estimation is performed using the prediction-error, all-pole, all-zero, and lattice models.

The estimation of frequency spectra for both deterministic as well as random signals is accomplished using the SIG *spectral estimation commands*, consisting of the standard FFT and FFT-based classical spectral estimators, as well as many of the modern approaches, including the all-pole and lattice methods. Besides a full suite of window functions to smooth spectral estimates, SIG also enables the signal processor to estimate coherence functions for both stationary and transient signals, as well as providing the FFT-based correlation/covariance estimator from [1].

Digital filtering is accomplished using the SIG *filter commands*, which consist of a full suite of analog prototype filters, including Butterworth, Bessel, and Chebychev I and II designs for low-pass, high-pass, band-pass and band-reject filters. The optimization-based Parks-McClellan FIR filter, as well as window-based designs, are also available. Random noise is also removed using the adaptive noise canceller, as well as the median filter for random "spike" removal.

Time signals can be manipulated using the SIG *manipulation commands*, consisting of signal alignment, decimation/interpolation, resampling, and normalize commands, while other features include convolution, differentiation, integration, as well as ensemble operations. The system impulse response and polynomial filtering are also available, as well as sample statistics.

We can operate on signals much the same as on ordinary numbers, using the SIG *arithmetic commands*, which include such operations as add, subtract, multiply, and divide for real as well as complex signals. Trigonometric functions such as sine, cosine, tangent, and their inverses, as well as exponential, square, logarithms, reciprocal, radians-to-degrees, and their associated inverses, are also included in these commands.

SIG also enables users to define and list both the SIG commands and their own custom commands, using the SIG *operation commands*.

The SIG command processor can be exploited using the *processor commands* to clear, continue, execute (command files), call for on-line help, display menus or the stack, or just read commands from the terminal.

The parameter file, which contains a list both local and global of all of the current SIG parameters can be written to or from using the SIG *parameter manipulation commands*. These commands include copy, delete, insert, list, replace, and writing to the file.

The user can interact with basic operations using the SIG *interaction commands*, which can set defaults, echoing, log files, messages, news, or pauses.

Finally, SIG can interact with the computer operating system through the use of SIG external commands which exit, communicate with the system, or execute an external program.

D.5 SIGNAL-PROCESSING ALGORITHMS

In this section, we briefly discuss the SIG signal-processing algorithms employed. When available, algorithms from the IEEE Digital Signal Processing Package were selected primarily because of their popularity and existing documentation [1]. Here we will highlight the algorithms of most interest from the signal-processing viewpoint.

The digital filter designs implemented in SIG follow the standard analog prototype filter designs [5] using the bilinear transformation. The Parks-McClellan design uses the Remez exchange algorithm and was selected from [1]. The adaptive noise canceller is the Widrow [6] LMS gradient algorithm, which provides the heart of the adaptive signal processing capability within SIG (see Chapter 8 for details).

The spectral estimation algorithms in SIG are again those selected from [1]. The classical correlation (or Blackman-Tukey method) and periodogram (or Welch's modified periodogram method) are employed. The modern all-pole spectral estimator employs the Levinson recursion (auto-correlation method), while the lattice spectral estimator uses Makhoul's covariance lattice estimator; both are available in [1]. Coherence function estimation is accomplished using the Welch algorithm as well. The covariance/correlation estimator uses the FFT-based method of calculating the sample covariance (cross or auto) function.

The linear system simulation commands include the discrete-time pulse transfer function and random ARMAX(N_a, N_b, N_c, N_d) simulators, as well as a specialized impulse response simulator. FFT-based fast convolution is also available [1]. System identification/parameter estimation commands used for parametric signal processing (see Chapter 7) are also available using the all-pole (Levinson) and lattice (Makhoul) estimators, as well as the ARMAX-based prediction-error model, estimated using the recursive extended least-squares (RELS) algorithm ([7] and Chapter 7). This same command can be used to perform the RLS adaptive FIR or IIR algorithms (Chapter 8) for adaptive signal processing. Finally, the deconvolution algorithms are essentially the all-pole (Levinson) and lattice (Makhoul) techniques, while the FIR method incorporates the Levinson-Wiggins-Robinson (LWR) recursion [8].

The time-signal manipulation commands include the FFT-based "fast" convolution technique available in [1], as well as the corresponding decimation/interpolation command, which incorporates the Parks-McClellan FIR filter to minimize aliasing. The resampler uses the Wiggins method of interpolation. Finally, the impulse-response estimator employs the LWR to provide the Wiener solution for both deterministic as well as random signals.

D.6   AVAILABILITY

SIG has been designed to be easily transportable. It is presently available in versions for VAX (VMS, Berkeley 4.2 UNIX) and SUN (Berkeley 4.2 UNIX). It can produce graphics on more than 20 different terminals and graphics devices.

SIG is a production code[†] and is available along with other related products (e.g., SIG tools development environment, short courses, etc.) from Techni-Soft, P.O. Box 2525, Livermore, CA 94550; phone (415) 443-7213. Technical details are available in the user's manual [2].

[†]Version 1.2, dated June, 1987, has approximately 40 new commands (over 200 total).

REFERENCES

1. Digital Signal Processing Committee of IEEE Acoustics, Speech, and Signal Processing Society, *Programs for Digital Signal Processing* (New York: IEEE Press, 1979).
2. D. Lager and S. Azevedo, *SIG—A General-Purpose Signal Processing Program*, Lawrence Livermore National Laboratory Report UCID-19912 Rev. 1, May 1985. Also available from Techni-Soft.
3. E. Yourdon and L. Constantine, *Structured Design—Fundamentals of a Discipline of Computer Program and Systems Design* (Yourdon Press, 1978).
4. H. Brand, *DIGLIB—Device Independent Graphics Library*, Lawrence Livermore National Laboratory, August, 1985.
5. A. Oppenheim and R. Shafer, *Digital Signal Processing* (Englewood Cliffs, N.J.: Prentice-Hall, 1975).
6. B. Widrow and S. Stearns, *Adaptive Signal Processing* (Englewood Cliffs, N.J.: Prentice-Hall, 1984).
7. L. Ljung and T. Soderstrom, *Theory and Practice of Recursive Identification* (Boston: MIT Press, 1983).
8. E. Robinson and S. Treitel, *Geophysical Signal Analysis* (Englewood Cliffs, N.J.: Prentice-Hall, 1980).

APPENDIX
E

SSPACK^{TM}:
AN
INTERACTIVE
STATE-SPACE,
MODEL-BASED
PROCESSING
SOFTWARE
PACKAGE

The design and analysis of model-based processors using state-space techniques are often formidable tasks when models are complex and multichannel. In this appendix, we introduce the State-Space, Model-Based Processing Software Package (SSPACK), an interactive software package for the analysis, design, and display of model-based processors. The package contains algorithms for model-based signal processing, simulation, estimation, and identification integrated into an easy-to-use package. It enables the user to quickly access algorithms and analyze their performance. A preprocessor designed to simplify model specifications allows even complex systems to easily be defined and processed. SSPACK enables easy interfacing to powerful model-based algorithms with display and analysis capability.

368

E.1   INTRODUCTION

SSPACK offers an integrated approach to various problems. It eases the burden of a priori knowledge by utilizing a menu-oriented supervisor, so the casual or new user can immediately begin operation. The package is designed to enable the serious user to run individual processors and algorithms separately as well as to interface his or her own particular algorithms. SSPACK utilizes step-by-step procedures that enable the user to monitor various operations in full detail.

The package consists of a supervisor, which controls the operation of the software, preprocessors, postprocessors, and sets of individual algorithms (or modules) that communicate using disk files.

The supervisor controls the package by enabling the user to perform various operations, ranging from specifying a problem title to preparing and running a particular problem. The user need not leave the supervisor to accomplish any of the required operations in SSPACK.

The processors aid the user in entering the problem of interest as well as displaying the results. The state-space preprocessor (SSPREP) algorithm program aids the user in preparing files for the individual algorithm programs, while the state-space postprocessor (SSPOST) program displays and analyzes the output from the algorithms. SSPREP prompts the user with a series of questions in a menu format. The initial menu lets the user choose the desired algorithm. Subsequent menus help specify all the necessary parameters for the selected algorithm. SSPOST is an interactive, command-driven processor. It is designed to interpret the output of the various SSPACK algorithms and display the results.

Besides the supervisor and processors, SSPACK currently consists of the following modules: a linear discrete-time state estimator (SSDEST) and simulator (SSDSIM), a nonlinear discrete-time state estimator (SSNEST) and simulator (SSNSIM), continuous-to-discrete-time conversion (SSCTOD), and a discrete-time parameter estimator/identifier (SSDID). All the algorithms are based on the Kalman filter/estimator [2,6] implemented using efficient factorized matrix techniques [3].

E.2   SUPERVISOR

The focus of SSPACK interactivity lies in its system supervisor. It is designed to enable the user to perform all the necessary operations from simulation to performance analysis, without the burden of understanding the detailed linkage between individual modules that comprise the package. Owing to its modular design, the user can interface his or her own particular algorithms to the package easily.

The control flow in the SSPACK supervisor is best described by the diagram shown in Fig. E–1. The supervisor enables the user to select the desired operation (simulation, estimation, or identification) through a menu format and then to perform the various steps necessary to accomplish it. Once selected, the user's choices are identical for each operation: RENAME the problem title,

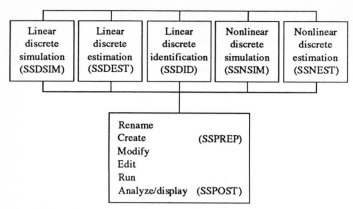

FIGURE E–1
SSPACK interaction diagram

CREATE the specification file for the particular algorithm, MODIFY the problem models, EDIT the specification file, RUN the selected algorithm, and ANALYZE and DISPLAY the results. Most of these operations are performed by specific SSPACK algorithms discussed in detail in the user's manual [2]. Note that some menu choices operate differently for different algorithms. For example, the MODIFY selection for the linear-discrete simulation enables the user to convert from a continuous-time model to a discrete-time model, while this selection for the nonlinear-discrete simulation actually enables the user to enter in a new set of nonlinear difference equations. These tasks are all accomplished within the SSPACK supervisor, thus simplifying the design procedure.

E.3   PREPROCESSOR

The preprocessor creates a problem-specification file, incorporating all of the necessary information to execute the desired program. This file is a simple text file in free format, making it easy for the user to read, print, or edit. Once the file is created, alterations in problem parameters are easily modified by selecting the EDIT operation, in which the actual problem-specification file is displayed for editing. Note that the preprocessor uses a question-and-answer format to enter parameters and a menu processor to alter these parameters after a particular category or model has been completely entered.

Once the specification file is completed, the user selects the RUN option of the supervisor to actually perform the simulation using selected algorithms. After optionally echoing the problem specifications back to the user's terminal, the algorithm executes the problem and creates a postprocessor file for display and performance analysis. Following the execution, the user selects the ANA-LYZE/DISPLAY option in the supervisor menu, which executes the postpro-cessor.

E.4 POSTPROCESSOR

SSPOST is a command-driven processor that displays the outputs of the various simulation and estimation algorithms in SSPACK and performs statistical tests to evaluate the performance of each particular algorithm. It reads commands and operates on a postprocessor (*pp*) file created by other SSPACK routines. The user issues commands to SSPOST interactively, from a terminal or from a command file for automatic execution. The commands are directives to the SSPOST program for reading data, setting scales, producing plots, etc. SSPOST also has the capability of selecting single or multiple plots on the display device by choosing various "viewports" available (e.g., four plots per screen). Along with this viewport capability, SSPOST enables the user to display data from multiple ppfiles as families of curves. This feature is particularly useful when attempting to analyze the effect of a particular adjustment on the performance of an algorithm, such as tuning an algorithm. SSPOST also offers the user an on-line help package that includes a description of each command for easy reference.

The following list summarizes the capability and important features of SSPOST. It can

1. *Display* the time history of various SSPACK algorithm variables.
2. *Test* statistically the performance of SSPACK algorithms.
3. *Execute* interactively from the keyboard or command files.
4. *Viewport* displays.
5. *Help* through an on-line list of commands.
6. *Plot* multiple ppfiles for comparison.

All of the SSPOST commands are succinctly summarized in Table E–1.

E.5 ALGORITHMS

SSPACK is primarily aimed at discrete-time, model-based signal processing, since the majority of real-time applications are discrete-time problems implemented on micro- or mini-computers. Problems formulated in continuous time are converted to discrete time and then executed by the various algorithms.

The conversion from continuous to discrete time is accomplished by first using the Paynter algorithm, to determine the number of terms required in the matrix exponential series, and then calculating the series using numerical methods [3].

There are two simulators available in SSPACK: a linear and a nonlinear algorithm. The simulators are stochastic and produce Gauss-Markov and approximately Gauss-Markov outputs for the state and noisy measurements [4]. The associated statistics are also calculated (mean and variance) and are used to determine confidence limits, as well as to evaluate the performance of the algorithm. Of course, deterministic models are special cases of these stochastic models and are easily simulated as well.

TABLE E–1
Command summary

Display commands

EP	I J	Plot innovation sequence
GK	I J K L	Plot gain
PC	I J K L	Plot corrected covariance
PP	I J K L	Plot predicted covariance
PS	I J K L	Plot simulated covariance
RE	I J K L	Plot innovation covariance
RZ	I J K L	Plot measurement covariance
U	I J	Plot forcing function
XC	I J	Plot corrected state
XP	I J	Plot predicted state
XT	I J	Plot true (mean) state
Z	I J	Plot measurement
ZEST	I J	Plot predicted measurement
ZT	I J	Plot the true (mean) measurement
TEXT		Scan opened ppfiles and copy text

Performance commands

PSDEP	I J ISTART	Plot innovations power spectrum
WTEST	I J ISTART	Plot innovations whiteness/zero-mean
WSSR	N	Plot the WSSR with window N
XCXP or XPXC	I J	Plot zig-zag estimated states
XERR	I J	Plot estimation error
XS	I J	Plot simulated state
ZS	I J	Plot simulated measurement

Input-output commands

COMFILE	NAME	Read next command from file NAME
DEVICE	NAME	Display plots on names device
END or EXIT		User is done—terminate
EXTRACT	DES1 DES2. . .	Copy data under DES1, etc.
PPFILE	NAME1 NAME2. . .	Get data from the names ppfiles

Parameter commands

FSCALE	FMIN FMAX	Set X-axis scale for frequency plots
HELP		Obtain on-line help
INFO	LINENO LABEL	Put LABEL on plots at LINENO
ITER	ISTART ISTOP	Only use data from ISTART to ISTOP
MAP	MAPNO	Set plot mapping according to MAPNO
NOQUICK		Change to normal plot mode
PTITLE	TITLE	Change each plot title to TITLE
QUICK		Change to "quick" plot mode
TIME	TSTART TSTOP	Only use data from TSTART to TSTOP
TSCALE	TMIN TMAX	Set time scaling for plots
TU	I TITLE	Set forcing function Y-axis title
TX	I TITLE	Set state Y-axis title
TZ	I TITLE	Set measurement Y-axis title
YSCALE	YMIN YMAX	Set Y-axis scale plots
YSCALEFD	YMIN YMAX	Set Y-axis scale of frequency plots
YTITLE	TITLE	Change the Y-axis title to TITLE

Two model-based processors are available in SSPACK. The linear discrete-time estimator uses the factorized matrix, or U-D form, of the recursive Kalman filter estimation algorithm, a superior numerical technique for small word-length machines or large-scale problems [1]. The nonlinear discrete-time algorithm is also the U-D form of the extended Kalman filter algorithm. Both algorithms produce postprocessor files, with estimates as well as associated statistics, to enable analysis and display of the overall statistical performance. For instance, such features as whiteness testing, spectral estimation, tracking errors, etc., are provided as some of the available analysis procedures (see Chapter 9.)

Two identification algorithms are also available. The first is a linear Kalman filter or, equivalently, a recursive least-squares identifier, capable of performing parameter estimation for a scalar discrete-time transfer function (ARX model) or state-space model in observer canonical form [5]. Nonlinear or multichannel models can be identified using the extended Kalman filter algorithm, mentioned earlier, by augmenting the parameters into the state-space models (see [4] for details).

All the algorithms communicate with the preprocessors and postprocessors through text files; these files are readable by the user.

E.6 AVAILABILITY

SSPACK is a production code and is available for purchase from Techni-Soft, P.O. Box 2525, Livermore, CA 94550; phone (415) 443-7213. Technical details are available in the user's manual [2].

REFERENCES

1. G. Bierman, *Factorization Methods for Discrete Sequential Estimation* (New York: Academic Press, 1977).
2. Technical Software Systems, *SSPACK: State-Space Systems Software Package*, 1983.
3. A. Gelb, *Optimal Estimation* (Boston: MIT Press, 1975).
4. J. Candy, *Signal Processing: The Model-Based Approach* (New York: McGraw-Hill, 1986).
5. L. Ljung and T. Soderstrom, *Theory and Practice of Recursive Identification* (Boston: MIT Press, 1983).
6. S. Azevedo, J. Candy, and D. Lager, "SSPACK: An Interactive State-Space Systems Software Package," *Proc. of CACSD '85 Confr.* (Santa Barbara, 1985).
7. S. Azevedo, J. Candy, and D. Lager, "SSPACK: A Multi-channel Signal Processing Package," *Proc. of IEEE Circuits, Systems Confr.* (San Jose, 1986).

INDEX

A posteriori mass, 153, 154
A priori mass, 154
Acoustic wave, 108, 126, 234
Adaptive:
  algorithm, 230, 263, 264, 266, 268, 276, 279, 280, 281, 291, 292, 309
    average trajectory, 274
    block, 288, 289, 291
    computations, 279
    convergence, 275, 278, 280, 283
    forgetting factor, 279
    misadjustment, 275, 279
    speed, 272, 279, 280
    stability, 279, 283
    step-size, 271, 272, 274, 275, 280, 284, 290, 292, 294, 298
    time constant, 275
    weighting, 287
  all-pole filter, 284, 286, 310
  all-zero filter, 267, 269, 275–277, 279, 281, 283
  approach, 264
  array, 312
  hyperstable adaptive recursive filter (HARF), 284
  identifier, 310
  lattice, 286, 290–292, 295
  line enhancer, 294
  mean-squared error, 274
  noise canceller, 294, 298
  nulling, 312
  pole-zero filter, 283–285
  predictor, 294, 298–301, 312

  processor, 263, 268, 281, 282, 288, 291, 297
  signal estimation, 264
  signal processing, 225, 237, 263, 264, 267, 286, 301, 308
Akaike H., 186, 189, 212, 241–244
  information criterion (AIC), 187, 188, 224, 235, 242, 244, 248, 251, 255, 291
Aliasing, 35–38, 40, 69, 71, 74, 75, 79, 255
All-pole model, 220, 284 (*see also* IIR filter)
All-zero model, 224, 267 (*see also* FIR filter)
Analog:
  prototype filter, 43, 44, 46
  signals, 33
Analytic function, 17
Anderson, B., 320, 321, 339, 340
Anderson, T., 175
Anti-aliasing filter, 36, 37, 52, 253
Aperiodic signal, 3, 7, 15, 56, 59, 62, 76
Applications:
  electromagnetic, 244
  transient pulse, 295, 302
Astrom, K., 6, 143, 214, 236, 241, 257
Auto-correlation method, 182, 184
  post-windowed, 183, 184
  pre-windowed, 183, 184
  un-windowed, 184
  windowed, 182, 184
Autoregressive (AR) model, 116, 120, 123, 124, 178–185, 188, 195, 196, 307, 352, 353
  exogenous input (ARX) model, 116, 132, 178, 179, 254, 283, 331, 333
  moving average (ARMA) model, 116, 119,

375

Autoregressive (AR) model (*cont.*)
    121, 132, 136, 145, 178, 179, 191–
    197, 205, 206, 209
  spectral estimation, 191
moving average with exogenous input
    (ARMAX) model, 25, 83, 115, 117–
    123, 128–132, 136–139, 147, 148,
    179, 215, 224, 225, 229, 230, 232,
    235, 237–239, 247, 249, 251, 254,
    280, 282, 283, 291, 293, 311, 343,
    354, 357, 358
  equivalence, 137
spectral estimation, 186, 187, 203, 209, 210,
    220, 221, 224, 260–262, 287
Azevedo, S., xii, 13, 14, 137, 329, 334, 337,
    340, 369, 373

Backward prediction error, 261, 262
Backward-shift operator, 25, 115–117, 126,
    180, 188, 191, 238, 247, 331
Bandlimited, 34, 38, 40, 52, 69
Bandpass filter, 73, 74, 173
Bandpass sinusoid example, 178, 185–188, 195,
    199, 200, 203, 204
Bartlett window, 164, 165, 167
Basewidth, 163
Bauer, F., 114, 143
Baum, C., 245, 258
Bayes' rule, 154
Bendat, J., 171, 175
Bergland, G., 71, 80
Bessel, 43, 49
Bevensee, R., 245, 258
Bias:
  error, 250
  estimator, 150, 151
Bierman, G., 218, 258, 322, 341, 373
Bilinear transformation, 41, 43, 53, 54
Bitmead, R., 274, 275, 308
Black-box, 246, 247
Blackman:
  Tukey method, 161 (*see also* Correlation
      method)
  window, 47, 49, 80
Block processor, 288–293
Box, G., 214, 257, 320, 321, 340
Brand, H., 361, 364, 367
Broadband, 300
Brigham, E., 71, 80, 349
Burg, 179, 196, 198, 199, 211, 234, 286, 350,
    353

Butterworth, 43, 44, 49, 54, 165, 212, 213, 311
Cadzow, J., 191, 212
Canceller, 296–300, 302, 303, 306
Candy, J., xii, xiii, 9, 14, 133, 151, 152, 174,
    214, 238, 257, 297, 302, 308, 310,
    320, 321, 327, 328, 331, 340, 341,
    373
Canonical form, 136
  observer, 139, 140
Capon, J., 200, 212
Causal, 19, 24, 27
Central limit theorem, 97, 103, 163
Chain rule, 95, 96, 151, 152, 175, 202, 268,
    355
Channel equalizer, 311
Characteristic equation, 206, 208, 210, 270
Chebyshev filter, 43, 44, 49
Childers, D., 179, 211, 214, 257
Chi-squared distribution, 162, 163, 168
Cholesky decompostion, 224, 262
Churchill, R., 17, 48
Circular shift, 58, 59
Coherence, 99, 110, 149, 171–174
  linearity, 172
  output spectrum, 174
  transfer function, 172
Conditional:
  expectation, 95, 144, 327
    properties, 144
  gaussian, 98
  mean, 152
  probability, 95, 155
Confidence interval, 104, 171, 317
Continuous:
  response, 40
  signal, 3, 29, 33, 35–38, 52, 68, 69
  spectrum, 35–38
  time, 32, 33
    Fourier, 7, 33, 34, 62, 65–71, 80, 82, 83,
        100, 159
Convolution, 9, 10, 20, 24, 34, 69, 83, 84, 86,
    111, 113, 119, 158, 159
Cooley, J., 79, 80, 160, 175, 345
Correlation, 86, 112, 300
  cancelling, 216, 300
  method, 161, 162, 166, 168, 190, 201
  time, 300
Covariance, 10, 11, 14, 86, 87, 92, 93, 99, 106,
    108, 109, 111–114, 117, 119, 121,
    122, 146, 147, 156, 158, 159, 160,
    175, 193, 204, 261, 350, 351, 353
DFT calculation, 160

estimation, 155, 156, 167, 177, 186, 196, 203
exponential, 223, 228, 259, 260, 300
matrix, 185, 252, 259, 272, 309, 323, 324, 326
method of spectral estimation, 183–185, 209
nonparametric, 156, 160
parametric, 156
periodic, 106
power spectrum pairs, 103, 145
propagation, 120, 123, 124
properties, 104, 106, 109
  maximum, 104
  symmetry, 104
sample, 157
sequence, 84-86, 161
sinusoidal, 205
stationary, 96
Cowen, C., 258, 279, 283, 284
Cramer Rao Bound (CRB), 151, 152, 175, 176
Criteria for AR transfer function (CAT), 244
Cross-covariance, 104, 106, 108–110, 112, 146, 161, 252, 309, 326
Cutoff frequency, 43, 46

Deconvolution, 230, 231, 252, 253, 289, 311, 314, 334–338, 343
adaptive, 301
Kalman, 334–338, 343
Wiener, 230, 231, 252
Delay, 108–110
Desired signal, 269, 283, 310
Deterministic signal, 3, 10, 13, 14, 68, 74, 83, 84, 112, 116, 159, 214
Deutsch, R., 217, 258
Diamagnetic loop, 302, 303
Difference equations, 9, 13, 20, 25–27, 29, 31, 62, 63, 81, 82, 112, 115, 116, 121, 128, 139, 146, 180, 188, 191, 205, 207, 224, 280
state, 133
Digital filter, 9, 40–42, 51, 52, 54, 84, 112, 116, 117, 123, 125, 146, 147, 301
Direction vector, 265
Discrete:
Fourier transform, 14, 30, 55–62, 66–71, 73–82, 159–162, 169, 170, 180, 195, 196, 300, 345–349
  coefficients, 56, 60, 65, 78
  properties, 59-62
frequency, 56
random signal, 9, 14, 83, 99, 106

time:
Fourier transform, 7, 8, 10, 14, 15, 29–31, 55, 56, 59-63, 65-68, 74, 78, 80, 82, 99, 101, 106, 109, 111
properties, 63, 64
signal, 1, 29, 55
system, 23, 25, 27, 29, 31, 40, 62
Distortionless response, 202
D-step predictor, 298–301, 312, 313
Dynamic system, 113

Eigen-equation, 206, 207
Eigenvalue, 206–208, 270, 272, 275, 309, 310
matrix, 271
minimum, 206–208
Eigenvector, 206, 207, 270, 272, 309
Electromagnetic:
signal processing, 214, 244
wave, 108, 126, 234
Elliptic filter, 43
Ensemble, 90–92, 94, 99, 100, 104, 247
average, 97, 198, 234, 287
statistics, 93
Entropy, 94, 95, 243, 350–353
Equalizer, 311
Ergodic, 97, 145
Error:
backward, 196, 197, 261, 286, 287
covariance, 152, 323
estimation, 151, 181, 316, 327, 328
forward, 196, 197, 286, 287
gradient, 198
mean-squared, 153, 191, 215, 283, 287, 309, 310
prediction, (*see* Prediction error)
sequence, 215
squared, 153, 209, 277, 278
variance, 155, 156, 180, 189, 198, 234, 317
Estimation, 214, 314
algorithm, 315, 316
error, 151, 316, 331
  Cramer-Rao bound, 151, 152
filter, 11, 12, 84, 86, 125
linear, 225
optimal, 215
problem, 182, 323
procedure, 316
Estimator, 11–13, 84, 149, 150, 157
adaptive, 263, 269, 279, 283, 286, 288, 289, 291, 298
all-pole, 284, 286

Estimator (*cont.*)
  all-zero, 225, 227, 269, 281
  batch, 223, 224, 229, 324, 342
  biased, 150, 157, 167, 177
  consistent, 150, 155, 157
  converges in probability, 150
  covariance, 177, 186, 279
  efficient, 150, 152, 155, 156
  extended least squares (ELS), 193, 195, 240,
      242, 247, 254
  gaussian, 155
  Gauss-Newton, 239, 355, 356
  gradient adaptive lattice (GAL), 291, 292,
      294, 295
  hyperstable adaptive recursive filter (HARF),
      284
  Kalman, 12–14, 140, 193, 214, 320–323,
      328, 329, 331–340, 342
  lattice, 212, 215, 258, 286, 288–295
  least mean-squared (LMS), 273–276, 279–
      282, 284, 286, 291, 297, 300, 304,
      307, 309, 310–313
  least-squares (LS), 153, 176, 194, 210, 241,
      277, 317
  Levinson, 223–225, 234–236, 280
    Wiggins-Robinson (LWR), 225–232, 252,
      260, 304
  line enhancer, 292, 300
  linear, 219, 225
  maximum:
    a posteriori (MAP), 153–156, 176, 315
    entropy method (MEM), 196, 315
    likelihood (ML), 152, 154–156, 176, 243,
      315
  mean-squared convergent, 150
  minimum variance, 153, 176, 315, 324, 326,
      333
  Newton, 254, 266, 267, 275, 276, 278, 308,
      309
  noise, 296, 297, 304, 305
  nonlinear least-squares (NLS), 254, 255
  normalized least mean-squared (NLMS), 275,
      276
  one-step predicted, 181, 284
  optimal, 215–217, 219, 231, 296, 297
  order, 241
  orthogonality, 152, 327
  parametric, 173, 186, 196, 219, 237–239,
      243, 264, 269
  performance, 150
  pole-zero, 283–285
  prediction error, 237, 241, 279
  predictor, 298–301
  pulse, 263, 306

  quality, 151, 152, 316
  recursive, 25, 27, 117, 123, 320, 321, 324
    extended least squares (RELS), 247, 254
    least-squares (RLS), 193, 255, 276, 278–
      282, 307, 309, 334
    prediction error method (RPEM), 236–242,
      276, 283, 284, 307, 354, 355, 357,
      358
    repeated least squares, 241
  sample mean, 320
  state, 314, 321–323, 327
  steady-state, 268, 280, 297
  stochastic gradient, 266, 268, 269, 271–276,
      283–285, 290, 291
  sufficient, 150
  unbiased, 150, 157, 158, 201, 273
    conditionally, 150, 152
    unconditionally, 150, 152
  uncorrelated, 152
  unknown constant, 155
Euler, 4, 6, 8, 61
Event, 88
Expectation, 10, 93, 100, 104, 151, 266, 273
Expected value, 86, 92–94
Experiment, 88
Exponential, 4, 5, 56, 57, 63, 74, 201, 207,
      208, 280, 347
  colored noise, 146, 223, 228, 259, 300
  matrix, 329
Extended least squares (ELS), 193, 195
Eykhoff, P., 263, 301, 308

Faurre, P., 340, 341
Filter, 11, 40, 84, 85, 102, 253
Final prediction error (FPE), 186–188, 244, 247,
      249, 251, 255
Finite:
  duration signal, 58, 64, 69, 79, 99, 160
  energy, 99
  impulse response (FIR), 26, 27, 42, 45-48,
      116, 128, 201, 267, 268, 282, 292,
      294, 295
Folding frequency, 36, 71
Forgetting factor, 279
Forward:
  prediction error, 126
  signal, 126
Fourier:
  series, 6, 14, 45, 55, 57, 58, 65, 66
  spectrum, 84, 161, 297
  transform, 7, 33, 34, 62, 65–71, 80, 82, 83,
      100, 159

Frequency:
  domain, 7, 29, 33–35, 45, 86, 115, 345
  interval, 60, 73, 74, 76, 77
  overlap, 35
  resolution, 76
  response, 7, 8, 14, 29, 30–33, 36, 37, 39, 45,
       46, 48, 50, 52, 53, 62, 82
  sampling, 68, 70–72
    design, 45
  scaling, 20, 64
  warping, 41
Friedlander, B., 125, 144, 195, 199, 212, 283,
       297
Fundamental period, 5
Fusion, 302

Gauss:
  Markov model, 12, 13, 83, 133, 134, 136–
       138, 140, 142, 147, 329, 330, 334,
       336, 342
  Newton, 238, 239, 355, 356
Gaussian, 97, 98, 133, 134, 140, 145, 146, 151,
       155–157, 162, 167, 188, 351
  conditional, 145
  independent, 162
  noise, 146, 175
  orthogonal, 145
  random variable, 145
Gelb, A., 369, 371, 373
Geometric series, 8, 17, 56, 61, 65, 67, 113
  finite sum, 56, 61, 65
Gersch, W., 242, 258
Giordano, A., 234, 258
Goodman, D., 254, 255, 258
Goodwin, G., xii, xiii, 115, 144, 214, 237, 257,
       263, 283, 301, 307, 320, 334, 340
Gradient, 189, 192, 198, 233, 238, 239, 254,
       264–269, 271–276, 280, 283–285,
       287, 289, 290, 293–295, 298, 308,
       309, 312, 355–357
  adaptive lattice (GAL), 290, 291
  chain rule, 151, 175
  vector, 151
Gram-Schmidt orthogonalization, 217, 218,
       259
Gray, A., 125, 128, 131, 144
Griffiths, L., 291, 308

Hamming, 46, 47, 49, 74, 80, 166, 170, 171
Hankel matrix, 211

Hanning, 47–49, 80
Harris, F., 45, 49
Haykin, S., xii, xiii, 199, 212, 220, 232, 234,
       263, 278, 293, 301, 307
Hessian, 238, 239, 266, 275, 280, 355, 356
Hogg, R., 143, 150, 174
Honig, M., 232, 258
Hyperstable adaptive recursive filter (HARF),
       284
Hypothesis test, 104

Ideal filter, 42, 45, 48
Identification, 245–247, 249, 251, 253,
       255–257, 263, 297, 314, 331, 333,
       334, 343
  adaptive, 301, 310
  off-line, 254, 255
Impulse, 3, 4, 71, 102, 103, 120, 162
  function, 89
  invariant transformation, 39–41, 43, 53, 54,
       255
  propagation, 120–122
  response, 9, 15, 24–26, 30, 31, 40, 42,
       49-51, 53, 81, 82, 112, 113, 121, 122,
       145, 146, 193, 195, 224, 225, 229,
       230, 236, 262, 281, 335
  sampler, 34, 68, 69, 70
  train, 34
Incoherent, 172
Independence, 93, 97, 157, 163
Infinite:
  impulse response (IIR), 24, 42, 44–46, 116,
       128, 132, 284, 285
  series, 17
Information, 94, 95, 242, 243
  matrix, 157
  theory, 188, 242
Innovations:
  covariance, 148, 322, 323, 326, 328, 335,
       340
  gradient, 239, 355
  model, 140, 143, 339, 343
  sequence, 216, 217, 219–221, 259, 320–328,
       331, 333–335, 337, 342, 343
  statistics, 217, 328
  vector, 218
  weighted sum-squared residual (WSSR), 331
  whiteness, 323, 327, 328, 331, 334, 335
Input vector, 133, 139
Instantaneous gradient, 267, 274, 284, 291, 294,
       297
Interpolation, 36, 65, 76–78

Inverse filter, 239, 252, 357
Inversion integral, 22, 23, 108, 113
Iterative, 114
Jategaonkar, R., 241, 244, 258
Jazwinski, A., 88, 99, 143, 320, 321, 336, 340
Jenkins, G., 157, 164, 167, 175
Joint:
  distribution, 94
  entropy, 96
  mass function, 93, 94
  process estimation, 226, 297
jw-axis, 7, 29, 33, 37, 39, 41, 62

Kailath, T., 114, 133, 136, 141, 143, 205, 212, 216, 257, 323, 325, 327, 340
Kaiser window, 46–49, 80
Kalman filter, 12–14, 140, 193, 214, 320–323, 328, 329, 331–340, 342
  algorithm, 322
  deconvolver, 334–338, 343
  frequency domain, 337
  gain, 327, 333
  identifier, 331, 332, 334, 335, 343
  power spectrum, 337, 339
  prediction form, 342
  Riccati equation, 340, 342
  Schmidt filter, 336, 337
  steady-state, 337, 339, 342
  tuning, 328, 337
  U-D form, 322
  Wiener filter, 337
Kalman, R., 214, 257, 321, 339, 340
Kay, S., 186, 190, 191, 203, 211, 212
Kesler, S., 179, 211
Kirchoff's law, 319, 328
Kronecker delta, 139
Kullback-Leibler measure, 242

Lager D., xiii, 14, 25, 31, 36, 44, 46, 49, 57, 62, 68, 74, 80, 84, 86, 104, 109, 117, 121, 122, 125, 143, 165, 171, 173, 175, 186, 188, 195, 199, 207, 212, 224, 229, 231, 236, 240, 249, 281, 291, 300, 360, 366, 367
Lagrange multiplier, 202
Laplace transform, 7, 27, 29, 33, 62, 107, 115
Lattice:
  adaptive, 286, 290–292
  block processor, 235, 288–293
  equivalence, 137
  autoregressive, 198, 199, 237
  state-space, 140, 141
  feedback, 130
  feedforward, 127–129
  filter, 125, 127, 214, 231, 234, 235, 237, 261, 262
  inverse transfer function, 127–128
  Levinson recursion, 130, 198, 212, 234
  model, 126, 129, 131, 132, 136–138, 147, 215, 232, 234, 235, 264, 287, 289, 291
  orthogonality, 286
  rational, 131, 132
  recursion, 130, 140, 196, 198, 232, 233, 287
  reflection coefficient, 126, 129, 132, 197–199, 222, 233–236, 261, 280, 287, 288, 290
  reverse polynomial, 128
  spectral estimation, 196, 199
  stability, 286
  stage transfer function, 127
  transfer function, 127
  two-port network, 127
Laurent series, 17
Layered medium, 126, 234, 235
Leakage, 46, 69, 72, 74
Least Mean-Squared (LMS) algorithm, 273–276, 279–282, 284, 286, 297, 298, 304, 307, 310–313
  complex, 311
Least-squares, 153, 176, 210
  extended (ELS), 193
Lee, D., 140, 141, 144, 308
Left
  half plane, 41
  sided signal, 18
Levinson
  all-pole filter, 220, 235, 262
  all-zero filter, 224, 225, 228–231, 262
  Durbin recursion, 128, 130, 198, 212, 214, 220, 222–225, 227, 228, 234, 235, 260, 261
  linear predictor, 220, 222, 227, 235
  Wiggins-Robinson (LWR) recursion, 225–232, 252, 304
Line enhancer, 294
Linear:
  combiner, 312
  discrete-time system, 24, 33
  system, 84, 111, 112, 115, 172–174, 179, 245, 335, 336, 339
    random, 112
  systems theory, 9, 10, 14, 29, 83–87, 96, 205, 207

time invariant (LTI), 15, 24, 25, 27, 31, 33, 41, 42, 50, 81, 82, 103, 146
Linearity, 19, 93, 95
Ljung, L., xii, xiii, 179, 191, 211, 214, 236, 237, 240, 241, 254, 257, 263, 278, 283, 297, 301, 303, 331, 334, 340, 354, 358, 366, 367, 373
Log-likelihood, 154, 155
Low pass filter, 35, 36, 43, 47, 48, 84
Luenberger, D., 264, 266, 308
Lyapunov equation, 135

Magnitude, 45, 46
Mainlobe, 46, 72
Makhoul, J., 186, 190, 191, 203, 211, 212
Markel, J., 126, 144, 214, 231, 234, 257
Markov process, 96, 103, 134
Marple, L., 211, 212, 233
Mason's rule, 130
Matched filter, 114
Matrix:
    covariance, 185, 193, 218, 324
    data, 183
    eigenvalue, 271
    exponential, 329
    Hankel, 211
    inversion lemma, 240, 278, 358
    lower triangular, 217, 262
    modal, 270
    normal form, 270
    Toeplitz, 182, 184, 206, 221, 225, 226, 228, 231, 252
    Vandermonde, 210
Maximum:
    a posteriori (MAP) estimator, 153–156, 315
    entropy method (MEM), 196, 315, 350–353
    likelihood (ML) estimator, 152, 154, 156, 187, 200, 243, 254, 315
Maybeck, P., 337, 341
Mean, 91, 92, 94, 97, 101, 117, 123, 137, 150, 157, 179, 320
    propagation, 117, 120–122, 124
    squared, 93, 106
        error, 153, 157, 215, 242, 251, 267–269, 272, 274, 283, 287, 309, 310
Measurement, 84, 316, 319, 326
    accuracy, 316
    actual, 320, 322
    bias, 316
    covariance, 136, 147, 148, 324, 339
        propagation, 135, 136, 329, 330
    filter, 321, 334

ideal, 316, 317
instrument, 317, 318
mean vector, 134, 329, 330
model 133, 155, 179, 188, 191, 237, 280, 283, 287, 294, 295, 303, 315, 318, 323, 324, 331
noise, 140
noisy, 84, 86, 172, 173, 316, 341
power spectrum, 136, 147
precision, 316, 319, 329
predicted, 320, 322
propagation, 135, 329, 330
quality, 317
system, 317
variance, 316, 317
vector, 139
Mendel, J., 334, 340
Miller, E., 245, 258
Minimum:
    Akaike information criterion (MAICE), 188, 244
    description length (MDL) criterion, 244
    variance, 152, 175, 180, 189, 202, 304
        distortionless response (MVDR) estimator, 178, 200–204
        estimator, 153, 175, 181, 259, 315, 324, 326, 333, 336
Misadjustment, 275
Mitra, S., 125, 127, 144
Model:
    all-pole, 220, 284
    all-zero, 224, 267
    antenna, 251
    autoregressive (AR), 116 (*see also* Autoregressive)
    based signal processing, 12, 14, 314, 316, 317, 328
        approach, 9
        processor, 12–14, 214, 215, 314, 316–321, 340, 341
        technique, 350
    covariance, 320
    dipole, 253, 255
    FIR, 26 (*see also* FIR)
    Gauss-Markov, 133 (*see also* Gauss-Markov)
    IIR, 24 (*see also* IIR)
    innovations, 140 (*see also* Innovations)
    input-output, 115
    inverse, 239, 252
    lattice, 125 (*see also* Lattice)
    layered medium, 126
    measurement, 137, 155, 315, 316, 319, 326
    moving average (MA), 116, 189
    noise, 315

Model (*cont.*)
  output error, 254, 255
  pole-zero, 283–285
  probe, 251
  process, 133, 315
  radar, 343
  RC circuit, 28, 318, 319, 334
  RLC circuit, 328–330, 332
  state-space, 133 (*see also* State)
  stochastic, 128, 320
  transfer function, 27 (*see also* Transfer
      function)
  validation, 250
  wave, 126
Mohanty, N., 241, 258
Monzingo, R., 263, 301, 307
Morf, M., 308
Moving average (MA) model, 116, 123, 128,
      141, 146, 178, 188, 189, 195, 260
Multichannel, 114, 127, 133, 136, 225
Multivariable system, 136
Mutual information, 95, 242
Narrowband, 201, 300
Natural mode, 253
Newton's method, 254, 266, 267, 275, 276,
      278, 308, 309
Noble, B., 217, 257, 280
Noise, 86, 117
  cancelling, 263, 294, 296–299, 302–304,
      306, 311
  model, 238, 311
  random, 173
  reference, 294, 295, 304
Nonlinear, 172
  algebraic equation, 114
  least-squares (NLS), 254, 255
  state space, 133
Nonparametric, 26
Nonrecursive, 26
Nonstationary, 263
Normal equations, 181, 184, 185, 221, 226, 231
Normalized least mean-square (NLMS), 275,
      298
Notch filter, 312
Nyquist frequency, 36, 38, 52, 71, 72

Observer canonical form, 139, 140
Oppenheim, A., xi, xiii, 5, 6, 14, 18–22, 24,
      33, 41–43, 45, 48, 49, 59, 80, 167,
      175, 206, 212, 348, 349, 365, 367
Optimal:

design, 45, 46
  estimation, 215, 216, 221, 226, 227, 231,
      232, 252, 263, 266, 269, 271, 292,
      297, 304, 305, 312
Optimization, 263, 264
  theory, 263
Order, 186, 187, 203, 241, 242, 247, 248
  estimation, 241
  tests, 244, 248, 249, 251
Orfanidis, S., xii, xiii, 132, 144, 216, 224, 257,
      272, 308
Orthogonal, 93, 98, 125, 145, 218, 244, 259,
      286, 289, 324, 325
  decomposition, 217
  eigenvectors, 270
  projection, 215, 216, 231
Orthogonality condition, 153, 181, 184, 189,
      192, 209, 216, 221, 226, 231, 232,
      261, 277, 324, 325, 327
Output:
  error, 254, 255
  vector, 133

Papoullis, A., 88, 89, 95, 104, 143
Parameter:
  random, 175, 219
  unknown, 175
Parametric, 26, 30, 119, 220
  design, 11, 13
  gradient, 189
  nonrandom, 151
  processor, 13
Parks-McClellan filter, 46, 48, 49
Parks, T., 45, 46, 49
Parseval, 21, 100, 163
Partial:
  correlation coefficient, 222 (*see also* Reflec-
      tion coefficient)
  fraction, 21–23, 29, 255
Parzen window, 47
Periodic, 29, 31, 37, 59, 63, 69, 76
  extension, 71
  signal, 3, 5, 6, 14, 20, 21, 29, 34, 36,
      55–58, 60, 63, 68, 71, 76, 345
Periodogram, 161, 166, 168–170, 201
Phase, 31, 45, 46
  distortion, 44
  wrapping, 45
Picket fence, 72–74, 79
Pisarenko harmonic decomposition (PHD)
      method, 140, 206, 207

Pisarenko, V., 206, 212
Pitfalls, 71, 74, 75
Pole, 18, 19, 22, 27, 32, 33, 39–41, 45, 113, 219, 241, 253, 256, 339
  dipole, 256, 257
Polynomial, 26, 29, 188
  operator, 25
  reverse, 128, 260
Power:
  average, 100, 106, 109
  series, 21, 22
  spectral density, 10, 14, 86, 100, 101, 103, 106, 113, 114, 122, 125, 146, 161, 162, 165–167, 171, 174, 179, 180, 196, 200, 201, 205, 210, 219, 350–353 (*see also* Spectral estimation)
    cross, 146
    measurement, 136
    pairs, 112
    properties, 106, 109
    rational, 191
    shaping, 116
    sinusoidal, 205
  spectrum, 10, 11, 14, 84–87, 99, 101, 102, 104, 105, 107, 108, 111, 112, 117, 121, 145 (*see also* Spectrum)
Predictable, 115
Prediction, 11, 222, 294, 312, 322
  distance, 298–300
  error, 181, 183, 187–189, 196, 197, 208, 221, 234, 237–239, 247, 250, 255, 261, 262, 279, 284–286, 288, 290, 298, 354, 358
    backward, 196, 197, 261, 286, 287
    filter, 214, 220, 236, 237
    forward, 196, 197, 286, 287
    gradient, 181
    test, 244, 247–252
    variance, 182, 188, 198, 234
Predictor, 298–301, 312, 313
  corrector, 322, 323
  polynomial, 261, 287
Primary strip, 38
Probability, 10, 88
  distribution function, 89, 90
  mass function, 89–92, 97
  theory, 88, 94
Process model, 133, 315, 318
Processor:
  model-based, 13, 14, 214, 215, 314, 316–321
Projection, 217
  orthogonal, 217
Prony method, 207, 209, 210, 211
Propagation equation, 117

Property, 18, 21, 89
Pulse, 14, 15, 263
Rabiner, L., 45, 77, 80, 214, 257
Radar, 343
Ramp, 4
Random:
  input, 111, 115, 179
  noise, 11, 173, 219, 302
    signals, 1, 3, 9, 10, 13, 14, 84–87, 92, 93, 95, 98, 99, 107, 108, 110, 115, 149, 178, 214, 260
  number, 104
    generator, 91, 104
  process, 91 (*see also* Stochastic process)
  variable, 88–92, 94–97, 100, 259
  vector, 152
    space, 217
  walk, 91, 92
Rational function, 17, 27, 31, 32
Rayleigh:
  distribution, 176
  quotient, 207
RC circuit, 28, 318, 319, 334
Realization, 89–92, 97, 99, 157
Received signal, 108
Reconstruction filter, 81
Rectangular, 47, 61, 62, 65, 69, 74, 76, 158, 159, 166, 170, 171
Recursive, 25, 27, 117, 123, 320, 321, 324
  extended least-squares (ELS), 247, 254, 255, 262
    off-line, 255
  form, 320, 321, 342
  in-order, 237
  in-time, 237, 239, 289, 356
  least-squares (RLS), 193, 255, 276, 278–282, 307, 309, 324
    U-D factorization, 279
  prediction error method (RPEM), 236–242, 307, 354, 355, 357, 358
Reflection coefficient, 126, 129, 132, 197–199, 222, 233–236, 261, 287, 289, 290
Region of convergence (ROC), 17–19, 29, 62
Representation theorem, 113, 114
Residual, 240, 250
Residue, 22, 40, 253, 255
  theorem, 22, 220
Resolution, 46, 76–79, 82, 166, 168, 305
Reverse polynomial, 128, 260
Rhodes, I., 341
Riccati equation, 340, 342
Right-sided sequence, 18
Ripples, 69
Rissanen, J., 114, 144

RLC circuit, 328–330, 332
Robinson, E., 119, 125, 126, 141, 199, 212, 367
Root mean-squared (RMS) error, 250

Sage, A., 153, 154, 174, 320, 339, 340, 341
Sample:
  covariance estimator, 157, 158, 161, 167, 184
  mean, 320
  space, 3, 9, 88, 98
Sampled:
  data system, 133
  signal, 3, 8, 33–35, 38, 59
  spectrum, 35, 38, 39
Sampling:
  frequency, 34–36, 38, 69
  function, 34, 70, 71
  interval, 34, 36, 39–41, 79, 82
  process, 34, 68, 71
  rate, 38
  theorem, 33, 35, 36, 38–40, 68, 71
Schmidt-Kalman filter, 336, 337
Schur-Cohn test, 129, 233
Schuster, A., 160, 175
Schwartz, M., xi, xii, 158, 175
Seismic, 108, 126, 234, 334
Self-tuning filter, 312
Shannon, C., 187, 212, 350, 353
Sidelobe, 46, 71, 72, 74, 76
SIG, 14, 25, 44, 46, 49, 57, 62, 68, 71, 76, 80, 84, 86, 104, 109, 117, 121, 122, 125, 143, 165, 171, 173, 175, 186, 188, 195, 199, 203, 212, 224, 229, 231, 235, 240, 249, 258, 280, 281, 291, 297, 300, 307, 359–367
  algorithms, 365–366
  availability, 366
  command processor, 360
  commands, 361–365
Signal, 1–3, 14, 33, 84, 88
  analysis, 55
  correlation-time, 300
  error test, 244, 248, 249, 251
  estimation, 11, 12, 149, 280, 314
  model, 220
  processing, 1, 3, 83, 155, 179, 245, 246, 253, 263, 267, 316, 334
  processor, 1, 7, 9, 33, 41, 45, 46, 48
  to-noise ratio (SNR), 165
Silvia, M., 225, 234, 258, 334, 340
Simulation, 117
Sinc function, 36, 61, 69, 71, 72, 74

Singularity expansion method (SEM), 245, 253, 257
Sinusoid, 4, 7, 14, 65, 69, 71, 77, 79, 84, 86, 101, 102, 173, 176, 178, 185, 188, 203–205, 207, 210, 213, 297, 302, 312
Smearing, 71, 79
Smoothing, 11, 162
Sonar, 108
Spectral:
  analysis, 55
  estimation, 76, 149, 156, 162, 163, 165, 167, 171, 172, 178, 179, 186, 189, 203, 213, 214
    autocorrelation method, 182, 184
    autoregressive (AR), 178, 182, 183, 185, 186, 203, 210, 212
      moving average (ARMA), 178, 191, 195–197, 212
      variance, 185
    bias, 164, 169, 177
    classical, 164
    confidence interval, 171
    consistent, 168
    correlation method, 161, 162, 166, 168, 172, 190, 195, 201
    covariance method, 183–185, 209
    discrete Fourier transform (DFT), 180
    extended least-squares (ELS), 195, 197
    harmonic, 178, 204, 210
    lattice, 196, 198, 199
    maximum entropy method (MEM), 196, 203, 350–353
    maximum likelihood method (MLM), 178, 200–204, 212
    mean-square, 164
    minimum variance distortionless response (MVDR), 178, 200–204, 212
    model-based, 179, 180
    modern, 179
    moving average (MA), 178, 188–190
    parametric, 178–180, 191
    periodogram, 168–170, 172, 201
    Pisarenko harmonic decomposition (PHD), 178, 206–208
    Prony method, 178, 207–211
    resolution, 168, 178, 186
    unbiased, 168
    variance, 164, 168, 169, 177
      reduction, 177
    Welch, 167, 168, 175, 212
  factorization, 113, 114, 170, 190, 195, 220, 339
    theorem, 113, 352

representation, 99
Spectrum, 29, 30, 34
  cross, 110, 146, 161
  inverse, 352
  matching, 305
  simulation procedure, 113
Speech, 126, 214, 334
S-plane, 33, 37, 39–41, 46, 253
SSPACK, 13, 14, 137, 143, 329, 331, 334, 337,
      338, 340, 368–373
  algorithms, 371
  availability, 373
  commands, 372
  post-processor, 371
  pre-processor, 370
  supervisor, 369
Stability test, 129
Stable, 18, 24, 27–33, 41, 49, 54, 113, 129,
      233, 271, 339
Standard error, 250
State, 133, 134
  covariance, 135, 147
  equivalence, 137, 141–143
    ARMAX, 137
    lattice, 141–143
  estimation, 314, 321–323, 327
  mean vector, 134, 135
  power spectrum, 147
  propagation, 135
  space, 9, 115, 132, 133, 137, 138, 143, 269,
      270, 286, 289, 314
    model, 133, 134, 136, 264, 321–324, 329,
      331, 333
  steady-state covariance, 147, 148
  transition matrix, 133, 270
  variance, 135
  vector, 133, 139
Stationary, 96, 113, 119, 137, 142, 147, 168
  order, 96
Steady-state, 136, 137, 280
  estimator, 268
Steepest descent, 265
Steinberg, B., 251, 253, 258
Stewart, G., 207, 212
Step, 4, 24, 49, 89
  size, 271
Stochastic:
  approximation, 266, 273, 287
  gradient, 266, 268, 269, 271–276, 280,
      283–285, 290, 291, 293, 294, 297,
      298, 309, 311, 312
  process, 9, 10, 13, 83, 87, 88, 90, 92, 94-
      96, 98–100, 103, 113, 117, 125, 132,
      142, 143, 156

Strip, 37, 38
Sum decomposition, 107–109, 113, 114, 145,
      219, 352
Superposition, 24
System, 9
  identification, 179

Taylor series, 264, 329, 343, 355
Tesche, F., 258
Time:
  average, 101, 182, 198, 234, 284, 287
  delay, 108–110
  invariant, 24
  reversal, 20
  series, 116, 214
  shift, 20, 27, 63
Toeplitz matrix, 182, 184, 206, 211, 221, 225,
      226, 228, 231, 252
Transfer function, 27, 28, 30–33, 39–42, 44,
      52, 53, 81, 113, 115, 116, 128 131,
      146, 147, 172, 173, 179, 245, 312,
      339
  matrix, 127, 133
  pole-zero form, 131, 132
  to lattice transform, 132
Transform equivalence, 64, 65, 67
Transformation, 10, 33, 34, 39–43, 46, 325
Transition matrix, 133
Transmission line, 126
Tretter, S., xi, xii, 22, 49, 106, 107, 139, 219,
      255, 258
Triangular, 47, 49, 80, 159, 297
Truncation, 69, 71, 72, 74, 75, 80
Tukey window, 47
Tuning analysis procedure, 328
Two-sided signal, 18

U-D factorization, 279, 322
Unbiased estimator, 150, 151
Uncorrelated, 93, 98, 103
Undersampling, 35–37
Uniform distribution, 97, 101, 102, 104, 112
Unit circle, 7, 29, 30, 33, 37–39, 41, 52, 60,
      65, 106, 107, 113, 129, 219, 271,
      339, 351

Van Trees, H., 151, 174
Vandermonde matrix, 210

Variance, 92-94, 97, 103, 112, 119, 120,
122–124, 132, 135–137, 147, 150,
156–158, 160, 163, 164, 167, 168,
177, 198, 201, 202, 268, 316, 342
  AR power spectrum, 185
  prediction error, 182, 234
  propagation, 120, 122, 123

Warping, 41, 43, 52
Wave model, 126, 234, 235
Weighted sum-squared residual (WSSR), 331,
332
Welch, P., 167, 169, 175
White:
  gaussian noise, 13, 103, 104, 162, 281
  light, 103
  noise, 103, 105, 109, 113, 114, 116, 119,
143, 146, 147, 173, 180, 199, 201,
226, 260, 280, 281, 323, 327, 328
Whiteness test, 104, 241, 248–252, 255, 331,
332
Whitening filter, 219, 260, 321
Wide-sense stationary, 96, 97, 101
Widrow, B., xii, xiii, 225, 230, 263, 268, 274,
284, 300, 301, 307, 367
Wiener:
  deconvolution, 230, 231, 252
  filter, 219, 220, 225, 230, 259, 260, 268,
277, 280–283, 292, 294, 297, 300,
308–310, 312, 314, 340, 343
  Hopf equation, 153
  Kalman filter equivalence, 337, 339
  Khintchine relation, 101, 106, 160, 161, 190,
351, 352
Wiggins, R., 225, 258
Window, 46, 47, 68–70, 72, 74, 80, 159, 161,
353
  bandwidth, 164, 165
  Bartlett, 164, 165, 167

  basewidth, 163, 164
  Blackman, 47
  characteristics, 47
  degrees of freedom, 164, 165
  design, 45
  Hamming, 47, 165, 166, 170, 171
  Hanning, 47, 165
  Kaiser, 47
  lag, 161–163, 165, 168
  mainlobe, 167
  Parzen, 165
  rectangular, 47, 165, 166, 167, 170, 171
  sidelobe, 167
  smoothing, 167, 168
  spectral, 161, 163–165
  triangular, 47, 169
  Tukey, 165
  variance reduction, 164, 165, 177
Wold decomposition, 113, 115

Yourdon, E., 360, 367
Yule-Walker equation, 181, 194
  extended method, 195

Zeros, 17, 18, 32, 33, 40, 113, 219, 240
  appended, 74
  order hold, 81
  padding, 74–77, 79
Z-plane, 33, 37, 39, 40, 41, 46
Z-transform, 7, 8, 14–22, 26–29, 33, 46, 49,
50, 58–60, 62, 63, 65–67, 80–82, 99,
101, 106–108, 112, 127, 133, 180,
189, 191, 205, 208, 219, 339, 352
  inverse, 7, 16, 21, 42, 49, 50, 112, 119
  one-sided, 27
  pairs, 19, 29